ONE WEEK LOAN

2005

'05

3 0 JAN 2003

Steels: Metallurgy and Applications

Third Edition

D.T. Llewellyn and R.C. Hudd

Butterworth-Heinemann
Linacre House, Jordan Hill, Oxford OX2 8DP
225 Wildwood Avenue, Woburn, MA 01801-2041
A division of Reed Educational and Professional Publishing Ltd

 A member of the Reed Elsevier plc group

OXFORD   BOSTON   JOHANNESBURG
MELBOURNE   NEW DELHI   SINGAPORE

First published 1992
Second edition 1994
Reprinted 1995
Third edition 1998

British Library Cataloguing in Publication Data
A catalogue record for this book is available from the British Library

ISBN 0 7506 3757 9

Library of Congress Cataloguing in Publication Data
A catalogue record for this book is available from the Library of Congress

Every effort has been made to trace holders of
copyright material. However, if any omissions
have been made, the authors will be pleased
to rectify them in future editions of the book.

Typeset by Laser Words, Madras, India
Printed and bound in Great Britain by MPG Books Ltd, Bodmin, Cornwall

Contents

Preface

The third edition of this book represents a significant change from the previous editions in that the chapter on *Low carbon strip steels* has been completely rewritten and expanded very considerably. This has been achieved through the input of Roger Hudd, who has had a long and distinguished career with British Steel in the strip steels sector, and I am delighted that he has agreed to be a co-author for this latest edition. Previous editions also contained a brief chapter entitled *Technological trends in the steelmaking industry* which dealt with steel production aspects. This chapter has been eliminated in the third edition because the authors are essentially product-orientated and their knowledge of process aspects is superficial. However, the remaining chapters on *Low carbon structural steels, Engineering steels* and *Stainless steels* are essentially unchanged and the text overall deals with the metallurgy of the mainstream commercial grades and the service requirements that govern their applications. As such, the text is again directed primarily towards the needs of undergraduates and steel users who have a basic knowledge of ferrous metallurgy. However, a minor innovation is the inclusion of brief sections, at the beginning of each chapter, on the *Underlying metallurgical principles* of the various steel types which may serve as a useful introduction to the basic concepts.

Since the second edition in 1994, major changes have taken place in the move from British to European specifications. These changes have been recorded to reflect the situation in spring 1997. However, this process is still incomplete and, in some instances, it has been necessary to refer to both British and European standards.

As far as possible, the data on steel prices have also been updated to reflect the current situation, but again with the proviso that such information can soon be out of date and should only be used as a guide to the relative costs of steel grades.

David T. Llewellyn
Department of Materials Engineering
University of Wales Swansea
May 1997

1   *Low-carbon strip steels*

Overview

Sheet iron was first rolled during the seventeeth century. It was hot rolled by hand as separate sheets and coated with tin to form tinplate. Cold rolling was later introduced initially to give an improved surface, but cold-rolled and annealed steel strip is now used for a number of reasons. It may be produced with good shape and flatness and with close control of gauge and width, and it may be made with a clean surface with a roughness that makes it very suitable for painting.

It has been found that formed components made of sheet steel may be easily welded to form complete structures such as motor cars. These structures may have a high degree of rigidity, partly as a result of good design, and partly also as a result of the high elastic modulus of the steel itself. Sheet steel may be coated with other metals, including zinc and aluminium, separately or together, to provide enhanced corrosion and oxidation resistance. Alloy coatings have also been developed to give enhanced properties. It is the ability of sheet steel, however, to be economically and satisfactorily formed into a wide range of complicated shapes without splitting, necking or wrinkling, as well as all its other advantages such as low cost, which is ensuring its continued use as a major engineering material.

The low cost of steel arises partly as a result of the nature of the extraction process and the abundance and cheapness of the raw materials, and also as a result of the continued development of the steelmaking process itself. The first continuous mill to produce steel strip in coil form was commissioned in the United States in 1923 and this was clearly cheaper than rolling individual sheets by hand. The first mill outside the United States was commissioned at Ebbw Vale in 1938. The introduction of oxygen steelmaking mainly in the 1960s also enabled the refinement of impure iron into steel to be achieved much more rapidly and cheaply.

The introduction of the continuous casting of slabs enabled the stage of hot-rolling ingots to slabs to be eliminated, and this combined with vacuum degassing enabled new and more consistent steel chemistries to be obtained. Close control of the continuous casting process itself has led to a reduction in the number of inclusions by several orders of magnitude. The result has been that steel may be used for many, very thin, tinplate applications that could not have been considered for ingot route-processed steel available previously. Other changes that have reduced costs have been the linking of pickle lines with tandem cold mills and the linking of cold mills to continuous annealing lines, both of which eliminate the between-process handling costs.

A further means of reducing costs has been the introduction of thin slab casting, which in one of its forms enables the roughing stage to be eliminated. Finally, the development of strip casting will enable the finish hot-rolling sequence to be removed. At the time of writing, this process has not yet reached the stage of routine production.

The motor industry, among others, has provided an important stimulus to the steel industry to further improve the mechanical attributes of strip steel, including gauge control, flatness, surface cleanliness and roughness. The motor industry, however, has also helped to stimulate the development of mild steel with improved formability and of some of the wide range of high-strength and coated strip steels that are now available. The building industry has also played its part and there has been a spectacular growth in the use of profiled, organic-coated, galvanized steel sheet for architectural roofing and cladding, particularly for industrial buildings. The galvanized steel lintel has also now substantially replaced the reinforced concrete lintel in domestic housing.

General processing considerations

Developments in steelmaking, the use of vacuum degassing and other secondary steel making techniques enable the steel to contain much more controlled levels of the important elements carbon, nitrogen, sulphur, aluminium and manganese than were available previously. Very low levels of the first three of these elements, down to below 0.003%, are now also easily obtained if required. In many cases, however, higher levels must be used to give the properties needed.

The most common of the existing process routes for the conversion of liquid steel into a usable sheet form are summarized in Figure 1.1. Prior to about 1970,

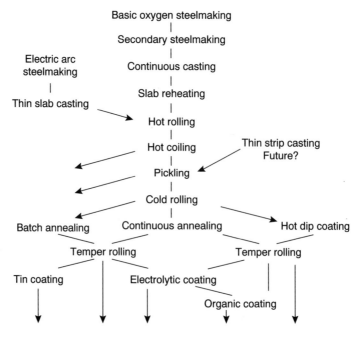

Figure 1.1   *Process route – low-carbon steel strip*

all strip steel was cast into ingots about 500 mm thick. After cooling and removal from the moulds the ingots were reheated to about 1250°C and rolled to slabs about 200–250 mm thick and allowed to cool. The surface was then scarfed to remove surface defects.

With the development of continuous casting, mainly in the 1970s, the ingot-rolling process was eliminated and the slabs were cast directly to a thickness of 200–250 mm as a continuous process and allowed to cool. The first hot-rolling stage then consisted of reheating to temperatures up to about 1250°C and rolling in two linked stages. The first stage, called *roughing*, reduced the thickness to an intermediate gauge, usually in the range 30–45 mm, and the second stage, called *finishing*, reduced the thickness to the final hot-rolled gauge required, often in the range 1–2 mm up to 5–12 mm, depending on details of the mill employed.

The roughing section may consist of a single reversing stand through which the steel passes backwards and forwards, usually five or seven times, or it may consist of several non-reversing stands through which the steel passes once. It may, however, consist of a combination of both reversing and non-reversing stands. Whichever combination is used, the steel completely leaves one stand before it enters the next one. Critical matching of rolling speeds is not, therefore, required.

A finishing train usually contains seven stands which are positioned close together. The front end of the strip exits the last stand well before the back end of the strip enters the first stand. Exact matching of the speeds of each stand is required, therefore, depending on the reductions in gauge in each stand. The steel usually exits the last stand at a finishing temperature up to and above 900°C depending on grade, which ensures that all the deformation takes place in the single-phase austenite region of the phase diagram. It is then cooled with water on a run-out table before coiling at a coiling temperature that is usually close to 600°C, but almost always in the range 200–750°C, depending on the metallurgical needs of the grade. Not all hot mills would, however, have the capability of covering this entire range. The finishing temperature may be controlled by using a suitable slab reheat temperature, usually in the range 1100–1250°C, by adjusting the speed of rolling through the mill and, if necessary, by introducing delays between roughing and finishing to increase temperature loss or by inter-stand cooling using water sprays. The temperature drop between roughing and finishing may, however, on certain mills be reduced by the use of radiation shields or by forming the steel into a coil to reduce the surface area for heat loss.

Some of the hot-rolled coil, covering the complete gauge range, is sold for direct use and some, usually in the gauge range up to about 5 mm, is cold rolled to thinner gauges.

An oxide-free surface is required for cold reduction and this is achieved by passing the steel through a pickle line. Traditionally, the acid used was sulphuric acid, but most pickle lines now use hydrochloric acid.

The cold rolling is usually carried out using a tandem mill containing five stands, with controlled front and back tension. Each stand usually contains four rolls consisting of two work rolls and two back-up rolls, but six high stands, each containing four back-up rolls, may be used in one or more positions if a particularly high cold reduction is required.

Cold-rolled gauges are usually in the range 0.4–3.0 mm and generally involve the use of 50–80% cold reduction, though tinplate mills provide up to 90% cold reduction to give thinner gauges. Clearly the thinner cold-rolled gauges are usually rolled from the thinner hot band gauges in order to limit the amount of work required from the cold-rolling mill. When the application requires a gauge that is common to both hot- and cold-rolled gauges, a cold-rolled and annealed material would only be used if a good surface and a high degree of formability were required. Otherwise, a hot-rolled material would be used for cost reasons. Tinplate and certain other packaging applications require gauges down to less than 0.2 mm and are rolled from the thinner hot-rolled gauges. Certain grades of tinplate, however, involve two stages of cold reduction.

Cold rolling causes the steel to become hard and strong and with very little ductility. For almost all applications, therefore, an annealing treatment is required to reduce the strength and to give the formability that is needed for the final application. Various metallurgical changes take place during annealing including recovery, recrystallization and grain growth, and the formation, growth or dissolution of precipitates or transformation products. It is the complex controlled interaction of these changes which provides the steel with its final properties.

Both batch or continuous annealing may be used for many types of steel, but certain steels may only be processed by continuous annealing. The traditional method was, however, batch annealing and this method in still in common use. With this method, several tightly wound coils are stacked on top of each other with their axes vertical. They are enclosed by a furnace cover which contains a protective atmosphere which is recirculated to promote heat transfer between the cover and the steel during both heating and cooling. Traditionally, the protective atmosphere has been HNX gas, which is nitrogen with up to 5% hydrogen, but furnaces designed for the use of 100% hydrogen were introduced during the 1980s and now provide an appreciable proportion of the batch-annealing capacity world-wide. Rapidly recirculated hydrogen has a higher heat transfer coefficient than the HNX gas and this enables faster heating and cooling rates to be achieved. The complete cycle may still, however, take up to several days. Hydrogen annealing gives improved control over mechanical properties and improved surface cleanliness.[1]

Continuous annealing for formable strip was first developed in Japan during the late 1960s and early 1970s, though continuous annealing for higher-strength steels for tinplate had been commonly used world-wide since the 1960s. The original cooling method for continuous annealing was by recirculated HNX gas, but other methods have now been developed. These include gas jet cooling with HNX gas, cold-water quenching, hot-water quenching (called HOWAQ), roll cooling, gas-assisted roll cooling, water-mist cooling and gas-jet cooling with high concentrations of hydrogen in nitrogen. A review of heating and cooling methods for continuous annealing has been given by Imose[2] and the importance of fast cooling rates for continuous annealing will become clear later. Continuous annealing cycles usually last for a few minutes depending on the type of steel being processed. A typical continuous annealing cycle, consisting of heating, holding, slow cooling, rapid cooling, very slow cooling or holding, slow cooling

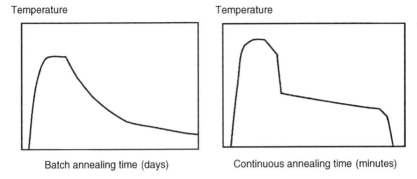

Figure 1.2   *Schematic annealing cycles for batch and continuous annealing*

and rapid cooling is illustrated in Figure 1.2. Clearly, as will be seen below, the details of the cycles to be used depend on the type of steel to be processed.

It is useful to note that top batch-annealing temperatures are rarely much above 700°C, whereas continuous annealing temperatures are usually above 700°C for strip grades and often above 800°C. These latter temperatures compensate for the large differences in time between batch and continuous annealing. Most continuously annealed steels for tinplate applications, however, are annealed below 700°C. It is usual but not essential to produce hot dip coatings on lines which also provide continuous annealing.

The last stage in the sequence for processing uncoated steel, prior to surface inspection, is called *temper rolling* or *skin passing*, which usually provides a cold reduction of 0.5–1.5%. This process is usually needed to remove the yield point in the tensile curve to ensure that stretcher-strain markings are not formed on pressing, but it also imprints a surface roughness that makes the steel surface suitable for its application and may improve the steel flatness.

Annealed and temper-rolled steel may be coated with a metallic coating using an electrolytic process on a separate coating line, but as indicated earlier, it is usual to combine a hot dip process with continuous annealing on a single combined line. Organic coatings may also be applied on top of either hot dip or electrolytic metallic coatings.

It is useful to note that when initially introduced, all the rolling stages in the processing of strip steel were intended merely to change the thickness of the steel and hot processing was used because it was easier to reduce the thickness of a thick material when it was hot and soft than when it was cold and hard. Each stage is now regarded, however, as a metallurgical process with closely controlled reductions, temperatures, cooling rates and times. The controlled processing sequence now enables desirable metallurgical structures and hence properties to be obtained at each stage that make the material either suitable for direct use or for proceeding to the next stage in the process.

As mentioned previously, an important recent development in steel processing has been the introduction of *thin slab casting*. Two main processes are involved, but in each case, the casting process is linked directly to a hot mill. With the CSP (compact strip production) process as practised at Nucor,[3] the slab is cast to a thickness of 50 mm, passes through a tunnel furnace to make the temperature uniform and then passes through a five-stand finishing train. With the ISP (inline strip production) process as practised at Arvedi,[4] the slab is squeezed while the core is liquid to a thickness of 43 mm. It passes through a three-stand roughing sequence to reduce the thickness to 15–20 mm and then passes through a four-stand finishing mill. An advantage of thin slab casting is the reduction in cost compared with the more conventional process, but at present, the surface quality of the product does not match that of the conventional process. So far, however, only a relatively small number of steel works have been equipped with the thin slab casting process and these are fed with steel from electric arc furnaces produced mainly by melting steel scrap. The steel may, therefore, contain relatively high residual elements which tend to limit quality.

Further developments are under way to enable the usual hot-rolled strip gauges to be cast directly. These processes are known as *thin strip casting* processes. So far, no steel works has been equipped with thin strip casting, but the expectation is that the elimination of the finish hot-rolling sequence will lead to a further reduction in cost. It is likely, however, that the material will not be suitable for direct use, but will require subsequent cold rolling and annealing for usable properties to be obtained.

Other means of reducing steel costs have been developed, including ferritic rolling, the combining of pickling and cold rolling into a single process, the combining of continuous annealing, temper rolling and inspection into a single process and the combining of cold rolling and continuous annealing. Only ferritic hot rolling, however, has a major impact on the metallurgy of the complete process.

Underlying metallurgical principles

Low-carbon strip steels are based primarily on ferrite microstructures and are almost invariably subjected to cold-forming operations in order to achieve specific shapes, e.g. automotive body panels. They are produced in the hot-rolled condition in thicknesses down to about 2 mm but they are used primarily in the cold-reduced and annealed state where the gauges can extend down to 0.16 mm for tinplate grades, e.g. for packaging applications. The basic consideration in the development of microstructure in strip and low-alloy steels is the Fe–C phase diagram, a portion of which is shown in Figure 1.3. At temperatures above about 900°C, a steel containing about 0.05% C will consist of a single phase called austenite or γ iron which has a face-centred cubic crystal structure, as illustrated in Figure 1.4(a), and all the carbon will be in solid solution. On cooling slowly, the material reaches a phase boundary at which a separate iron phase, called ferrite or α iron, begins to form. This phase contains a low carbon content and has a body-centred cubic structure, as illustrated in Figure 1.4(b). As cooling continues, the formation of further low-carbon ferrite leads to a carbon *enrichment* of the untransformed austenite. This process continues on cooling down to a temperature of 723°C, the eutectoid temperature, at which stage the remaining austenite will have been enriched in carbon to a level of about 0.8%.

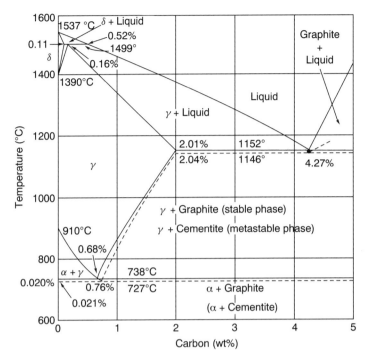

Figure 1.3 *The iron–carbon equilibrium diagram, for carbon contents up to 5 per cent* [5]

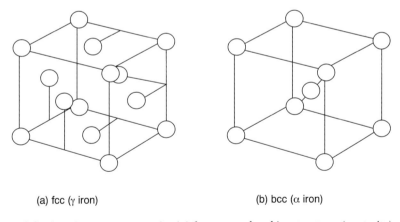

(a) fcc (γ iron)   (b) bcc (α iron)

Figure 1.4 *Atomic arrangement in (a) face-centred cubic structure (austenite) and (b) body-centred cubic structure (ferrite)*

A little below 723°C, the remaining austenite may transform to pearlite which consists of a lamelar structure of alternate laths of ferrite and cementite (Fe₃C). Below the eutectoid temperature, the solubility of carbon in ferrite decreases rapidly and is approximately given by the following equation[6]:

$$C_{soln} = 9.65 \cdot \exp -12\,100/RT \text{ wt\%}$$

where T is the absolute temperature and the gas constant $R = 1.987$ cal. degree^{-1}.mole^{-1}.

As the carbon content of a steel is increased, the γ to $\gamma + \alpha$ transformation temperature is depressed and the ratio of pearlite to ferrite in the microstructure is progressively increased, reaching 100% pearlite at about 0.8% carbon, the eutectic composition. Whereas the above sequence of events relates to slow cooling and an approach to equilibrium, faster cooling rates usually produce ferrite with globular carbides within the grains. Lower-temperature transformation products such as bainite or martensite may, however, be produced particularly when a suitable alloy addition has been made. These will be discussed later.

The ferrite grain size has a very important effect on the properties of low-carbon strip steels, as indicated by the Hall–Petch equation[7,8]:

$$\sigma_y = \sigma_i + k_y d^{-1/2}$$

where σ_y = yield stress
σ_i = the friction stress opposing dislocation movement
k_y = a constant (dislocation locking term) and
d = the ferrite grain size

Thus the yield stress increases with decreasing ferrite grain size. Pickering[9] has indicated that the value of k_y often lies in the range 15–18 MPa mm$^{1/2}$.

The strength is also influenced by other factors such as solid solution and precipitation strengthening. In the former, the strengthening is often related to the square root of the atomic concentration of the solute atoms, but at low concentrations, the strengthening effect may be regarded as linearly dependent. The magnitude of the effect depends on the atomic size difference between the iron and the solute element, the largest effects being produced by small elements such as carbon and nitrogen which go into interstitial solid solution. Elements such as phosphorus, manganese and silicon are often added to low-carbon strip to provide solid solution strengthening.

Both strip and structural grades of steel may be precipitation strengthened by the addition of elements such as titanium, niobium and vanadium and these elements may have an additional effect on strength by leading to a finer grain size. These elements are strong carbide and nitride formers which may be partially or completely dissolved at the slab reheating stage prior to hot rolling and which may then be reprecipitated into a fine form on subsequent cooling and transformation to ferrite. The degree of strengthening is dependent on both the volume fraction and size of the precipitates, finer precipitates producing the greater effect. Coarse precipitates which are not dissolved at the slab reheating stage are ineffective as strengthening agents. The solubility product for the precipitate is important, therefore, since it determines the amount of precipitate that can be taken into solution at any temperature and, consequently, the volume fraction that may be subsequently reprecipitated in a fine form. The temperature dependence of the solubility product is generally represented by an equation of the form:

$$\log[X][Y] = -A/T + B$$

Table 1.1 *Temperature dependence of solubility products (Wt%) for carbides, nitrides, carbonitrides,[10] sulphides and carbosulphides[11]*

Solubility product	$log_{10}k$	Solubility product	$log_{10}k$
[B][N]	$-13970/T + 5.24$	[V][N]	$-7700/T + 2.86$
[Nb][N]	$-10150/T + 3.79$	[V][C]$^{0.75}$	$-6500/T + 4.45$
[Nb][C]$^{0.87}$	$-7020/T + 2.81$	[Ti][S]	$-16550/T + 6.92$
[Nb][C]$^{0.7}$[N]$^{0.2}$	$-9450/T + 4.12$	[Ti][C]$^{1/2}$[S]$^{0.5}$	$-15350/T + 6.32$
[Ti][N]	$-15790/T + 5.40$	[Mn][S]	$-9020/T + 2.93$
[Ti][C]	$-7000/T + 2.75$	–	–

where $[X]$ and $[Y]$ are the weight percentages of elements in solution, such as titanium and carbon, T is the temperature in degrees Kelvin and A and B are constants. Table 1.1, prepared mainly by Turkdogan,[10] gives the solubility products for a number of compounds in austenite, but additional data for the sulphides have been added from a separate source.[11]

It is useful to note that a number of the precipitates are solid soluble in each other and that the precise composition of such a precipitate depends on the composition of the austenite matrix with which it is in equilibrium, as well as on the temperature. Niobium carbonitride, for example, has a wide range of solid solubility and an early theory[12] showed how the ratio of carbon to nitrogen in the precipitate in equilibrium at a given temperature could be calculated for different amounts of carbon, nitrogen and niobium in the steel.

In this model, a precipitate with formula $NbC_xN_{(1-x)}$ was assumed to be in equilibrium with a matrix according to the reaction:

$$Nb_{sol} + x \cdot C_{sol} + (1 - x) \cdot N = NbC_xN_{(1-x)ppt}$$

The solubility product K for this reaction was, therefore, written as

$$[Nb] \cdot [C]^x \cdot [N]^{(1-x)}/[NbC_xN_{(1-x)}] = K$$

where $[NbC_xN_{(1-x)}]$ was the activity of the carbonitride which was taken as unity. It was assumed, however, that the effective activity of NbC in the precipitate was x and the effective activity of NbN in the precipitate was $(1 - x)$ and that separate solubility product equations for the matrix would apply for NbC and NbN as follows:

$$[Nb] \cdot [C]/x = K_1$$

and $[Nb] \cdot [N]/(1 - x) = K_2$

where K_1 and K_2 are the solubility products for the pure carbide and nitride respectively at the desired temperature. For a given steel composition and temperature, these equations could be solved to give a value of x. The predictions from the model were in good agreement with the compositions of the niobium carbonitride precipitates that had been reported in the literature.

The model was later developed further to involve more than one precipitate[13] and a number of computer programs able to predict the equilibrium conditions for a series of multicomponent systems have now been developed. These programs can be used to predict the amounts of titanium, niobium, carbon, nitrogen and sulphur, etc, retained in solution in equilibrium at any temperature as well as the amounts and compositions of the precipitates.

In addition to producing precipitation reactions on transformation to ferrite, elements such as titanium or niobium also play an important part in controlling the recrystallization kinetics during the hot-rolling stage. Elements in solid solution may retard recrystallization through the process of *solute drag* but the strain-induced precipitation of niobium carbonitride is also very effective in retarding or stopping recrystallization. The final effect depends on the initial austenite grain size, the temperature, the amount of deformation given after the last previous recrystallization and the level of alloy addition.[14] As an example, Figure 1.5 shows how the critical strain for austenite recrystallization varies with prior austenite grain size and rolling temperature for a steel containing 0.03% niobium. On cooling of a recrystallized austenite structure, ferrite grains may nucleate at the austenite grain boundaries. With an unrecrystallized austenite, the austenite grain boundary area per unit volume is increased due to grain elongation. The nucleation frequency per unit grain boundary area is higher and nucleation may also occur on deformation bands.[15] The overall effect is that a finer ferrite grain size is produced from a deformed unrecrystallized austenite than from a recrystallized austenite. Figure 1.6 shows that for a given grain boundary area per unit volume S_V, the ferrite grain size also decreases with increasing cooling rate.[16] These effects are utilized in the processing of micro-alloyed steels and are discussed further in the section on high-strength steels.

There is a limit to the strengthening that can be achieved by a combination of grain refinement, solid solution and precipitation effects. Transformation strengthening is, therefore, employed to obtain values of tensile strength above

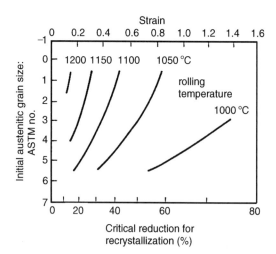

Figure 1.5   *Effect of initial grain size on the critical rolling reduction needed for austenite recrystallization of a 0.03 wt% Nb steel reheated to 1250°C for 20 min (Kozasu et al.[15]) (The initial grain size was varied by rolling at high temperatures.)*

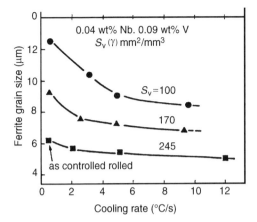

Figure 1.6 *The change in the transformed microstructure with increasing cooling rate after controlled rolling, for different austenite grain boundary surface area S_v per unit volume for a 0.10 wt% C. 1.50 wt% Mn, 0.04 wt% Nb. 0.09 wt% V steel (Ouchi[16])*

about 500 MPa, depending on whether the steel is in the hot-rolled condition or whether it has also been cold rolled and annealed. Thus, if a steel containing about 1% manganese is cooled rapidly from an austenite state to a temperature close to 450°C, the ferrite and pearlite reactions are suppressed and the austenite will transform to a lower temperature transformation product called *bainite*.

Bainite consists of small packets of lath-like ferrite grains with low misorientation between the grains and high-angle boundaries between the packets.[9] Carbides are present at the lath and packet boundaries and there is a variable dislocation density within the laths. Upper bainite formed at higher temperatures in the bainitic range has a coarser structure than lower bainite formed at lower temperatures and the finer lower bainite may also contain carbides within the laths.[9] Acicular ferrite may be formed at temperatures between the pearlitic and bainitic regions and may be considered to be a form of very low-carbon upper bainite. The formation of both pearlite and bainite involve the diffusion of carbon to form separate ferrite and carbide phases. Bainite may form over a range of temperatures starting with the bainite start temperature (B_s) and ending at the bainite finish temperature (B_F). These temperatures are given approximately by the following equations,[17] where the symbols in brackets refer to weight percentages.

$$B_s(C°) = 830 - 270[C] - 90[Mn] - 37[Ni] - 70[Cr] - 83[Mo]$$

and $B_F(C°) = B_s - 120[C]$

High-alloy contents and rapid cooling to below the bainite finish temperature may lead to the formation of a separate structure called martensite by a diffusionless transformation. In this transformation, a large number of atoms shear cooperatively to form a fine plate-like structure with the carbon held substantially in solution.[18] Carbon, however, may be precipitated as carbide by tempering. The crystal structure of martensite is tetragonal and the ratio of the *c*-axis of the unit cell to the a-axis depends on the carbon content. Martensite is the hardest structure that can be produced in steel, the strength increasing with the carbon content. The temperature at which martensite can start to form is called the martensite start temperature (M_s) and depends on the carbon content and other alloy additions made to the steel. The following equation gives an estimate of the M_s temperature for certain steels where the symbols in brackets again refer to weight percentages.[19]

$$M_s(°C) = 539 - 423[C] - 30.3[Mn] - 12.1[Cr] - 17.7[Ni] - 7.5[Mo]$$

In steels, the martensite finish temperature M_f is usually about 200°C below the M_s temperature.

Further details on the effects of alloying elements and cooling rate on the formation of martensite are given in the chapter on *Engineering steels*.

Whereas each of the strengthening reactions described above is employed singly or together to produce high-strength grades, they will also decrease the ductility. However, a very important consideration in the production of cold-rolled and annealed steels is the development of the preferred crystallographic orientation texture. If grains of every possible orientation are present in a steel with equal volume fraction, the steel is said to possess a random orientation texture. Such a texture is difficult to achieve completely and it is usual to find that certain orientations are present to a greater extent than others. The steel is then said to possess a preferred orientation texture. The strength of any component in a texture is often given in 'times random' units which is the ratio of the strength of the component in the texture divided by the strength that would exist in a similar steel with a random texture.

Orientation textures may be studied using X-ray diffraction and may be expressed in several ways. The first, called a pole figure, gives a plot, on a stereographic projection, of the strength of a particular crystallographic plane in each direction. Pole figures may be plotted for each relevant plane including, for example, (100), (110) or (111) planes and an example is illustrated in Figure 1.7, which also gives the positions of planes for several ideal orientations. Pole figures have the limitation that they do not give precise information concerning the strength of any component nor the way in which directions within each plane are oriented.

The second method of expressing texture is called an inverse pole figure. This gives the intensity of planes parallel with the strip surface, again in times random units. This has the limitation that it gives no information concerning the orientation of directions within the rolling plane. Nevertheless, inverse pole figures are useful since they give the precise total intensity of planes that are parallel to the surface. These intensities largely determine the mean r value (see next section).

A complete characterization of any texture may be given by means of an orientation distribution function which is calculated from several pole figures. An orientation distribution function gives a plot of intensity of each orientation in Euler space using angles that relate each orientation to the orientation of the strip. The method was originally described independently by Roe[21] and Bunge[22] using different but related angles. In the Bunge notation, the Euler angles ϕ_1, Φ

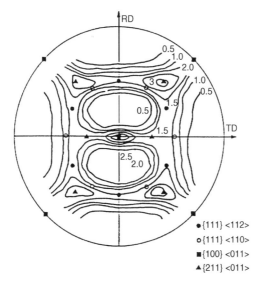

Figure 1.7 *(200) pole figure for cold-rolled steel showing important orientations (Hutchinson[20])*

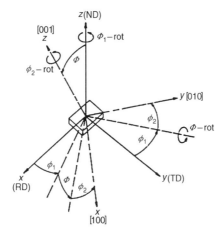

Figure 1.8 *Diagram illustrating the relationship between the crystal coordinate system [100], [010], [001] and the rolling, transverse and normal directions by means of the Euler angles ϕ_1, Φ, ϕ_2*

and ϕ_2 relate the crystal coordinate system [100], [010] and [001] to the physical reference frame RD, TD, ND (the rolling transverse and normal directions) by three rotations taken in order. These rotations are illustrated in Figure 1.8 and are further illustrated on a stereographic projection in Figure 1.9. This figure uses a (111)[121] orientation as a particular example.

An outline orientation distribution function giving the positions of some important orientations is given in Figure 1.10. It is useful to note that any

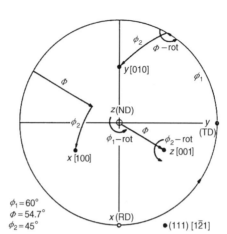

Figure 1.9 *Stereographic projection illustrating the relationship between the crystal coordinate system [100],[010],[001] and the specimen coordinate system, the rolling, transverse and normal directions, by rotations through the Euler angles, ϕ_1, ϕ, ϕ_2 using a (111)[1$\bar{2}$1] orientation as an example (Hu[23])*

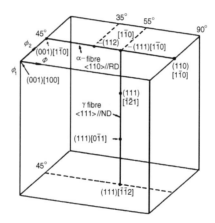

Figure 1.10 *Diagram of orientation space showing the important section at $\phi_2 = 45°$ containing the α fibre orientations ($\langle 110 \rangle$//RD) and the γ fibre orientations ($\langle 111 \rangle$//ND (Emren et al.[24])*

set of orientations represented by a straight line parallel to one of the axes in orientation space is known as a fibre texture. Figure 1.10 gives the position of the γ fibre texture, which includes all orientations with a [111] direction normal to the strip surface and the α fibre texture, which includes all orientations with a [110] direction in the rolling direction. These two fibre textures are particularly important since, as mentioned previously, they have a dominant influence on the r values and hence deep drawability of a ferritic steel.

A limitation of all the above three methods of representing texture is that the methods characterize the average texture in the volume of sample examined, but give no indication of the spatial distribution of grains with a particular orientation. The spatial distribution of texture is important because what happens to a particular subgrain in a material during recrystallization is strongly influenced by the orientation and structure of the material around it. The orientation of groups of neighbouring grains may be laboriously studied using Laue diffraction photographs, but the recently developed technique of electron backscatter diffraction[25] enables the orientation of neighbouring grains to be studied more easily.

During cold rolling, the texture, which is often close to random in the hot band structure, develops progressively with increasing cold reduction, as illustrated by means of inverse pole figure data in Figure 1.11. An equivalent plot, by means of two fibre lines through an orientation distribution function, is shown in Figure 1.12. This figure gives the intensities along both the alpha and gamma fibres. Clearly both fibres become stronger with increasing cold reduction due to the progressive rotation of the orientation of grains from other orientations into these two fibre orientations. The precise intensities along the fibre groups of components depend on the type and details of the steel and there is also a marked effect of hot band grain size.

The other important effect that occurs during rolling is the development of a deformation structure. As deformation proceeds, the dislocation density increases, but there is a tendency for dislocations to cluster together. The final result after sufficient deformation is that a subgrain structure is formed with a high dislocation density in the subgrain boundaries and a relatively low density within the subgrains.[28] The total stored energy of deformation depends on the total mean dislocation density. It depends, therefore, on the subgrain size and the orientation difference between adjacent subgrains since the latter is determined by the dislocation density within the subgrain boundaries.

It was indicated previously that iron may have a face-centred cubic or a body-centred cubic form. When either of these two crystalline structures is deformed, the deformation can take place by dislocation movement only on particular slip

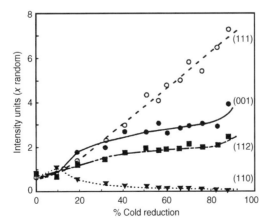

Figure 1.11 *Variation of the intensity of the components of the deformation texture of aluminium-killed steel with cold reduction (Held[26])*

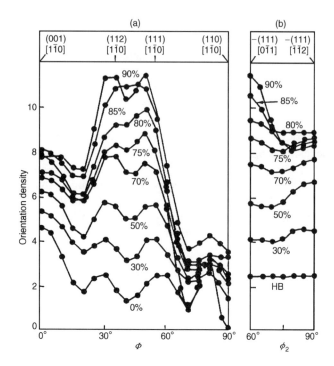

Figure 1.12 *Development of the rolling texture of an aluminium-killed 0.007 wt% C steel with cold reduction, (a) α fibre (b) skeleton line of γ fibre (Schlippenback and Lucke.[27]) The skeleton line is the line close to the γ fibre exhibiting the maximum orientation density*

planes in particular slip directions. For a body-centred cubic structure, such as ferrite, slip can occur in the four [111] directions on 12 slip planes divided between (011), (112) and (123). This gives 48 possible slip systems.[29] For a face-centred cubic-structure, such as austenite, slip can occur in the three [011] slip directions on the four (111) slip planes giving 12 possible slip systems.[29]

In any sample containing grains with a range of orientations, grains with certain orientations will deform with greater difficulty than other grains because their possible slip systems are oriented less favourably than others to the direction of the applied stress. The result is that grains with certain orientations develop a higher stored energy of deformation than other orientations. Dillamore *et al.*[30] showed that after 70% cold reduction the subgrain size decreased from (100) through (112) to (111) and (110) planes, as illustrated in Figure 1.13. In addition the subgrain misorientation increased in the same order. The result was that the total stored energy of deformation increased in the order (100), (112), (111) and (110). Other crystallite size work[31] and X-ray line broadening studies[32] have confirmed that the stored energy of deformation varies with orientation.

The stored energy of deformation is important because it can lead to selective nucleation of new grains during recrystallization when the steel is heated. Decker and Harker[33] suggested that the highest energy regions would be able to nucleate recrystallized grains first and that they would then be able to consume regions with low initial energy that would not nucleate so easily. The dislocation densities

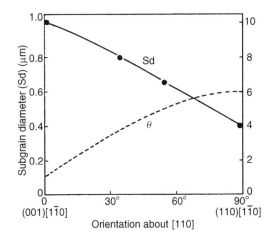

Figure 1.13 *Variation of subgrain diameter (sd) and subgrain misorientation (θ) with orientation from (001) [1̄1̄0] to (110) [1̄1̄0]. (Dillamore* et al.[30])

close to the original grain boundaries are usually higher than within the grains because the deformation must accommodate the deformation on differently orientated slip systems on either side of the boundary. The deformed grain boundaries are, therefore, preferred sites for the nucleation of recrystallized grains.[29] When recrystallization takes place after cold rolling, the recrystallized texture is clearly influenced by the deformation texture that is developed as well as by oriented nucleation and selective growth effects[34] that arise from the deformation structure itself and from other effects described below.

An example of a texture change that occurs during recrystallization, using data for a vacuum-degassed steel, is given in Figure 1.14. This gives the intensities along the α and γ fibre components as a function of time through the annealing process. It is seen that overall, the γ fibre components increase in intensity and that many of the α fibre components decrease except those that are common to the γ fibre components. In the γ fibre, the components close to (111)[112] clearly develop more strongly than components close to (111)[011].

The percentage cold reduction has an important influence on the grain size developed during recrystallization as well as on the texture. Increasing cold reduction generally leads to a finer grain size, but both the grain size and the recrystallization texture are also strongly influenced by the presence of alloying elements in solution and by the presence of precipitates. The hot band grain size also has an influence on the recrystallized grain size as well as its influence on texture, as illustrated for an aluminium-killed steel in Figure 1.15. It is also useful to note that the textures in many steels may be improved as a result of grain growth after recrystallization is complete, but grain growth may be inhibited by the presence of precipitates.

Many workers have shown that carbon and nitrogen in interstitial solid solution during annealing have a detrimental effect on the texture developed during recrystallization. With higher interstitial carbon contents there is a gradual decrease in the (111) components during recrystallization, whereas with low-solute carbon,

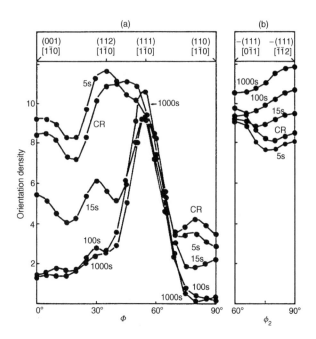

Figure 1.14   *Development of orientation density during the recrystallization of a vacuum-degassed steel annealed at 700°C (a) α fibre and (b) γ fibre (skeleton line) (Emren et al.[35])*

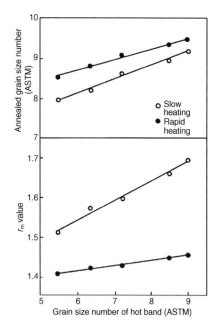

Figure 1.15   *Effect of hot band grain size on the recrystallized grain size and r_m value of aluminium-killed steel heated to 700°C at 100°C/h and held for 1 hour and at 50°C/s and held for 1.5 min (Ono and Nishimoto[36])*

the initial decrease in the (111) components is compensated for by a subsequent increase especially in the (111)[112] component during the later stages of recrystallization.[37] The influence of the dissolved carbon is mainly to increase the proportion of the minor components with a consequent decrease in the proportion of the important (111) γ fibre components that lead to good drawability. The higher carbon content is also accompanied by a finer grain size.[37] Nitrogen in solution also leads to low r values a fine grain size. The effect of nitrogen in solution is, therefore, similar to the effect of carbon in solution.[20] It is usually necessary to ensure that the solute carbon and nitrogen is reduced to a low level in the hot band structure and during recrystallization itself, if high r values are to be obtained by cold rolling and annealing. The development of high r values in batch-annealed, aluminium-killed steel, however, as further discussed below, has a different requirement.

Cold-forming behaviour

Cold formability and strength, as indicated above, represent the two most important requirements for low-carbon strip grade steels. For many applications, the main requirement is to be able to form the part without splitting, necking or wrinkling. The most suitable steel, therefore, is one with a low strength and high formability. For structural applications, the strength of the steel is more important and must be above a given minimum value. It is found, however, that there is a general tendency for the cold formability of any type of steel to reduce as the strength increases. The reduced formability of higher-strength steels tends, therefore, to limit their use to those applications which do not require the very highest formability. Many of the developments of higher-strength steels, therefore, have been specifically aimed at providing higher strength while minimizing the loss in formability that would otherwise have taken place.

It is now generally accepted that, for many applications, the cold formability of sheet steel may be resolved into two separate but related components, namely its drawability and its stretchability. The drawability of a steel is its ability to be drawn in to make a component without local necking or splitting whereas the stretchability of a steel is its ability to be stretched to form a component, again without local necking or splitting. Stretching involves major and minor strains in the plane of the sheet that are both positive, whereas drawing involves major and minor strains, one of which is positive and one of which is negative. Many applications, however, involve both drawing and stretching. Success in forming for such applications involves, therefore, the factors which influence both drawing and stretching.

A simple measure of the deep drawability of a steel may be obtained by forming flat-bottomed cylindrical cups from circular blanks, as illustrated in Figure 1.16(a). The maximum ratio between the circular blank diameter and the punch diameter that may be drawn in a single stage to form a cylindrical cup without necking or splitting is a measure of the drawability of the steel. This ratio is called the *limiting drawing ratio* and may vary up to about 2.5 for very good deep-drawing steels.

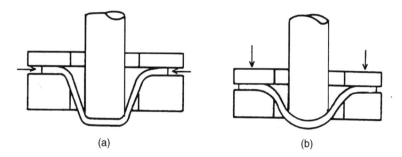

Figure 1.16   *(a) Deep drawing; (b) stretch forming*

A simple measure of the stretchability of a steel may be obtained by using a hemispherical punch to form a circular dome, as indicated in Figure 1.16(b). The flange of the circular blank is prevented from being pulled in by the use of a draw bead or by using sufficient blank holder pressure. The maximum ratio of dome height to dome diameter at the moment of necking or splitting is a measure of the stretchability of the steel. A similar test, called a *hydraulic bulge test*, may be used, employing oil under pressure to form the dome. In this case, the result would not be influenced by friction. In these tests in which the shape of the dome is circular, the strain at the top of the dome is almost the same in two directions at right angles. The strain, therefore, is said to be balanced biaxial strain.

Work-hardening coefficients and normal anisotropy

The formability of sheet steels may also be assessed using parameters that may be measured directly from a conventional tensile test, provided suitable length and width extensometers are available. The first parameter is the strain ratio which was originally devised by Lankford and others[38] and is usually called the *r* value. It gives a measure of the drawability of the steel but also gives a measure of the resistance to thinning resulting from the orientation of the slip systems that are active during drawing. The second parameter is the work-hardening coefficient, designated the *n* value which is closely related to the stretchability of the steel. The uniform and total elongation measured in a tensile test are also related to the stretchability of a steel, but it is necessary to introduce the concepts of true stress and true strain before the *r* value and the *n* value may be defined precisely.

True stress and true strain

The stress used in a conventional tensile test, often called the *engineering stress*, is defined as the load at any moment during the test divided by the original cross-sectional area of the test piece. During the test, the load increases up to the point of maximum load which defines the tensile strength of the material and then decreases as the specimen undergoes local necking. The true stress, however, is defined as the load at any moment during the test divided by the current cross-sectional area at the same moment. Thus, as illustrated in Figure 1.17, the true

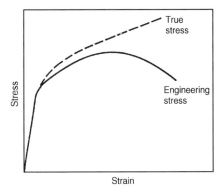

Figure 1.17   *Stress–strain diagrams*

stress continues to increase with increasing strain and represents the actual stress on the specimen at every moment during the test.

The engineering strain during a test is defined as the change in length divided by the original length. The concept of true strain has been developed by regarding any overall strain as being made up of a series of small elements of strain each related to the length of the specimen at the start of each element. Thus if an overall engineering strain $(L_3 - L_0)/L_0$ were to be regarded as made up of three elements, the true strain ε would be given by the sum of the three elements. Thus:

$$\varepsilon = (L_1 - L_0)/L_0 + (L_2 - L_1)/L_1 + (L_3 - L_2)/L_2$$

In the limit as each element becomes very small and the number of elements becomes very large:

$$\varepsilon = \int_{L_0}^{L} \frac{\delta L}{L}$$

and

$$= \ln L/L_0$$

Thus, the true strain is equal to the natural logarithm of the ratio of the final length to the original length. With this concept, any material that is extended to give a particular true strain would need the same numerical value of true strain but of opposite sign to restore it to its original length. Clearly, the engineering strains to achieve the same result would not have had the same numerical value. It is also useful to note that if, for example, any steel is given several separate strains to make up a complete strain, the total true strain would be equal to the sum of the separate true strains, whereas the total engineering strain would not be equal to the sum of the individual engineering strains.

The strain ratio r or Lankford value

The definition of the strain ratio r or Lankford value[38] depends on the observation that, as a tensile test proceeds, the gauge length of the sample becomes both thinner and narrower. The strain ratio r is defined, therefore, as the ratio of the true strain, ε_w, in the width direction to the true strain, ε_t, in the thickness direction at a particular moment during the test, usually when the engineering strain is either 15 or 20%. Thus:

$$r = \varepsilon_w/\varepsilon_t$$

In practice, it is usual to assume that the sample retains its original volume during the test which means that the sum of the true width, thickness and length strains would be equal to zero. Only the width w_f, therefore, needs to be measured if the length-engineering strain is fixed at either 15 or 20%, corresponding to true length strains of either 0.1398 or 0.1823. The final formula then becomes, for 20% engineering strain:

$$r = \frac{\ln w_f/w_0}{(\ln w_0/w_f - 0.1823)}$$

The strain ratio usually has different values for different directions in the plane of any sheet. A mean r value, r_m or $\bar{r}$, may, therefore, be defined by measuring a value in the rolling, transverse and diagonal directions and calculating a mean giving the diagonal value double weight as follows:

$$r_m = (r_0 + 2r_{45} + r_{90})/4$$

where the subscripts refer to the angle of the test direction to the rolling direction.

The variation in r value around the rolling plane is also important and a a Δr value is defined as:

$$\Delta r = (r_0 - 2r_{45} + r_{90})/4$$

The main structural feature influencing the r value is the crystallographic texture, but the relationship between r value and crystallographic texture is a complicated one. It is useful to note here, however, that a high proportion of grains with (111) planes parallel to the surface, which is the γ fibre texture component mentioned in the previous section, leads to high r values. A high proportion of grains with (100) planes parallel to the surface, which is one component of the α fibre texture, tends to lead to a low r value. Various empirical relationships between the strength of texture components and r value have been established. One of the most simple relationships is illustrated in Figure 1.18, which shows a substantially linear relationship between the r_m value and the ratio of the strengths of the (111) and (100) texture components.

The r_m value is important because it affects the distribution of strain in a pressing by influencing the thinning tendency and gives a measure of the deep drawability of the steel. This is illustrated in Figure 1.19, which gives a plot of r_m value versus limiting drawing ratio for a number of steels drawn to form cylindrical cups. There is a linear correlation between the r_m value and the limiting drawing ratio of the steel. This arises partly because the flange being drawn in to form the wall of a cup exhibits less thickening for a high r_m value steel than for a low r_m value steel, as illustrated in Figure 1.20. The r_m value also influences the strain distribution during stretching, as illustrated in Figure 1.21,

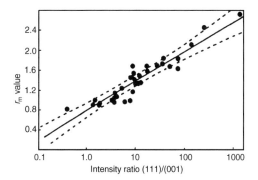

Figure 1.18 *Relation between the ratio of the intensity of the (111) component to the intensity of the (001) component and the r_m value of low-carbon steel sheets (After Held[39])*

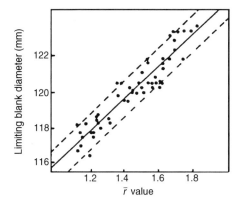

Figure 1.19 *Effect of $\bar{r}$ on limiting blank diameter for a range of low-carbon steels for Swift cups drawn using polythene sheet lubrication (After Atkinson and Maclean[40])*

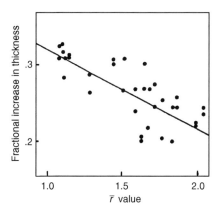

Figure 1.20 *Relationship between mean fractional increase in thickness at the top rim of a Swift cup and $\bar{r}$ value, for a range of low-carbon steels, Blank diameter 63.5 mm – Punch diameter 32 mm (After Hudd and Lyons[41])*

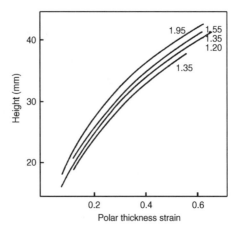

Figure 1.21   *Relationship between bulge height and polar thickness strain in sheets with different r̄ values. One 1.35 r̄ material had a lower uniform elongation than the remaining materials (After Horta et al.[42])*

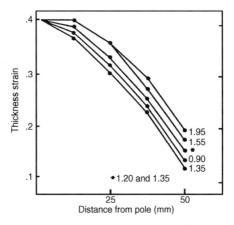

Figure 1.22   *Radial distribution of thickness strain, measured directly, in sheets with different r̄ values bulged to give a polar thickness strain of 0.4. One 1.35 r̄ material had a lower uniform elongation than the remaining materials (After Horta et al.[42])*

which gives a plot of the polar thickness strain developed in circular bulge tests of progressively increasing height using steels with different r_m value. It is seen that for a given bulge height, the polar thickness strain is lower for a high r_m value steel than for a low r_m value steel. The reason is, as shown in Figure 1.22, that the strain is more uniformly distributed across the surface of the bulge for the high r_m value steel than for the low r_m value steel. This type of influence of r_m on strain distribution is common to all pressings.

The Δr value is important because its ratio with the r_m value determines the height of ears that would be obtained in cups prepared using a cylindrical punch and a circular blank. The ear height is defined as the mean distance between the

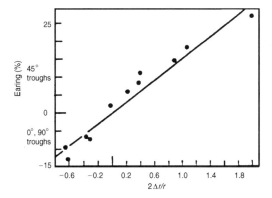

Figure 1.23   *The relationship between earing and Δr (After Wilson and Butler[43])*

high parts of the rim and the bottom of the troughs. The relationship between $\Delta r/r_m$ and earing is illustrated in Figure 1.23. For this figure, ears in the rolling and transverse directions are regarded as positive earing and ears in the diagonal direction are regarded as negative earing. The punch and blank diameters (P and B) used to make a cup also influence ear height and for a given material, the ear height is given with reasonable accuracy by the following equation:[44]

$$\text{Ear height} = M \cdot (B^2 - P^2)(B - P)^{1.5}/P^{2.5}$$

where M is a dimensionless earing parameter related mainly to the r and Δr values of the steel but also influenced a little by cup drawing conditions. Ears are clearly more noticeable on a simple cup than on a more complicated pressing, but the equivalent effect on strain distribution is obtained on all pressings, unless the steel is specially manufactured to give a low earing tendency.

The work-hardening coefficient n

Any stress/strain curve may be regarded as approximately conforming to an equation of the type:

$$\sigma = k\varepsilon^n$$

where σ and ε are the true stresses and the true strains measured throughout a tensile test as defined above and n is a constant. It follows that:

$$\log_{10} \sigma = n \cdot \log_{10} \varepsilon + \log_{10} K$$

The coefficient n is, therefore, defined as the slope of a true stress – true strain curve plotted on a logarithmic scale. It determines, therefore, how quickly the true stress builds up with increasing true strain and is, therefore, referred to as the *work-hardening coefficient*. In practice, it is found that the above logarithmic plot is not always a perfect straight line. The n value is then said to vary with strain and more than one value is sometimes quoted from a single tensile test, corresponding to different ranges of strain.

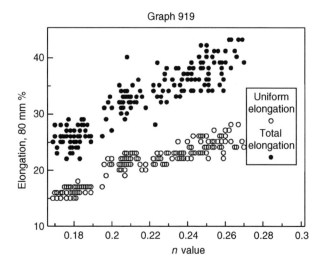

Figure 1.24   *A plot of total and uniform elongation versus* n *value for several cold-rolled, high-strength steels continuously annealed on an experimental line using a range of annealing cycles (without temper rolling)*

There is a close relationship between the n value of the steel and both the uniform and the total elongation measured in a tensile test. This relationship is illustrated for several types of high-strength steel in Figure 1.24. It is useful to note that if the n value were to be truly constant throughout a tensile test, the n value would be numerically equal to the true strain at maximum load which is the uniform elongation.

The most important factors that influence the n value of a steel are its strength and the strengthening mechanisms used to develop the strength. This is illustrated in Figure 1.25. It is seen that for each type of steel the n value decreases with increasing strength and that, for example, solid solution strengthening gives a lower loss in n value per unit increase in strength than strengthening by precipitation effects and grain refinement. It is clear that the highest possible n value is obtained when the strength of any type of steel is as low as possible and this is the reason why low-strength mild steels are always more formable than higher-strength steels. As mentioned previously, much emphasis has been given during the development of higher-strength steels to choosing or developing strengthening mechanisms that give the highest possible n value (and hence elongation) for the strength needed.

The n value is important because it is the second of the two main material parameters that determine the distribution of strain across any pressing. In general, a high n value for a steel leads to a more uniform strain distribution than a low n value steel. A higher n value enables the forming strains, therefore, to be distributed across a pressing with less likelihood of local necking and failure than for a low n value.

The strain distributions obtained across a simple hemispherical bulge may be used to illustrate the effects of n value on strain distribution. Figure 1.26 shows

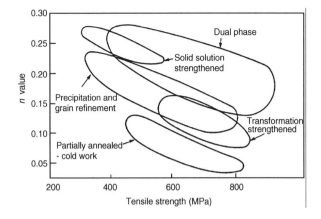

Figure 1.25 *Relationship between* n *value and tensile strength (After Dasarathy and Goodwin*[45]*)*

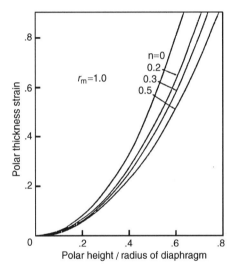

Figure 1.26 *Variation of calculated polar thickness strain versus polar height for different values of* n *for* $r_m = 1$ *(After Wang*[46]*)*

that the polar thickness strain for a given bulge height decreases as the *n* value increases. The implication is that the strain becomes more uniformly distributed as the *n* value increases. Figure 1.27(a) shows that the polar thickness strain at instability increases with increasing *n* value and the result is that the maximum bulge height at instability also increases with increasing *n* value (Figure 1.27(b)). As with r_m value, equivalent effects are obtained with the more complicated pressings that are needed across the engineering industry. In practice, it is found that success with any particular pressing may be dependent on a high value of one or other of the parameters *r* or *n*, whereas success with another pressing may depend on high values of both parameters.[47]

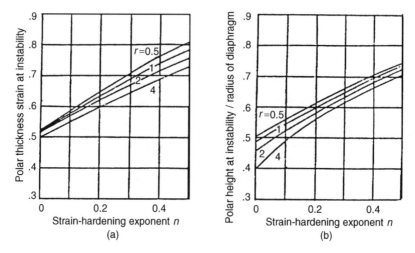

Figure 1.27   *Calculated (a) polar thickness strain (absolute value) and (b) polar height at instability versus strain-hardening exponent* n *for several values of* r *(After Wang[46])*

Forming limit diagrams

In 1963, Backhofen and Keeler[48] introduced the concept of a forming limit diagram (FLD) which gives a plot of the major and minor strains that may be obtained by deformation. This diagram enables the strains that are obtained across any commercial pressing to be compared with the maximum strains that could be sustained by the material without necking or splitting. These maximum strains are defined by a line called the forming limit curve which separates the strains that would give an unsatisfactory pressing from those that would give a satisfactory pressing. The forming limit line for any material may be measured experimentally, as indicated below, and its position on the diagram depends on the properties of the steel and its thickness. It usually has the general shape illustrated in Figure 1.28. The right-hand side of the diagram covers stretching when both the major and the minor strains are positive, whereas the left-hand side covers drawing for which, as mentioned previously, one of the strains is negative.

The strains to be plotted on a forming limit diagram may be measured by first etching a grid pattern, consisting of an array of circles about 2 mm in diameter, onto the steel surface prior to forming. This is done using a stencil and an electro-chemical or photographic marking technique. On forming, the circles change in size and shape and usually become elliptical, depending on the local strain at each point over the pressing. The circles and ellipses plotted in Figure 1.28 indicate the nature of the strain produced by deformation in different areas of the diagram. The major component of strain plotted vertically is obtained by measuring the long axis of the ellipse, whereas the other component is obtained from the minor axis. A comparison of strains on a component may be made with those for the forming limit curve. It is usually considered that if the plotted strains are well away from the forming limit line, the pressing is likely to be made without risk of necking or splitting. If the strains are close to the forming limit line, however,

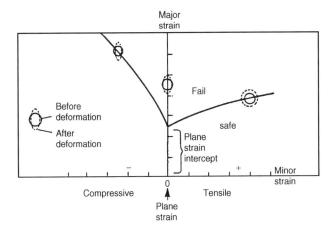

Figure 1.28 *Forming limit diagram*

there would be the risk that some pressings would be unsatisfactory. It would then be necessary to adjust the press, modify the lubrication, modify the blank shape, clean or modify the tools or use a more formable steel, to ensure that the pressings would become consistently satisfactory. Clearly modifying the tools would only be an option during the development and die try-out stages of a pressing before the design itself has become fixed or if the overall effect does not change panel shape.

The forming limit diagram itself may be established by subjecting a series of samples of the material to be used to a range of forming tests to give failure involving different ratios of major and minor strains, including negative ones. The strains close to the failure points are measured using the grid marking technique and plotted on the diagram in the normal way to establish the forming limit line itself.

It is useful to note, as illustrated in Figure 1.29, that the plane strain intercept on the major strain axis of a forming limit diagram is a function of the sheet thickness and the work-hardening coefficient n. Thus higher-strength steels with lower n values have a lower forming limit curve than lower-strength steels. It has also been found that the presence of macro inclusions reduces the forming limit.[49] In addition, fine elongated sulphide inclusions may also reduce the forming limit, but a chemical addition to modify the inclusion shape to make them globular may restore the forming limit to a higher level.[50]

Other forming effects

The above remarks have concentrated primarily on the factors that influence the possibility of local necking or splitting during the formation of any pressing. It should be remembered, however, that there are other factors that may lead to a pressing being unsatisfactory. These include wrinkling, hole expansion effects and constancy of shape, sometimes known as shape fixability which is related to springback.

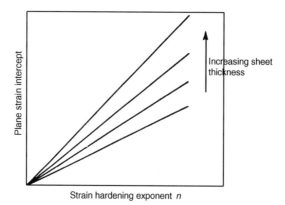

Figure 1.29   *Relationships between the plane strain intercept on a forming limit diagram and the strain hardening exponent as a function of sheet thickness*

Springback

Springback occurs on forming whenever the component is withdrawn from the tool set. The forces used to make the pressing are then no longer applied and the pressing relaxes in an elastic way to minimize its residual internal stresses. In general, the plastic strains introduced into many pressings are not uniform over the pressing. The result is that the elastic relaxation is not uniform and the shape of the pressing will not be exactly the same as the shape of the punch that has been used in its manufacture. The change in shape is known as springback and it clearly varies widely for different pressings. The most important material factor affecting springback is the yield stress, but gauge and work-hardening coefficient also have an influence. It is clearly important that all examples of the same pressing should have a uniform shape in order that they fit together easily with other pressings when they are assembled into complete structures. It is important, therefore, that each sheet of the material used should have closely the same yield stress, gauge and work-hardening coefficient. Press conditions may, however, be adjusted to minimize the effect of any property variation.

Buckling, distortion and wrinkling

Sometimes, when there are large strain gradients across a pressing or when there is some other critical condition, the transfer of stress across a pressing may cause buckling, or the relaxation of the elastic stresses introduced during forming may cause distortion. The extent of these effects is very dependent on the nature of the pressing.

Wrinkling consists of corrugations produced mainly near the flanges of a pressing when the material is subject to in-plane compression. Wrinkling would make a pressing unsuitable for any exposed applications, but the existence of wrinkling on internal flanges makes the process of spot welding difficult. Wrinkling may be controlled by adjusting press conditions but the main material parameters that affect the tendency are yield stress and gauge. The tendency for wrinkling increases with increasing yield stress and decreasing gauge.

Hole expansion effects

Many pressings are made with holes to accommodate screws for fixing and the forming operation often involves an increase in the size of the punched hole after it is made. An example would be a wheel centre for a motor vehicle which usually has four holes for securing the wheel in position. It is found that different steels have a tendency to split as the hole is expanded. They are said, therefore, to have different edge ductility. A hole expansion index has, therefore, been devised to give a measure of the quality of the steel in this respect and represents the percentage increase in the size of a hole at the moment that a crack forms.

It is found that the hole expansion index is often influenced by the presence of elongated sulphide inclusions, but the index may easily be improved by reducing the sulphur content and/or by making additions to the steel to modify the sulphide shape. Steels with certain types of microstructure are, however, more prone to cracking at cut edges than steels with other types of microstructure. Thus, for a given strength, a ferrite – martensite structure gives a lower hole expansion value than a ferrite – bainite structure.

Bending

Many components are made by a bending operation, for example simple brackets or tubes and the minimum bend radius that may be produced without cracking on the outside of the bend are often regarded as measures of the bendability of the steel. Low-strength steels with high *n* values are clearly more bendable than high-strength steels. It is found, however, that for a given strength, the minimum bend radius is again influenced by the presence of elongated sulphide inclusions.

Recently, new applications have required that the bendability of ultra-high-strength steel is optimized for tube-making processes. It was found[51] for three steels that the elongation as measured in a tensile test increased, as expected, with decreasing gauge length but that the steel with the lowest elongation, measured over a gauge length of 50 mm, had the highest elongation measured over a gauge length of 2 mm, as illustrated in Figure 1.30. The important observation was that the minimum bend radius correlated well with the 2 mm elongation (the local elongation) rather than with the 50 mm elongation. Other work showed that a homogeneity index, defined as the standard deviation of hardness over a specimen, correlated well with the local elongation (Figure 1.31) and confirmed that the minimum bend radius correlated with the local elongation (Figure 1.32). It was clear, therefore, that the best bendability is obtained from steels with the most homogeneous microstructure with the least variation in hardness throughout the steel. For the steels studied, there was no correlation between either the *n* value or the total elongation measured using a 50 mm gauge length and the minimum bend radius that would just avoid splitting.

Delayed fracture

Another phenomenon which may be important for certain applications is known as delayed fracture.[53] This is caused by hydrogen absorbed during corrosion and increases when the tensile strength is above 1170 N/mm². Delayed fracture

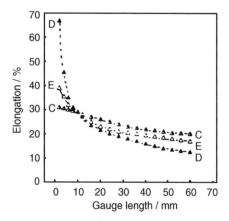

Figure 1.30 *Variations of elongation with gauge length for three ultra-high-strength steels (After Iwaya et al.[51])*

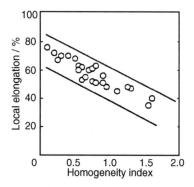

Figure 1.31 *Relationship between homogeneity index and local elongation for ultra-high-strength steel (After Yamazaki et al.[52])*

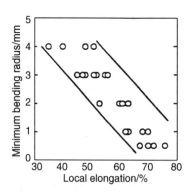

Figure 1.32 *Relationship between local elongation and minimum bending radius for ultra-high-strength steel (After Yamazaki et al.[52])*

resistance may be improved by reducing the carbon content and by controlling the size and morphology of carbides.[54]

Stretcher strain markings

Temper rolling serves several functions. As mentioned previously it imprints a surface roughness onto the steel surface and may improve the steel flatness, but it also removes the yield point elongation. It is found, however, that a number of steels exhibit a phenomenon called strain ageing, whereby the yield point returns on storage at room temperature due to the diffusion of carbon or nitrogen in solution onto free dislocations. Normally, it is considered that a visible marking would occur on the surface of a pressed component if the yield point elongation increases above about 0.3%. These markings are called *stretcher strain markings* and would lead to the rejection of the pressing if the markings are on the exposed surface of, for example, a motor vehicle body component. The markings are avoided if the strain ageing rate is sufficiently low or if the steel is not stored for very long periods before pressing. It will be seen below that most low-strength strip steel grades are guaranteed not to exhibit stretcher strain markings on forming for at least six months after delivery.

Surface roughness

The final surface roughness is imprinted onto the steel surface by the temper mill rolls as indicated in the previous section. The roughness is usually defined by the average peak height and the peak count per inch across the surface and typical values to use would be $1-2$ μm and 200 peaks per inch, depending on the application. The surface of the temper mill rolls may be roughened using one of four different methods,[55] namely shot blasting, spark erosion, using a pulsed laser (the lasertex process) or electron beam roughening, but each method produces a different type of surface.

The roughness is important because it influences the way in which forming lubricant is held on the surface and has, therefore, a direct influence on formability as a result of frictional effects. The roughness also has an influence on the appearance of a steel surface after painting.

Use of laser-welded tailored blanks

Traditionally, each body panel of a motor vehicle is press formed from a single sheet of steel cut to a suitable blank size and shape prior to forming. The implication was that each part of a pressing was made from the same grade of steel with the same strength and thickness. The result has been that certain parts of a pressing may have been stronger than is strictly necessary or additional strengthening members may have been needed to provide strength in parts of a pressing that may otherwise have been too weak.

This problem may now be overcome by making blanks from two or more pieces which have been laser welded together before forming to make what have become known as *tailored blanks*. The separate pieces may have different strength, formability, thickness or coating, depending on the different requirements for

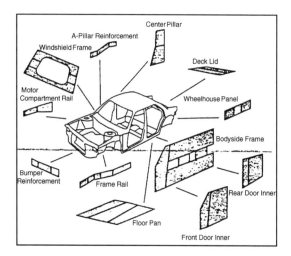

Figure 1.33   *Typical tailored blank automotive applications (After HSS Bulletin[56])*

the different areas of the pressing. With this technique, it is possible to reduce the weight and cost of a complete structure since each area of a pressing may be designed specifically and economically in relation to the required performance and the need for additional reinforcement or stiffening members is reduced. A number of typical automotive applications are illustrated in Figure 1.33.

Hydroforming

This is a technique for producing elongated hollow components from a steel tube. A tube, which may be either straight or bent, is first placed in a forming die. The tube is then filled with a liquid (usually a water-based emulsion) and then pressurized which causes the tube to expand and conform to the internal shape of the die.

The technique may be used to produce complicated hollow shapes that could not previously be made in one piece by a pressforming technique. Weld seams are, therefore, reduced in number. The method provides high precision in shape and may give reduced weight or higher torsional rigidity compared with a similar component manufactured by welding several subcomponents together.[57]

The method is starting to be used in the automotive industry for the manufacture of structural components that would previously have been made as box sections welded together from pressformed components. Applications so far have included engine cradles, lower and upper longitudinal body rails and various body crossmembers.[58]

The use of the hydroforming technique is now being extended to cover the forming of flat sheets as well as tubes.

Roll forming

Roll forming is a technique for continuously converting flat steel into a final profile by passing it through a series of suitably designed and contoured rolls

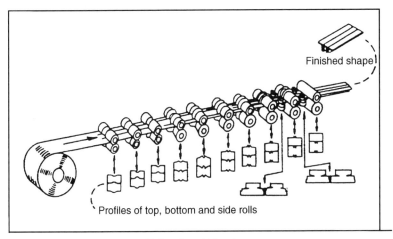

Roller die cold forming

Figure 1.34 *A schematic diagram showing the formation of a shape from a coil of raw material to finished configuration (After Ferry[59])*

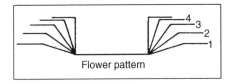

Flower pattern

Figure 1.35 *A flower pattern showing the progressive change of shape by rolling to form a top hat section (After Cain[60])*

which incrementally change the shape until the final shape is obtained. Most of the deformation is by bending, with the result that very little change in thickness occurs. Narrow strips may be converted into structural sections, whereas wide strips may be converted into profiled sheets for cladding applications mainly for building applications for which the profile imparts increased rigidity and load-bearing capability.

Figure 1.34 gives an example of the profiles of top bottom and side rolls that are required to give a particular structural section, whereas Figure 1.35 illustrates the progressive change in shape through each roll as a strip is roll formed to give a top hat section. This type of diagram is known as a *flower pattern*. Some of the profiles that are used for building applications are illustrated in Figure 1.122 at the end of this chapter.

Finite element modelling

Although the FLD provides a valuable guide to forming behaviour and can assist in component design, the approach is still empirical. Additionally, it has limitations by virtue of the fact that the strain paths in commercial pressings may be

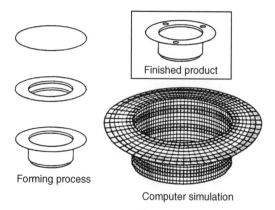

Finished product

Forming process

Computer simulation

Figure 1.36   *Computer simulation of deep-drawing operations (After Miles[61])*

significantly different from those encountered in the simple laboratory tests that are used to construct the FLD. There has, therefore, been a move to eliminate the empiricism involved in the approaches described above through the use of finite element modelling. It is now possible to simulate a wide variety of different forming processes by computer-aided engineering and Miles[61] has described the use of this technique in drawing operations on low-carbon steel. This is illustrated in Figure 1.36 which shows the computer simulation of a deep-drawn cup. The purpose of this particular experiment was to predict the tool forces required to form the component, the hold-down pressure on the blank and the tearing tendency of the sheet if the friction at the blank holder stopped the material from drawing correctly. Using this approach, it is also possible to identify areas of thinning and thickening in the sheet and where excessive compression might lead to wrinkling.

The computer programs that have now been generated for this type of work are extremely powerful and versatile and permit the analysis of a wide variety of three-dimensional shapes. The effects of planar and normal isotropy can also be accommodated.

Strip steel manufacture

The strip steels supplied for cold-forming applications may be divided into two main groups, namely those for which the main requirement is the provision of a defined minimum level of formability and those for which the main requirement is a defined minimum level of strength with a more limited level of formability. The former may be called *mild steels* whereas the latter may be called *high-strength steels* though in a few cases there may be little difference in strength between the two. Many of the steels available are covered by standards, including coated steels, but some of the most recently developed materials are not yet covered. It is also common, particularly with a formability requirement, for a specification that is tighter than the standard specification to be agreed between a customer and a supplier in order that a particularly difficult customer requirement can be satisfied.

It is convenient to present information on mild steels and higher-strength steels in separate sections even though they may be closely related. Metallic-coated steels and organic-coated steels are also covered in separate sections.

Mild or low-strength steels

BS 1449 was the definitive specification in the UK for carbon and carbon–manganese steels in the form of plate sheet and strip. The last version of this specification – BS 1449 Section 1.1: 1991 – is still in force and provides brief details of both hot-rolled (HR) and cold-reduced (CR) grades. A summary of this information is given in Table 1.2. It is understood, however, that BS 1449 should only be used to specify hot-rolled grades and this specification will eventually be withdrawn in favour of European standards.

The cold-rolled and annealed carbon steel flat products supplied for cold forming in Europe are covered by European standard EN 10130: 1991 and designated Fe P01 to Fe P06 but with Fe P02 missing. The most important feature of these grades is their formability, with grade Fe P06 having the highest formability and Fe P01 the lowest formability as indicated in Table 1.3. The higher grades, therefore, have stringent requirements for r value and n value. The other main difference between the grades concerns the acceptable strain ageing limit. The Fe P06 grade is regarded as being completely non-ageing, whereas a small degree of ageing is regarded as acceptable for grades Fe P03 to 5, provided the steel remains suitable for use after storage for up to six months. An even higher degree of ageing is acceptable for grade Fe P01.

Each grade may be obtained from either continuous or batch annealing, but the chemistry and prior processing needed to provide the necessary properties from the two methods are usually, but not always, different.

There are two main types of mild steel. The first is the more traditional steel described as *aluminium-killed* (AK) steel in which the carbon is relatively free

Table 1.2   *BS 1449: 1991 Section 1.1 (General specification)*

Grade	Rolled condition	Quality	C% max.	Mn% max
1	HR	EDDAK	0.08	0.45
	CR			
2	HR	EDD	0.08	0.45
	CR			
3	HR	DD	0.1	0.5
	CR			
4	HR	D or F	0.12	0.6
	CR			
14	HR	FL	0.15	0.6
15	HR	C	0.2	0.9

Key: EDDAK, extra deep drawing, Al-killed; EDD, extra deep drawing; DD, deep drawing; D or F, drawing or forming; FL, flanging; C, commercial.

Table 1.3 BS EN 10130: 1991 Cold-rolled carbon steel flat products for cold forming

Grade	Definition and classification according to EN 10020	Deoxidation	Validity of mechanical properties [1]	Surface appearance	Absence of stretcher strain marks	R_e N/mm²	R_m N/mm²	A_{80} % min. [3]	r_{90} min. [4,5]	n_{90} min. [4]	Chemical composition (ladle analysis % max.)				
											C	P	S	Mn	Ti
Fe P01 [6]	Non-alloy quality steel[7]	Manufacturer's discretion	–	A	–	–/280[8,10]	270/410	28			0,12	0,045	0,045	0,60	
			–	B	3 months										
Fe P03	Non-alloy quality steel[7]	Fully killed	6 months	A	6 months	–/240[8]	270/370	34	1,3		0,10	0,035	0,035	0,45	
			6 months	B	6 months										
Fe P04	Non-alloy quality steel[7]	Fully killed	6 months	A	6 months	–/210[8]	270/350	38	1,6	0,180	0,08	0,030	0,030	0,40	
			6 months	B	6 months										
Fe P05	Non-alloy quality steel[7]	Fully killed	6 months	A	6 months	–/180[8]	270/330	40	1,9	0,200	0,06	0,025	0,025	0,35	
			6 months	B	6 months										
Fe P06	Alloy quality steel	Fully killed	6 months	A	no limit	–/180[9]	270/350	38	$\bar{r}$min.[4,5] 1,8	$\bar{n}$ min.[4] 0,220	0,02	0,020	0,020	0,25[11]	0,3
			6 months	B	no limit										

Notes:

1 The mechanical properties apply only to skin-passed products.

2 The values of yield stress are the 0.2% proof stress for products which do not present a definite yield point and the lower yield stress R_{eL} for the others. When the thickness is less than or equal to 0,7 mm and greater than 0,5 mm the value for yield stress is increased by 20 N/mm². For thickness less than or equal to 0,5 mm the value is increased by 40 N/mm².

3 When the thickness is less than or equal to 0,7 mm and greater than 0,5 mm the minimum value for elongation is reduced by 2 units. For thickness less than or equal to 0,5 mm the minimum value is reduced by 4 units.

4 The values of r_{90} and n_{90} or $\bar{r}$ and $\bar{n}$ (see annexes A and B) only apply to products of thickness equal to or greater than 0.5 mm.

5 When the thickness is over 2 mm the value for r_{90} or $\bar{r}$ is reduced by 0,2.

6 It is recommended that products in grade Fe P01 should be formed within 6 weeks from the time of their availability.

7 Unless otherwise agreed at the time of the enquiry and order Fe P01, Fe P03, Fe P04 and Fe P05 may be supplied as alloy steels (for example with boron or titanium).

8 For design purposes the lower limit of R_e for grade Fe P01, Fe P03, Fe P04 and Fe P05 may be assumed to be 140 N/mm².

9 For design purposes the lower limit of R_e for grade Fe P06 may be assumed to be 120 N/mm².

10 The upper limit of R_e of 280 N/mm² for grade Fe P01 is valid only for 8 days from the time of the availability of the product.

11 Titanium may be replaced by niobium. Carbon and nitrogen shall be completely bound.

After BS EN 10130: 1991.

in the structure and may form iron carbides. The second is the newer type of steel called *interstitial-free* (IF) steel in which a strong carbide-forming element is present, such as titanium or niobium, to combine with some or all of the carbon usually to leave very little carbon in solid solution. The IF steels usually have an ultra-low-carbon content well below 0.005%, whereas the AK steels usually have a carbon content above 0.015%. The general chemical limits are given in Table 1.3. The more usual IF steels and the AK steels when batch annealed are almost completely non-ageing, whereas an AK steel may exhibit varying degrees of ageing after continuous annealing, depending on the chemistry and annealing conditions used. Some batch-annealed AK steels, however, will exhibit a small amount of room temperature strain ageing if they are intended to be bake-hardening steels.

An IF steel is the only type of steel that can satisfy the formability requirements of the Fe P06 grade, whereas a batch-annealed AK steel may be used to satisfy the formability requirements of the Fe P05 grade. A continuously annealed AK steel would only be expected to satisfy the formability requirements of this grade if very specific chemistry, hot rolling conditions and continuous annealing cycles are used. It would be more usual to use an AK steel only for the Fe P03 and P01 grades. It would also be more usual to use an IF steel to satisfy the requirements of the Fe P05 grade, if continuous annealing is the only method that is available.

Clearly, it is possible to use the chemistry and prior processing necessary for a higher grade to satisfy the requirements of a lower grade. This may be done either to reduce the number of independent steel chemistries that are needed, to be able to reduce the complexity of the annealing cycle, or to be sure that the property requirements would be easily met. Further information on the effects of chemistry and processing on the properties of mild steel are given in the following sections.

Low-strength, interstitial-free (IF) steels

Early work[62,63] showed that very low-carbon steels could develop high r_m and elongation values when sufficient titanium was added to tie up all the carbon and nitrogen and that the r_m value increased with increasing titanium content. Later work showed that it was also possible to obtain good r values by the addition of niobium,[64] and steels alloyed with titanium and niobium either separately or together are now in common use. Recent work has shown, however, that there may be some potential for IF steels containing vanadium.[65]

From the early 1970s, carbon contents close to 0.01% became readily available and it was then necessary to add sufficient alloying addition to be able to combine with all the carbon and nitrogen in the steel and to leave a small surplus in order that very good r values were obtained. This effect is illustrated in Figure 1.37(a) and (b) for both titanium and niobium additions.

More recently, with improvements in vacuum degassing techniques, ultra-low-carbon (ULC) contents below 0.003% have become easily available and with these ultra-low-carbon contents, it has been possible to lower the alloy addition while still achieving high mean r values. The original sharp rise in r value,

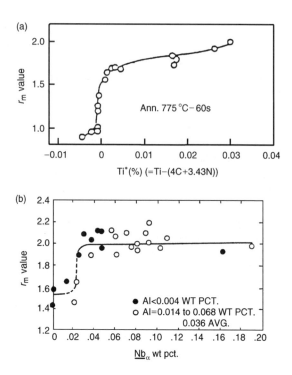

Figure 1.37 *Variation of r_m value with excess (a) titanium content and (b) niobium content (After Takahashi et al.[66] and Tokunaga et al.[67])*

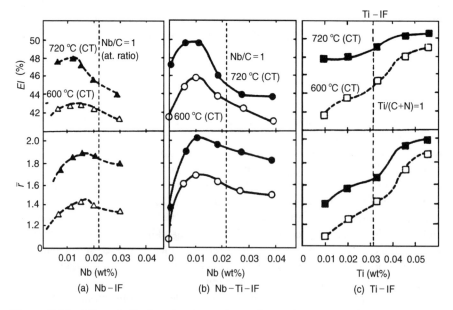

Figure 1.38 *Effect of alloying element on the mechanical properties of Nb-, Nb–Ti- and Ti–IF steels containing 20–30 ppm carbon and 26–35 ppm N. (After Tokunaga et al.[68])*

just above the stoichiometric composition for titanium steels, has become more gradual and the highest r values in certain niobium-bearing steels may be obtained with niobium contents close to 0.01% which is below the stoichiometric composition (Figure 1.38). A number of different combinations of niobium and titanium are also, however, in common use. Figure 1.38 also shows that the relationships between elongation and niobium or titanium content are similar to those relating r value to niobium or titanium content and that coiling temperature also has a marked effect on properties. There is, however, a strong relationship between hot band grain size and mean r value with a finer grain size leading to higher values (Figure 1.39). This is consistent with the effect of hot band grain size on final texture mentioned in a previous section.

Slab reheat temperature has a marked influence on elongation and r value with higher values coming from lower slab reheat temperatures, as illustrated in Figure 1.40. A high cold reduction (Figure 1.41) and a high annealing temperature (Figure 1.42) also lead to higher r values, but the properties are relatively insensitive to heating rate. This is the reason why ULC IF steels may be satisfactorily processed by both batch and continuous annealing.

Recent work[73] has highlighted the importance of sulphur with respect to the properties of steels containing titanium. This is because titanium carbo-sulphide is more stable than titanium carbide, with the result that the former compound can

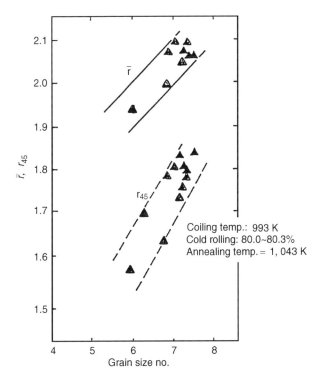

Figure 1.39   *Effect of grain size on the r values of IF steel sheets (After Kino et al.[69])*

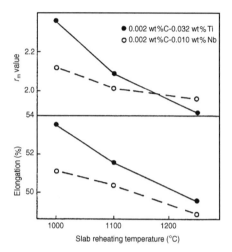

Figure 1.40   *Effect of slab reheating temperature on the mechanical properties of 0.002 wt% C steel containing 0.01 wt% Nb or 0.032 wt% Ti (Satoh et al.[70])*

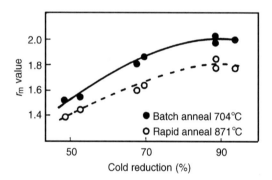

Figure 1.41   *Effect of cold recuction on the r_m values of titanium-stabilized steels for batch and continuous type annealing cycles (Goodenow and Held[71])*

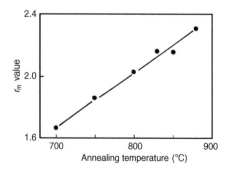

Figure 1.42   *Variation of r_m value of 0.002 wt% carbon–niobium steel with annealing temperature after annealing for 40 seconds (Satoh et al.[72])*

form during the hot-rolling sequence and influence the development of the hot-rolled texture.[74] The work showed that at the slab reheat temperature, titanium sulphide (TiS) forms and may transform into titanium carbo-sulphide ($Ti_4C_2S_2$) on cooling in the hot-rolling temperature range. This is accompanied by a much greater drop in the solute carbon content than would be possible in the absence of sulphur. It was considered[74] that the low-solute carbon content would promote the development of a desirable (111) texture in the hot band which would, in turn, lead to a more beneficial texture in the final cold-rolled and annealed product.

Batch-annealed, aluminium-killed steel

These steels are batch annealed using annealing temperatures close to 700°C.

The main elements which influence the properties are aluminium, nitrogen and carbon, with a minor influence of manganese. The optimum nitrogen and soluble aluminium contents are 0.005–0.01 wt% and 0.025–0.04 wt% respectively, although acceptable r values may be obtained just outside these limits, as illustrated in Figure 1.43. An increase in carbon content above and below the range causes an increase in strength. Normally with carbon contents close to 0.04%, the carbon that is taken into solution is reprecipitated completely during cooling mainly by diffusion onto the undissolved carbides. The nitrogen is combined as aluminium nitride and the steel becomes perfectly non-ageing as indicated previously. At lower carbon contents, the number of undissolved

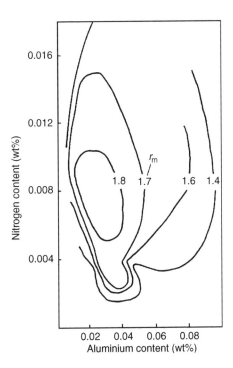

Figure 1.43 *Effect of finished sheet nitrogen and aluminium on the r_m values of batch-annealed, aluminium-killed steel (Gawne[75])*

carbides reduces and eventually becomes zero at about 0.02% carbon depending on the annealing temperature. Below this carbon content, the reprecipitation process becomes progressively more difficult since it involves a nucleation as well as a diffusion process and the reprecipitation process may remain incomplete. The result is a rise in strength and the steel ceases to be completely non-ageing. Such a steel exhibits bake hardening as discussed further below.

The most important metallurgical feature of the processing of an aluminium-killed steel to give high r_m values concerns the distribution of aluminium nitride. The main requirement is that the aluminium nitride is held in solution in the hot band structure prior to cold reduction. It can then lead to an enhancement of the (111) γ fibre texture components and to a decrease in the deleterious (100) texture components during annealing. At the same time a pancake grain structure develops. It is generally agreed that the beneficial effect is caused by the precipitation of aluminium nitride or the clustering of aluminium and nitrogen on sub-boundary sites during slow heating to the annealing temperature. This enables the beneficial (111) components to be strong components in the recrystallized texture. The effect is associated with an inhibition of recovery and subgrain growth and an increase in recrystallization temperature compared with a similar steel treated to precipitate the aluminium nitride prior to cold reduction. The maximum amount of inhibition of recrystallization does not, however, correlate with the optimum texture. The best texture, giving the highest r_m value, is obtained when there is just sufficient inhibition of recrystallization and when the drag on grain boundary movement as a result of aluminium nitride particles decays slowly as a result of particle coarsening.[76] The slow coarsening of the aluminium nitride particles leads to the preferential development of the (111) components and a coarser grain size due to restricted nucleation of other components.

Clearly, the optimum aluminium and nitrogen contents identified in Figure 1.43 provide the optimum amount of inhibition of recrystallization which decays at the optimum rate for normal processing conditions. It is generally agreed, as indicated by Hutchinson,[44] that the pancake grain structure is caused by the preferential precipitation of aluminium nitride on the prior cold-worked grain boundaries which provide an anisotropic barrier to grain boundary movement.

The aluminium nitride must be taken into solution at the slab reheat stage in order that it may be held in solution in the hot band structure prior to coiling and cold rolling.

For practical purposes, the solubility of aluminium nitride in austenite is given by:[77]

$$\log\,[Al][N] = -6770/T + 1.033$$

where T is the absolute temperature.

It may be shown, therefore, that relatively high slab reheat temperatures are required to take the aluminium and nitrogen into solution and the most commonly used temperature is 1250°C.

Relatively low coiling temperatures below 600°C are used to avoid substantial precipitation of aluminium nitride in the hot band, otherwise a deterioration in r_m

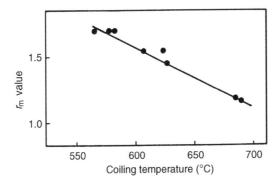

Figure 1.44   *Variation of r_m value of batch-annealed, aluminium-killed steel with coiling temperature, aluminium content 0.036% (Whiteley and Wise[78])*

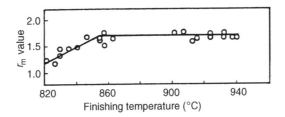

Figure 1.45   *Variation of r_m value with finishing temperature for batch-annealed, aluminium-killed steel coiled at 525–565°C (Parayil and Gupta[79])*

value would be obtained (Figure 1.44). Finishing temperatures must also be in the single-phase austenite region to avoid a similar fall in r_m value (Figure 1.45)

Manganese has a detrimental effect on r_m value, but the magnitude depends markedly on the carbon content (Figure 1.46). The highest r_m values tend to be provided by about 70% cold reduction and the heating rate during annealing also has an effect. Figure 1.47, for example, shows that there is a general tendency for the r_m value to decrease with increasing heating rate but that there are a series of optimum heating rates that provide higher r_m values than the general trend, depending on the aluminium content. These optimum heating rates also provide maxima in grain size and coincide with minima in yield stress. The optimum heating rate was shown to depend approximately linearly on the aluminium content,[81] but later work showed that it depended also on the nitrogen and manganese contents and on the cold reduction.[82] The equation below was found to give the optimum heating rate in °C/hour where [Al], [N] and [Mn] represent the weight percentages of these elements in solid solution and *CR* is the cold reduction:

$$\log (\text{optimum heating rate}) = 18.3 + 2.7 \log \, [\text{Al}][\text{N}][\text{Mn}]/CR$$

It is well known that an increase in temperature and/or time of annealing leads to a decrease in yield stress and an increase in elongation. Over the temperature range up to 700°C, the softening may be described by an apparent activation

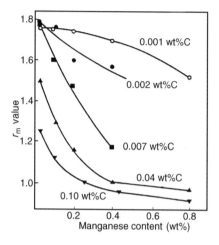

Figure 1.46   *Effect of manganese content on the r_m values of cold-rolled and annealed steels with various carbon contents, heating rate 100° C/h, held at 700° C for 1 hour (Osawa et al.[80])*

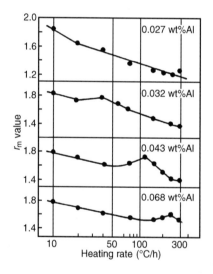

Figure 1.47   *Variation of r_m value with heating rate for batch-annealed, aluminium-killed steels containing different aluminium contents (Shimizu et al.[81])*

energy, Q, equal to 250 KJ/mole which corresponds approximately with that for the self-diffusion of iron.[83]

Continuously annealed, aluminium-killed steel

The developments in continuously annealed, aluminium-killed mild steels have been concerned primarily with matching the properties previously available from batch-annealed steels. It has been necessary, therefore, to compensate for the lack of time on a continuous line by changing the steel chemistry, prior processing and annealing temperature.

The general problem of matching the properties was resolved into three components:

1. Obtaining a suitably large grain size.
2. Obtaining a suitable orientation texture.
3. Minimizing the carbon and nitrogen in interstitial solid solution at the end of the process.

As with batch annealing, the steel must undergo recrystallization and grain growth during the annealing cycle. Carbon is inevitably taken into solution unless it is already combined as a stable carbide. It must, therefore, be substantially reprecipitated during cooling in order that the steel undergoes only relatively slow room temperature strain ageing to ensure that stretcher strain markings are avoided on subsequent pressing.

The grain size problem was solved by using a number of scavenging reactions involving mainly aluminium, nitrogen, carbon, manganese and sulphur to purify the ferrite matrix and by using a suitably high annealing temperature.[84] The texture problem was solved by the same means as the grain size problem and the carbon in solution problem was solved mainly by the incorporation of a 'so-called' overageing section into the cooling part of the annealing cycle. This usually consists of holding the steel at temperatures in the range 350–450°C for up to a few minutes or allowing the steel to cool slowly from such temperatures as illustrated previously in Figure 1.2.

Early work by Toda et al.[85] showed that controlling the manganese in relation to the sulphur content had a marked effect on grain size and hence properties. They defined a parameter denoted K by the equation:

$$K = [Mn] - 55/32\,[S] - 55/16[O]$$

where the symbols in square brackets represent the weight percentages of these elements in the steel. They observed that for rimming steel, there was a peak in grain size, elongation and r_m value when the k value was in the range 0–0.15%, as illustrated in Figure 1.48. In addition, there was a corresponding minimum in yield stress. Further work confirmed that a similar type of relationship applied to aluminium-killed steel, provided the term involving oxygen was removed since no oxygen would be available in such a steel to combine with manganese.

The effect of manganese was also dependent on the carbon content, as illustrated in Figure 1.49 which also shows that the detrimental effect of manganese was greatly reduced when the carbon in the steel was present as coarse carbides, arising from the use of a high coiling temperature such as 750°C.

The effect of carbon content is further illustrated in Figure 1.50, which shows that there is a minimum in yield stress at about 0.01–0.02% carbon which corresponds with a maximum in elongation. The maximum in yield stress below 0.01% carbon coincided with a maximum in ageing index and was clearly associated with additional carbon being retained in interstitial solid solution. The reason is that the tendency for reprecipitation during overageing depends on the

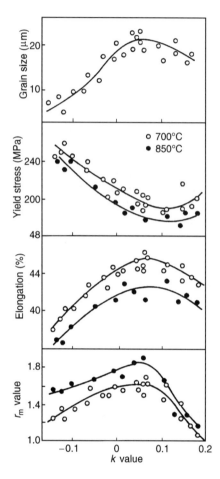

Figure 1.48   *Variation of yield stress, elongation, r_m value and grain size with K value for rimming steel subcritically and intercritically annealed (Toda et al.[85])*

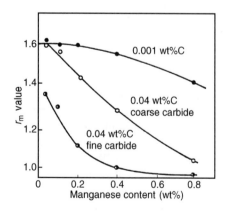

Figure 1.49   *Variation of r_m value with manganese content for steels with different carbon contents heated at 50°C/s (Osawa et al.[86])*

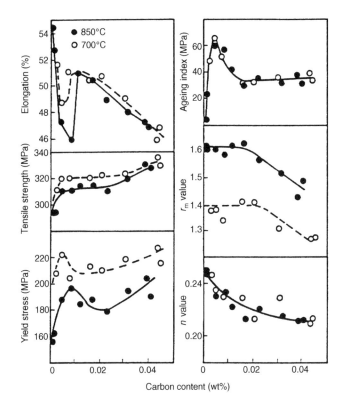

Figure 1.50   *Variation of the mechanical properties of aluminium-killed steel with carbon content, continuously annealed at 700 and 850°C (Ono et al.[87])*

carbon content at the start of overageing and will be discussed in more detail later. The lowest yield stress values and highest elongation values were obtained with ultra-low-carbon contents below 0.002% which also corresponded with the lowest ageing index. This was clearly due to the ultra-low total amount of carbon present in the steel, regardless of its position in the structure.

The grain size of continuously annealed, aluminium-killed steel increases with the coarsening of precipitates in the hot band which are mainly carbides and aluminium nitride. This may be achieved using high coiling temperatures, depending on the carbon content. Matsudo et al.[88] and Hutchinson[20] showed that higher r_m values are obtained with increasing carbide size due to the slower dissolution rate of coarse carbides during heating. This enables recrystallization to occur in a matrix that is relatively free from interstitial carbon.

The effect of nitrogen content on the properties of a rimming steel without an aluminium addition is illustrated in Figure 1.51, which shows an increase in yield stress with increasing nitrogen content. There is also a decrease in r_m value and grain size with increasing nitrogen content in aluminium-killed steel and a corresponding influence on yield stress depending on the coiling temperature used (Figure 1.52). High coiling temperatures lead to the combining of aluminium and nitrogen. The combined nitrogen is then no longer able to have any further

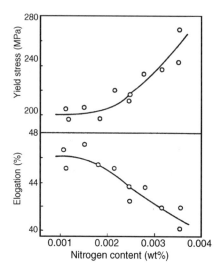

Figure 1.51   *Effect of nitrogen content on the yield stress and elongation of subcritically annealed rimming steel (Toda* et al.[85]*)*

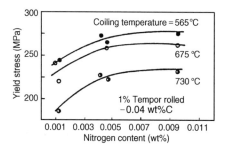

Figure 1.52   *Effect of nitrogen content on the yield stress of aluminium-killed steel continuously annealed at 760°C for different coiling temperatures (Pradhan[89])*

major deleterious influence on the annealing process. Low total nitrogen contents however, are still important if the highest formability is to be obtained.

An addition of boron is often used to improve the formability of continuously annealed steel by using an amount that is just able to combine with the nitrogen, the reaction taking place during hot rolling. It is not necessary, therefore, to use a high coiling temperature as is the case if the nitrogen is to be combined with aluminium. The deleterious effects of using a high coiling temperature (higher yield loss and more difficult pickling) can, therefore, be avoided. The formability of a boron-bearing, low coiling temperature steel is, however, lower than that of a high coiling temperature, non-boron steel due to the presence of relatively fine carbides which dissolve quickly on annealing. Boron-bearing continuously annealed AK steels are, therefore, only suitable for use for Fe P01 applications.

It is mentioned above that high slab reheat temperatures should be used for aluminium-killed steel for batch annealing in order that aluminium nitride is taken into solution and retained in solution in the hot band structure by means of a low coiling temperature. The ideal requirements for continuous annealing are the exact opposite. The use of a low slab reheat temperature leaves undissolved aluminium nitride particles present in the structure prior to hot rolling and these particles act as nucleation sites for precipitation after hot coiling. There is, therefore, a progressive decrease in yield stress with decreasing slab reheat temperature, as illustrated in Figure 1.53.

Finishing in the single-phase austenite region is preferred for continuous annealing if good formability is required, since lower r_m values are obtained when lower finishing temperatures are used. A practice has been developed, however, as mentioned previously, which utilizes hot rolling in the ferrite region. This will be discussed further at the end of this section.

The overall effect of coiling temperature on the properties of two aluminium-killed steels rolled in the austenite region is illustrated in Figure 1.54. The effect is clearly greater for the higher carbon steel than for the lower carbon steel.

Figure 1.55 shows that for a 0.044% carbon steel, the r_m value increases progressively with cold reduction up to about 80%, whereas the increase continues to at least 90% for the 0.018% carbon steel. This behaviour is similar to that of an IF steel but is in contrast to the behaviour of an aluminium-killed steel, processed by batch annealing. Clearly a very high cold reduction should be used for continuously annealed steels if the highest formability is required.

Each part of the annealing cycle needs to be carefully designed if the most favourable formability is required. A suitable grain size and texture is obtained, as mentioned previously by the appropriate scavenging effects, but this must be combined with a suitable annealing temperature. As shown in Figure 1.56, an annealing temperature close to 850°C gives a higher r_m value and a lower yield stress than annealing at 700°C, but the cooling part of the cycle is equally important. It is used to reprecipitate the carbon that is inevitably taken into solution, down to a sufficiently low level to give an adequately low level of room temperature strain ageing.

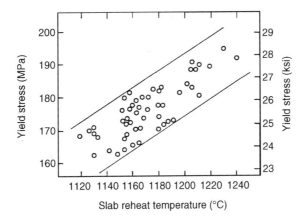

Figure 1.53 *Variation of YS with the reheating furnace temperature for continuously annealed, extra-low-carbon, aluminium-killed steel (After Prum et al.[90])*

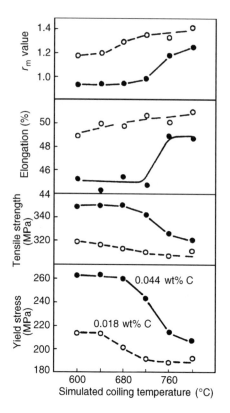

Figure 1.54   *Effect of coiling temperature on the mechanical properties of alumi-nium-killed steel containing 0.018 or 0.044 wt% C annealed at 700°C for 1.5 min (After Ono et al.[87])*

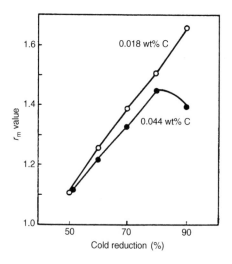

Figure 1.55   *Variation of r_m value with cold reduction for aluminium-killed steel containing 0.018 or 0.044 wt% C annealed at 700°C for 1.5 min (After Ono et al.[87])*

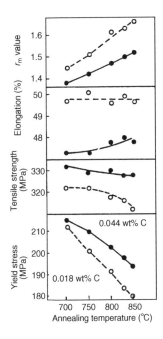

Figure 1.56 *Variation of mechanical properties with annealing temperature for aluminium-killed steel annealed for 1.5 min after 71% cold reduction (After Ono et al.[87])*

When the carbon in solution at the start of overageing is low, the carbon precipitates by diffusion to the ferrite grain boundaries. This is a long-range diffusion process and takes place, therefore, relatively slowly. With higher solute carbon contents at this stage, there is the possibility of the nucleation of new, fine precipitates within the grains. Reprecipitation then involves short range diffusion and can proceed, therefore, more rapidly. The objective of the part of the annealing cycle immediately prior to the overageing section is, therefore, to retain a high level of carbon in solution in order to maximize the possibility of the nucleation of a fine array of closely spaced carbides. This may be done by cooling rapidly from a temperature close to the temperature (723°C) of maximum solubility. In practice, however, it is usual to commence the rapid cooling from a temperature well below 700°C to minimize the risk of strip distortion and to avoid the formation of too many fine carbides.

It has been found that the carbides also tend to precipitate on fine manganese sulphide precipitates. The provision of such particles by control of the manganese content and the slab reheating temperature provides a means, therefore, of promoting the overageing process.

The effect of cooling rate itself on final properties is illustrated in Figure 1.57. This figure shows that there is a tendency for the yield stress to decrease as the cooling rate increases to an optimum value of about 100°C/s, but that higher cooling rates lead to higher values of yield stress. Additionally, lower overageing temperatures lead to a higher yield stress in spite of the short precipitation diffusion distances due to the closely spaced carbides. The reason is clear from

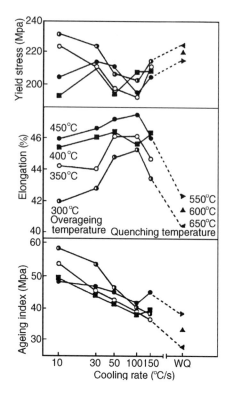

Figure 1.57 *Effect of cooling rate and overageing temperature on the mechanical properties of an aluminium-killed steel annealed at 800°C (After Obara et al.[91])*

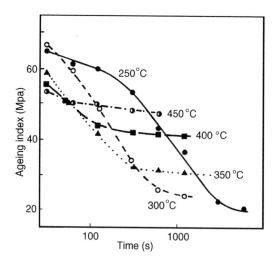

Figure 1.58 *Variation of ageing index with time for aluminium-killed steel for different overageing temperatures after isothermal overageing (After Katoh et al.[92])*

Figure 1.58, which shows how the carbon content, as reflected in the ageing index, varies with time for different overageing temperatures. The reprecipitation proceeds quickly to a final high level at temperatures close to 450°C due to a high diffusion coefficient but proceeds very slowly to a lower level at temperatures close to 250°C due to a low diffusion coefficient. Certain relatively short over-ageing processes, therefore, involve cooling to below 300°C to take advantage of the nucleation of closely spaced carbides and then include reheating to a higher temperature such as 350°C, to take advantage of a higher diffusion rate.[92]

The early continuous annealing lines for strip gauge steels used gas cooling, gas jet cooling or cold water quenching to cool the strip down to the overageing temperature. These cooling methods did not, however, provide the optimum cooling rate. The methods mentioned previously of hot-water quenching, roll cooling, gas-roll cooling, water-mist cooling and high-hydrogen gas-jet cooling were, therefore, developed to give cooling rates closer to the optimum.

Ferritic rolling

It is clear from the above sections that the most formable properties from both interstitial-free and aluminium-killed steels are usually obtained when the finishing temperature is in the single-phase austenite region. For many applications, however, the best properties are not required and this gives scope for the use of cheaper processes. One such opportunity currently being developed is *ferritic rolling*, whereby the roughing sequence is carried out at a lower temperature than previously, but still usually in the single-phase austenite region, and the finish rolling is carried out in the single-phase ferrite region. This process takes advantage of the fact that the resistance to hot deformation is similar in the ferrite region below 800°C to the resistance in the austenite region above 900°C, as illustrated in Figure 1.59.

When extra-low-carbon steel was ferritically rolled in the laboratory, using a finishing temperature of 757°C and a high coiling temperature of 650°C, a yield stress of 166 N/mm^2 was obtained for the as-hot rolled condition compared

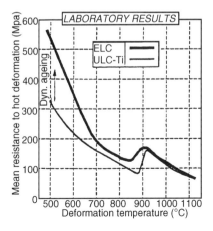

Figure 1.59 *Mean hot deformation resistance of ELC and ULC-Ti steels (After Herman et al.[93])*

with 238 N/mm^2 for conventional rolling.[93] The difference was due to a grain size effect. Under production conditions, ferritically hot-rolled steel developed similar properties and was considered to be suitable for direct application or for cold rolling and annealing to give an Fe P01-type product. It was also found, however, that a finishing temperature in the range 700–750°C, combined with a low coiling temperature at least below 600°C, led to a hot band structure that retained a deformed structure that could be directly annealed without an intermediate cold reduction. For this process, a yield stress in the final product of either 250 or 130 N/mm^2 was obtained depending on whether the steel was an extra-low-carbon steel or an IF steel. The r_m value of the IF steel was, however, low at 1.1. It was considered that this low r_m value was due to the formation of a deleterious shear texture close to the surface of the strip, but this shear texture could be minimized by the use of lubrication on the hot mill.[93]

High-strength steels

The low-strength steels considered in the previous section have been in use for many years, but changes mainly in the automobile market have prompted the development of higher-strength cold-reduced steels which have penetrated the market for the more traditional steels. The first higher-strength cold-rolled steels considered were the previous mild steels but with a higher degree of temper rolling to increase the yield strength.[94] The main problem, however, was the low formability which restricted use. Nevertheless, strengthening by cold reduction is an effective way of strengthening when a very low degree of formability is acceptable. Cold-reduced steels continue to be used, therefore, for many strapping applications and in the zinc coated condition for corrugated roofing panels.

Other early cold-rolled high-strength steels were obtained by cold rolling and annealing the hot rolled type of micro-alloyed steels already available but with some modification to chemistry to provide specific strength requirements after annealing. These steels, with yield stresses up to about 400 N/mm^2, were strengthened mainly by grain refinement and were more formable than the steel with a high degree of temper rolling, but they had low r values.

The main emphasis subsequently was the development of steels for which the loss in formability with increasing strength was minimized. Substitutional solid solution-strengthened steels were developed and the main reason, as illustrated in Figure 1.60, was that the loss in elongation per unit strength increase is less for a solid solution-strengthened steel than for a micro-alloyed steel. These relationships are similar to those for n value given in Figure 1.25. The equivalent relationships between r value and strength are given in Figure 1.61.

The strength increase that could be obtained with a rephosphorized addition was, however, limited by the detrimental effect of phosphorus on welding. These steels were restricted, therefore, to relatively modest strength increases with minimum yield stresses up to but usually well below 300 N/mm^2. The steels were clearly, therefore, highly formable.

Dual-phase steels, based on a ferrite matrix but containing up to about 20% of dispersed martensite islands, have been developed which generate tensile

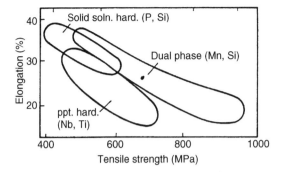

Figure 1.60 *Variation of tensile elongation with tensile strength for various steel types (After Hayami and Furukawa[95])*

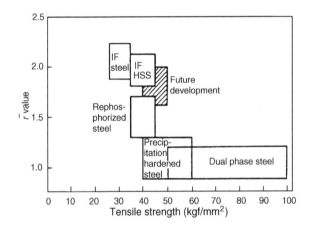

Figure 1.61 *Relationship between r value and tensile strength (After Takechi[96])*

strengths up to and above 600 N/mm². These steels have relatively high n values for their strength but the *r* values are low.

Bake-hardening steels, usually with strengthening by phosphorus, develop a substantial increase in yield stress during the paint stoving treatment after painting. These steels, with yield stresses up to about 300 N/mm² in the as-supplied condition, have the formability, therefore, of a relatively low-strength, solid solution-strengthened steel but the performance in use of a much higher-strength steel.

The development of highly formable low-strength IF steels subsequently led to the generation of solid solution-strengthened versions of these steels with values of tensile strength up to about 440 N/mm². It was found, however, that some of these steels could be potentially affected by a phenomenon called *secondary cold work embrittlement* when subjected to fairly severe drawing strains. This tendency has been reduced by the addition of boron and is considered further in a section below.

Steels with very high strength up to above 800 N/mm² have been developed using the 'so-called' TRIP mechanism. This refers to transformation-induced plasticity. The steels contain a ferrite matrix with islands of separate phases that usually contain bainite and metastable-retained austenite. Under the action of a forming strain, the austenite transforms to martensite and imparts a high work hardening coefficient and hence elongation to the steel.

Other steels may contain a substantially complete bainitic or martensitic microstructure to give very high strength but these steels have low ductility. A steel with a completely bainitic structure has, however, been used for certain strapping applications for many years. A small quantity of steel is, however, supplied in the relatively soft condition for forming and is then heat treated to give a substantially martensitic structure and a tensile strength up to about 1600 N/mm^2. Further details concerning all these steels are given in the remainder of this section.

It is useful to note that most higher-strength steels have yield stresses and tensile strengths that are noticeably higher, by 10 or 20 N/mm^2 or more, in the transverse direction than in the rolling direction. It is important, therefore, that the direction of testing should be specified if a minimum strength requirement is to be satisfied.

Micro-alloyed, high-strength, low-alloy (HSLA) Steel

Micro-alloyed steels are essentially low-carbon manganese steels alloyed with additions of the strong carbide- or nitride-forming elements niobium, titanium or vanadium, separately or together and are often known as HSLA steels. In the hot-rolled condition, they usually have values of yield stress in the range from 300 up to 500 or 600 N/mm^2, but the greater tonnage tends to lie towards the middle of this range. The upper limit of the potential yield stress range is usually lower for a cold-rolled and annealed product, depending on the processing given.

The alloying elements have widely differing effects[97] due to the different solubilities of their carbides and nitrides in both austenite and ferrite, and due to their different precipitation kinetics. They increase strength by grain refinement and precipitation effects when sufficient carbon (and nitrogen for vanadium steels) is present in the steel, but the grain refinement itself may arise from several mechanisms.

The addition of alloying elements may restrict austenite grain growth at the slab-soaking stage through the presence of undissolved particles such as niobium carbonitride or titanium nitride, as illustrated in Figure 1.62. It is seen that niobium is more effective in restricting grain growth than vanadium, but there can be an even more marked effect of titanium due to the formation of titanium nitride.

A proportion of any precipitates present is also taken into solution during slab reheating and this can lead to strain-induced reprecipitation on cooling during hot deformation. This, in turn, as mentioned previously, retards austenite recrystallization. When the final rolling is at a sufficiently low temperature, called the recrystallization stop temperature, the inhibition of recrystallization is complete. Subsequent transformation to ferrite then occurs from an unrecrystallized austenite which in turn leads to a very fine-grained ferrite. Figure 1.63 illustrates for steels with one base composition how the recrystallization stop temperature varies with increasing amounts of the several alloy additions. It is

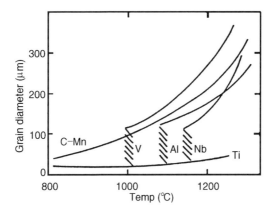

Figure 1.62   *Austenite grain-growth characteristics in steels contain various micro-alloy additions (After Cuddy and Roley[98])*

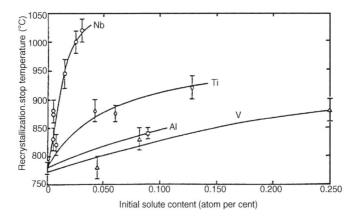

Figure 1.63   *The increase in recrystallization-stop temperature with increase in the level of micro-alloy solutes in a 0.07 C, 1.40 Mn, 0.25 Si steel (After Cuddy[99])*

clear that the addition of niobium has the greatest effect, but with a much smaller effect from titanium. Jonas[100] has studied the effect of all three alloy additions and has expressed the results in the form of the following equation:

$$T_{\text{stop}} = 887 + 464C + (6445Nb - 644Nb^{1/2}) + (732V - 230V^{1/2})$$
$$+ 890Ti + 363Al + 357Si$$

As mentioned previously, the recrystallization characteristics of any steel during hot rolling depend on the amount of deformation given and the strain rate, as well as on the composition and the deformation temperature.[101] Thus different recrystallization temperatures would be obtained for the different strains, strain rates and interpass times than would be obtained in the successive stands of a hot mill.

At temperatures above the recrystallization stop temperature, full or partial recrystallization may take place between each stand or before cooling, but the

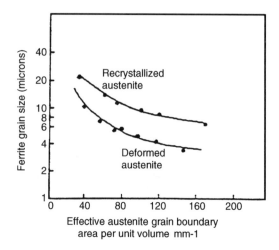

Figure 1.64   *Variation of ferrite grain size with austenite grain boundary area per unit volume for deformed and recrystallized austenite. (After Kasper etal.[102])*

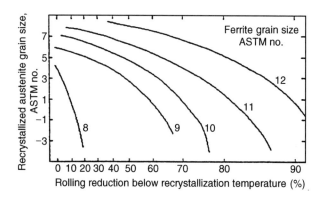

Figure 1.65   *Effect of recrystallized austenite grain size and total reduction below recrystallization temperature on ferrite grain size (After Sekine et al.[103])*

formation of strain-induced fine precipitates may still lead to a finer recrystallized austenite grain size than would otherwise have been the case. This again leads to a finer ferrite grain size on subsequent cooling and transformation. Figure 1.64 shows that the ferrite grain size produced on transformation depends on the austenite grain boundary surface area per unit volume before transformation and that a much finer ferrite grain size, for a given grain boundary surface area, is produced from an unrecrystallized austenite than from a recrystallized austenite. It is clear that austenite grain boundaries provide nucleation sites for the formation of ferrite grains. The grain boundary surface area per unit volume itself depends on the austenite grain size, but for the unrecrystallized structure, it also depends on the strain given after the last previous recrystallization since this largely determines the grain shape which in turn influences grain boundary area per unit volume. Figure 1.65 shows how the ferrite grain size varies with recrystallized austenite grain size and the amount of deformation given below the austenite recrystallization temperature.

It is also useful to note, as mentioned earlier, that the final ferrite grain size depends on the cooling rate after the steel leaves the last finishing stand due to the influence of cooling rate on transformation temperature. One consequence of this is that richer compositions are often needed to develop a given strength from a relatively thick steel that cools slowly compared with a relatively thin steel that may be cooled more quickly.

The formation of fine precipitates themselves may also lead to precipitation strengthening when the precipitation occurs during the transformation from austenite to ferrite or after the transformation is complete. These precipitates may interfere with ferrite grain growth during or after the transformation and may, therefore, have a further grain-refining effect. It is useful to note, however, as pointed out by Pickering,[104] that precipitates formed in the austenite do not contribute directly to precipitation strengthening in the final hot-rolled ferrite structure.

Hot-rolled, high-strength, micro-alloyed steels may be cold rolled and annealed using either batch or continuous annealing to give steels that also have high strength and it is generally found that for a given annealing method, there is a correlation between the strength of the hot band and the strength of the cold-rolled and annealed product. This is illustrated for batch-annealed niobium steels in Figure 1.66. Figure 1.67 shows that a fairly similar relationship applies to vanadium steels and also shows that the strength of a continuously annealed vanadium steel is higher than that of a batch-annealed steel of the same composition. A similar relationship applies to niobium steels, as illustrated in Figure 1.68, and comparison between Figures 1.67 and 1.68 shows that a given niobium addition may develop greater strength than the same vanadium addition.

Hayami and Furukawa[95] showed that the strength increase to be obtained from a niobium or titanium addition may be empirically related to the square root of the composition. Their results are in qualitative agreement with Figures 1.67 and

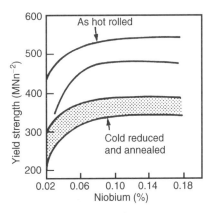

Figure 1.66 *Effect of niobium additions on yield strength of hot strip and cold-reduced strip after batch-annealing and temper rolling (After Bordignon et al.[105])*

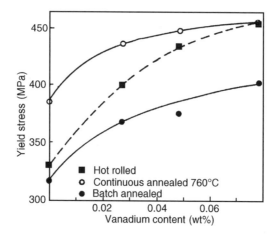

Figure 1.67   *Variation of yield stress with vanadium content of hot-rolled and batch- and continuously annealed steel also containing 0.06 wt% P and 0.05 wt% Si (After Pradhan[106])*

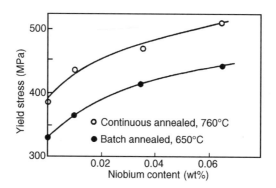

Figure 1.68   *Comparison of the effect of continuous and batch annealing on the yield stress of cold-rolled niobium steels also containing 0.06 wt% P and 0.5 wt% Si (After Pradhan[107])*

1.68 and imply that titanium gives more strengthening per unit addition than niobium.

It is sometimes necessary to supplement the strengthening from grain refinement by strengthening from solid solution elements if a relatively high strength is required in a cold-rolled and annealed product. The resulting steel, however, has the general forming characteristics of a micro-alloyed steel rather than of a solid solution-strengthened steel because the presence of the precipitates leads to low r values.

Cold reduction is an important variable for micro-alloyed steels because it influences both the temperature required for complete recrystallization and the ~rength. After continuous annealing for one minute at 760°C, for example, an ~ase in cold reduction from 50 to 70% for a niobium steel decreases the temperature for the completion of recrystallization by about 20°C and increases the yield stress by about 20 N/mm^2, with a larger increase for vanadium-bearing steels.[106] As expected, higher annealing temperatures lead to lower strength for both batch and continuous-type annealing.[106,108,109]

As with mild steel, temper rolling is the last stage in the process sequence that substantially affects the properties. Usually, it is found that a higher temper rolling reduction is needed to remove the yield point of a micro-alloyed steel than for a mild steel.

In summary, cold-rolled and annealed micro-alloyed steels are steels with low r values and moderate n values for their strength. They are used, therefore, for high-strength, structural applications with modest formability requirements.

Solid solution-strengthened steels

Each type of mild steel may be given a modest increase in strength by the addition of elements that remain in solution in the final product. The elements most commonly used are phosphorus, manganese and silicon, though boron may also be used in IF steels. Carbon and nitrogen are the most potent solid solution-strengthening elements but substantial quantities of these elements in solution are not normally used because they lead to deleterious room temperature strain ageing. Very small quantities in solution, however, are used to give bake hardening (see below) but the amounts involved do not lead to a major increase in strength in the as-supplied condition.

Solid solution-strengthened steels usually have a minimum yield stress in the range 220–300 N/mm^2, though higher strengths may be obtained depending on the processing (mainly annealing) facilities available. Phosphorus is the element most commonly used for fairly low increases in strength since it is relatively cheap and gives a higher increase in strength per unit addition than manganese or silicon. The phosphorus addition is usually restricted to well below 0.1%, however, to avoid problems with welding and because phosphorus may lead to secondary cold work embrittlement of IF steels (see below). For the higher-strength steels, the strengthening by phosphorus is often supplemented first by manganese and then by silicon.

An advantage of solid solution-strengthened steels is that they retain the general characteristics of the mild steel from which they are derived. Thus, they retain good formability, including fairly high r values. A further advantage over micro-alloyed steel is that there is less loss in strength from the hot-rolled to the cold-rolled condition. Cold mill loads are, therefore, correspondingly lower for a given strength in the final batch-annealed product.

Solid solution-strengthened AK steels

The following equation indicates approximately how the tensile strength of a batch-annealed product varies with the steel composition (wt%).[95]

$$TS_{BA} = 270 + 441[C] + 64[Mn] + 98[Si] + 930[P]$$

Another equation is as follows:[110]

$$TS_{BA} = 292 + 563[C] + 678[P] + 90[Si] + 18[Mn] - 1534[S]$$

An equivalent equation for a continuously annealed steel is as follows:[111]

$$TS_{CA} = 477 + 48[Mn] + 127[Si] + 918[P] - 0.019(AT°C)$$

where $AT°C$ is the annealing temperature. It is clear that the strengthening effect from each element is similar for each type of annealing but that, as expected, the precise effects depend on the different processing used.

The prior processing requirements for each type of steel are the same as for the equivalent type of mild steel. Thus a low coiling temperature below 600°C must be used for the batch-annealed product to retain nitrogen in solution following hot rolling, in order that satisfactory r values are obtained from the final product. For the same reason, the continuously annealed product should be processed using a high coiling temperature to precipitate nitrogen as aluminium nitride. Low manganese and extra-low-carbon contents should also be used if high r values are to be obtained.

Different workers have reported different effects of phosphorus on r value, but it is now generally agreed that an increase in strength is accompanied by a small drop in r value.

The effects of cold reduction and annealing conditions are generally similar to the effects for the equivalent type of mild steel. A variation in annealing temperature over the practical range, however, has a greater effect for the continuously annealed product than for the batch-annealed product. High annealing temperatures up to 800°C or more should be used for the continuously annealed steel if high r values are to be obtained.

Solid solution-strengthened IF steels

These steels are usually based on titanium, niobium or titanium plus niobium IF ultra-low-carbon steels but with sufficient phosphorus, manganese, silicon or boron to give the strength required. They are usually processed by continuous annealing but may also be processed by batch annealing. They are used in the uncoated condition when higher formability is required than can be obtained from an aluminium-killed steel, but provide the only way of achieving a highly formable high-strength product from a hot dip coating line (see below).

Figure 1.69 gives a plot of properties versus [Si] + [Mn] + 10[P] for niobium-bearing ULC IF steels. These curves imply that the increase in strength from manganese is equivalent to that from silicon and that the effect of phosphorus is ten times as great. The drop in r value with increasing strength is, however, greatest for manganese and least for phosphorus. Figure 1.70 gives an example of the effect of solid solution additions to a titanium steel. In this case the effect of silicon is greater than the effect of manganese as it is in an aluminium-killed steel. For the titanium steel, the decrease in elongation is greatest for a phosphorus addition and least for silicon (Figure 1.71). Although boron has a high affinity for nitrogen, it imparts solid solution strengthening to titanium-treated IF steel because the nitrogen present is precipitated as titanium nitride.

An important effect of phosphorus in IF steels is the greater tendency for secondary cold work embrittlement. This is discussed further in a later section.

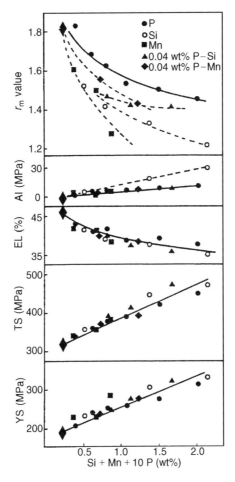

Figure 1.69 *Variation of the mechanical properties of ultra-low-carbon continuously annealed niobium-bearing steel with [Si] + [Mn] + 10[P] (After Ohashi et al.[112])*

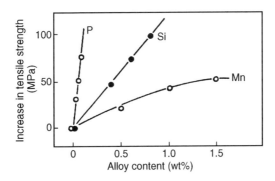

Figure 1.70 *Variation of tensile strength of ultra-low-carbon titanium-bearing steel with solid solution elements (After Katoh et al.[111])*

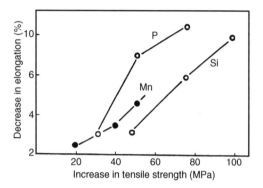

Figure 1.71   *Deterioration of elongation on strengthening by solid solution elements (After Katoh* et al.[111]*)*

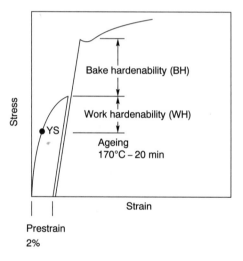

Figure 1.72   *The measurement of bake hardening (BH) and work hardening (WH) using a tensile test*

Bake-hardening steels

The essence of a bake-hardening steel is that it has the ability to increase its yield stress significantly during a paint-stoving process after forming by means of a type of strain-ageing process. In the as-delivered condition the steel has the formability, therefore, of a relatively low-strength steel but the performance in service of a steel with a much higher yield stress. Paint stoving is often carried out using a heat treatment at 170°C for about 20 minutes and bake hardening is most useful in the part of any pressing which is subject to a low degree of work hardening. A standard index to quantify bake hardening is used, therefore, which involves straining a sample by 2% in a standard tensile test, baking the sample at 170°C for 20 minutes and then continuing the tensile test. The difference between the lower yield stress after baking and the 2% flow stress before baking is taken as the bake-hardening index, as indicated in Figure 1.72.

The minimum levels of bake-hardening index that are usually considered to provide a useful benefit are 30 or 40 N/mm^2. An increase in yield stress of about 40 N/mm^2, for example, would compensate for a reduction in gauge of about 0.1 mm to give the same dent resistance.[113] The maximum levels of bake hardening that are commonly used are 50 or 60 N/mm^2. Higher indices than these would lead to excessive room temperature strain ageing with the result that pressings would be subject to stretcher strain defects on forming. Recently, it has been indicated,[114] however, that a sufficiently low and satisfactory degree of strain ageing is obtained for exposed parts when the bake-hardening index is below 50 N/mm^2. Bake-hardening indices above 50 N/mm^2 for exposed parts were, therefore, regarded as suitable only for *just-in-time* delivery before forming. Figure 1.73 gives an example of the return of yield point elongation for steel with different bake-hardening indices after an accelerated strain-ageing test for one hour at 100°C. A yield point elongation of just 0.2% was regarded as satisfactory if stretcher strain markings are to be avoided.[113]

The earliest type of bake-hardening steel available was rimming steel which acquired its bake-hardening tendency as a result of free nitrogen in solution. These steels were, however, subject to high levels of room temperature strain ageing with the result that forming without stretcher strain markings could only be guaranteed for a short period. With the widespread use of continuous casting, nitrogen is usually combined as stable aluminium nitride. The bake hardening must, therefore, be obtained from solute carbon. Consequently, the chemistry and processing of the steel must be manipulated to leave sufficient carbon in solution after annealing. It is usually accepted that the ideal level of solute of carbon is in the range 5–20 ppm, but certain workers regard a figure low in this range as essential whereas others indicate an optimum figure high in the range. The ideal figure depends on the type of steel and, in particular, its grain size since it is well known that bake hardening depends on grain size as well as on the solute carbon or nitrogen content.[115]

Bake-hardening steels may be processed by batch annealing or by continuous annealing including continuous annealing on galvanizing or aluminizing lines.

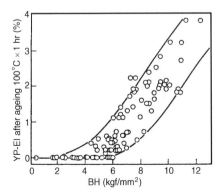

Figure 1.73 *Relationship between bake hardenability and YP-El caused by strain ageing (After Takechi, H.[113])*

The details of the chemistry and the processing that must be used, however, depend on the steelmaking capability and on the annealing method, as well as on the strength and r value of the steel required.

Batch-annealed, bake-hardening steels

Most batch-annealed, bake-hardening steels are based on a conventional aluminium-killed steel as indicated above, using hot-rolling conditions and aluminium and nitrogen contents necessary to give high r values. With normal annealing conditions with a top annealing temperature below 723°C and with a carbon content above about 0.03%, the carbon that is taken into solution at the annealing temperature reprecipitates almost completely during cooling and the steel becomes non-strain ageing and non-bake hardening. With lower carbon contents, reprecipitation becomes more difficult, a small amount of carbon remains in solution and a useful degree of bake hardening is obtained. Figure 1.74 shows how the bake hardening varies with total carbon content for tight coil annealing, but also shows that a higher bake-hardening index is obtained using the faster cooling rates and higher annealing temperatures that may be obtained from open-coil annealing. Such a method, however, is rarely used. It is common practice, however, to utilize the higher cooling rates that may be obtained from hydrogen annealing compared with the older HNX annealing to develop a higher degree of bake hardening.

The yield stress of a steel prepared as above is comparable with that of a conventional non-bake-hardening, batch-annealed, aluminium-killed steel, but

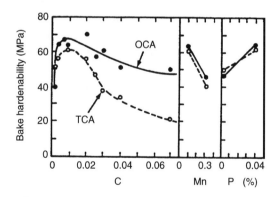

Figure 1.74 *Effect of chemical composition and batch-annealing furnace on bake-hardenability (base composition:*

0.01% C–0.11% Mn–0.07% P–0.05% Al
OCA: Open coil annealing
        soaked at 740°C
        cooled at 80°C/h
TCA: Tight coil annealing
        soaked at 710°C
        cooled at 20°C/h)

(After Mizui et al.[116])

higher strength may be obtained by alloying with phosphorus, manganese or silicon to provide solid solution strengthening. The addition of phosphorus also increases the bake hardening by refining the grain size, but manganese leads to a reduction in the bake hardening. Silicon increases the bake hardening since it delays carbon precipitation,[117] but it causes a higher yield point elongation. Silicon is only used, therefore, when sufficient strength cannot be obtained by the use of other elements.

A batch-annealed, bake-hardening steel with a yield stress below 150 N/mm^2 may be obtained using an ultra-low-carbon IF steel with a small titanium and niobium addition to combine with part of the carbon but to leave sufficient excess carbon to provide bake hardening.[117]

Continuously annealed, low-carbon, aluminium-killed, bake-hardening steel

Bake-hardening, aluminium-killed steels may be produced by continuous annealing using the method given above for the production of formable aluminium-killed steel. The retention of carbon in solution to provide the bake hardening is a natural consequence of the fast cooling rates that must be used on a continuous line. The rapid cooling and overageing sections on any line are designed to remove as much carbon as possible, and as with a batch-annealed bake-hardening steel, the strength may be increased using suitable solid solution-strengthening elements, such as phosphorus, manganese or silicon with similar overall effects.

The steels need to posses fairly high r values in order to provide formability suitable for the intended applications and this is achieved by the use of a low nitrogen content and a high coiling temperature using the same mechanisms as for low strength steel. An increase in annealing temperature causes a decrease in yield point elongation after ageing, an increase in r value and a decrease in strength.[118] Other work[119] has shown that the bake hardening tendency may be controlled at a satisfactory level by over cooling prior to overageing and reheating to an optimum overageing temperature.

The use of low carbon, aluminium-killed steel provides a method for producing bake-hardening steel that is suitable for use on any continuous annealing line which incorporates a relatively long overageing section. The method is not suitable for use on hot dip coating lines which do not incorporate such a section.

Ultra-low-carbon, interstitial-free, bake-hardening steels

Continuously annealed, bake-hardening steels, containing ultra-low-carbon contents, may be produced in several ways, depending on the r values required and on the steelmaking capability available, but most of these steels are alloyed with varying quantities of titanium and/or niobium. With the lower additions, some carbon is left uncombined and this solute carbon is available, therefore, to provide the bake hardening. With the higher additions, substantially all the carbon is combined with the titanium or niobium prior to annealing. The free carbon needed to provide the bake hardening is obtained, therefore, by annealing at a sufficiently high temperature such as 850°C to take carbon back into solution. The steel must then be cooled quickly to minimize the reformation of the precipitates on cooling. Detailed studies[120] showed that a minimum cooling rate from 850°C greater than 30°C/s is necessary to obtain sufficient bake hardening for a niobium-bearing steel with an Nb/C atomic ratio of unity, as illustrated in Figure 1.75.

Figure 1.76 illustrates the effect of annealing temperature on bake hardening and r value for several steels containing different amounts of either titanium or niobium. Figure 1.76(a) shows that the steel with an Nb/C atomic ratio of 0.3, a high degree of bake hardening is obtained for all annealing temperatures, whereas

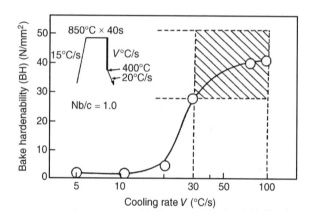

Figure 1.75 *Effect of cooling rate on the bake hardenability of Nb-added extra-low-C steel. The Specimens were annealed at 850°C for 40 s (After Sakata et al.[120])*

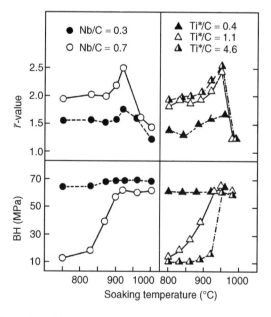

Figure 1.76 *Effect of soaking temperature on bake hardenability (BH) and average* r *value in Nb- and Ti-added extra-low-C steels (After Sakata et al.[120])*

the steel with an Nb/C ratio of 0.7 increases its bake-hardening effect up to an annealing temperature of 900°C. The r values of the 0.3 ratio steel are very much less than those of the 0.7 ratio steel.

Figure 1.76(b) also shows similar relationships for titanium-bearing steels where Ti* is the residual titanium content after the amount that is tied up with nitrogen or sulphur is taken into account. The titanium steel, however, leads to a lower bake hardening than the niobium steel for the same atomic ratio. In addition, the titanium steel would require a higher annealing temperature if a combination of good bake hardening and the highest possible r value is required. Ultra-low-carbon, bake-hardening steels are often produced using a combined titanium and niobium addition and, in principle, with no titanium or niobium addition. It is found, however, that the latter steels may develop relatively large grains in the hot band structure which lead to high levels of planar anisotropy.[120]

Recent work[65] has indicated that ultra-low-carbon steels containing titanium and vanadium could also be made suitable for bake-hardening applications. Further research, however, needs to be carried out on this system.

Transformation-strengthened steels

Transformation-strengthened steels are steels that develop high strength by containing a proportion of transformation products such as martensite or bainite and retained austenite in their microstructure. When these relatively hard phases are distributed throughout a matrix of ferrite, favourable combinations of strength and ductility may be obtained when the proportion of the hard phases is up to about 20% or more. The ultra-high-strength steels, however, contain a high proportion of bainite, martensite or retained austenite. The dual-phase steels exhibit continuous yielding which is followed by a high work-hardening rate. They possess, therefore, a low yield stress to tensile strength ratio and good cold formability for their strength. Steels containing ferrite, bainite and/or martensite, but also containing retained austenite (TRIP steels), provide even higher formability than dual-phase steels, as illustrated in Figure 1.77. The benefits of the TRIP effect were first recognized in steels containing high levels of chromium, nickel and molybdenum,[121] but it is now well known that steels containing manganese and silicon or aluminium are also suitable.

Both dual-phase and TRIP steels, with tensile strengths up to above 600 and 800 N/mm^2 respectively, may be manufactured in the hot-rolled condition by controlling the cooling conditions in relation to the chemistry, but they may also be obtained by cold rolling and annealing to give higher strength. The annealing must, however, be continuous because the long slow cool associated with batch annealing makes it impossible to generate the necessary microstructures with this method. Ultra-high-strength steels with a tensile strength of up to 1000 or 1200 N/mm^2 or more may only be manufactured at present in cold-rolled gauges, using continuous annealing.

Dual-phase steels

Williams and Davies[122] in 1963 were the first to reveal the beneficial effects of a ferrite–martensite structure, but it was not until the 1970s that practical interest

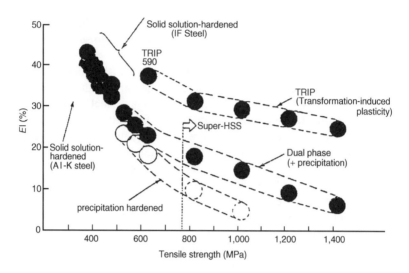

Figure 1.77   *Strength ductility balance of high-strength steel (After Shimada et al.[114])*

was taken in this type of steel. The steels are first annealed in the intercritical region of the phase diagram to produce a structure of ferrite and austenite with a relatively high carbon content in the austenite. On cooling through the inter-critical region, the austenite may become further enriched with carbon and the subsequent cooling rate must be sufficiently fast to avoid the formation of pearlite or bainite. Sufficient martensite must then form to be able to eliminate the yield point elongation. This critical cooling rate depends on the manganese content of the steel, as illustrated in Figure 1.78, but is also influenced by the chromium and molybdenum content. An equation was given[123] to calculate the equivalent manganese content to take into account the effect of molybdenum and chromium as follows:

$$[Mn]_{eq} = [Mn] + 1.3[Cr] + 2.67[Mo]$$

This equation shows that both chromium and molybdenum are more effective in reducing the critical cooling rate than manganese. Other work has shown that the critical cooling rate may depend on the silicon and phosphorus contents and on the holding time at temperature. It is clear that the fairly rapid cooling should continue at least to the M_s temperature if the formation of higher temperature transformation products, such as bainite, is to be avoided.

It is found that the tensile strength of a classical, ferrite–martensite dual-phase steel increases as the volume fraction of martensite increases, but that the yield stress first decreases and then increases, as illustrated in Figure 1.79. This is associated with the gradual removal of the yield point elongation at low martensite volume fractions. For a given volume fraction of martensite, Gladman[125] showed that a decrease in mean island diameter had little influence on tensile strength but has a marked effect in increasing the uniform elongation. Other workers,[126] however, observed an increase in tensile strength. The effect of the martensite volume fraction on the yield stress depends on the relative strengths

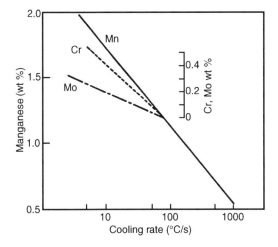

Figure 1.78 *Relation between critical cooling rate and amount of alloying element present (After Irie et al.[123])*

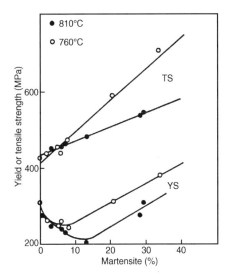

Figure 1.79 *Variation of yield and tensile strength with volume per cent of martensite for a dual-phase steel containing 0.063 wt% C and 1.29 wt% Mn annealed for 10 min at 760 or 810°C (After Lawson et al.[124])*

of the ferrite and the martensite which depends primarily on the martensite carbon content[127] and on any solid solution strengthening of the ferrite. The martensite carbon content depends on the annealing temperature and the initial cooling rate, but manganese and silicon additions also influence this carbon content.[128] The optimum combination of strength and formability is obtained by a very fine distribution of martensite islands and a very fine grain size.[126]

The chemical composition of the steel has a marked influence on the properties developed by influencing the volume fraction of martensite formed, with increasing carbon, manganese and silicon all leading to an increase in strength.[129] Phosphorus and vanadium[130] also increase strength: the phosphorus by solid solution strengthening and the vanadium by increasing the martensite volume fraction and by decreasing the ferrite grain size. Silicon also provides solid solution strengthening, increases the carbon content of the martensite, as mentioned previously, and suppresses pearlite formation.[131]

Variations in hot-rolling conditions may affect the strength of a cold-rolled and annealed dual-phase steel by influencing the hot band microstructure,[132] but no specific microstructure was reported to give a superior strength/ductility balance, as illustrated in Figure 1.80. A higher coiling temperature may lead to a coarser grain size and a lower volume fraction of martensite.[133] In many cases, however, hot-rolling conditions must be selected to give a low hot band strength to minimize cold mill loads rather than to optimize product properties.

Annealing temperature is important for the processing of dual-phase steel because it has a major influence in determining the volume fraction of martensite. Figure 1.81 shows, for a range of carbon, manganese and silicon contents, how the yield stress and tensile strength vary with annealing temperature for gas jet-cooled steels. The strength increases rapidly through the intercritical range which leads to an increase in martensite volume fraction, but then increases more slowly at higher temperatures. Sufficient time is needed at the annealing temperature for the formation of sufficient austenite and data for a 0.06% carbon, 1.23% manganese steel (Figure 1.82) show that the volume fraction continues to increase after holding for 10 minutes at 775°C. The yield point elongation is, however, eliminated after 1 minute at this temperature (Figure 1.83). The effect of cooling rate is illustrated in Figure 1.84 for steels containing 1.2% manganese and up to 0.5% chromium. The tensile strength increases with increasing cooling rate, whereas the yield stress first decreases due to the removal of the yield point elongation and then increases as a proof stress.

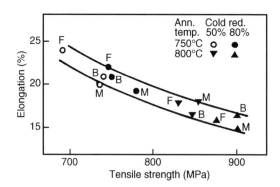

Figure 1.80 *Tensile strength-elongation balance for dual-phase steels showing the strength developed from different hot band structures and different cold reductions (After Shirasawa and Thomson[132]). F = Ferrite and carbide prior microstructure, B = bainite and M = martensite prior microstructure*

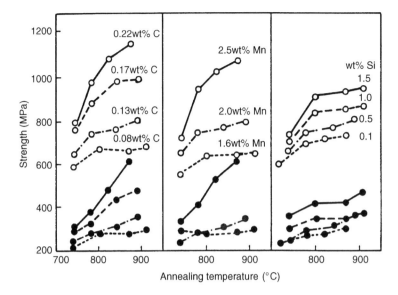

Figure 1.81 *Effect of annealing temperature and chemical composition on the yield (closed circles) and tensile strength (open circles) of gas jet-cooled steel (After Okamoto et al.[134])*

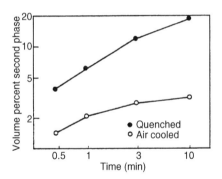

Figure 1.82 *Variation of volume per cent of second phase with time of annealing at 775°C (After Sudo and Kokubo[135])*

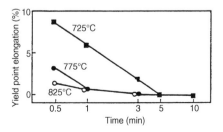

Figure 1.83 *Variation of yield point elongation with annealing time for a 0.06 wt% C, 1.23 wt% Mn Steel for different annealing temperatures (After Sudo and Kokubo[135])*

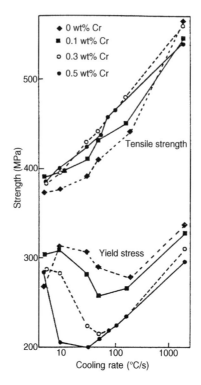

Figure 1.84   *Variation of yield and tensile strength with cooling rate for steels containing 1.2 wt% Mn and different chromium contents (After Irie et al.[123])*

It is useful to note that dual-phase steels tend to possess high *n* values for their strength at low degrees of strain in a tensile test, but that the *n* values reduce at higher strains. In addition, dual-phase steels usually possess poor hole expansion values. The hole expansion values may, however, be improved using a titanium addition to provide precipitation strengthening to the ferrite matrix which reduces the hardness difference between the two phases present.

TRIP or multiphase steels

These steels are based on carbon–manganese compositions, but usually also containing a high level of silicon or possibly aluminium to inhibit carbide precipitation. The early stages of annealing are similar to those of dual-phase steels, but the main difference between the annealing cycles is that TRIP steels are cooled to a temperature reasonably close to 400°C, called the *austempering temperature*, to develop the necessary transformation products. The details of the microstructural changes that occur and their effects on final properties depend critically on the precise chemistry and processing used. The information given in this section is, therefore, intended to give a broad illustration of the microstructural changes that may take place.

The steels are given an intercritical anneal on a continuous annealing line since, as for dual-phase steels, the cooling rates possible with batch annealing would be insufficient to develop the necessary structures. At the top annealing temperature, the two-phase structure of ferrite and austenite is developed with a decreasing proportion of ferrite as the temperature is increased.[136] The carbon clearly partitions between the two phases to give a much higher concentration of carbon in the austenite, but there may also be a smaller partitioning of manganese and silicon with the manganese content higher and the silicon content a little lower in the austenite than in the ferrite. During cooling, the volume fraction of the austenite may decrease and its carbon content may increase, particularly for slow cooling rates, as illustrated in Figure 1.85. In some cases, the carbon diffusing into the austenite may not reach equilibrium and may, therefore, develop a carbon-rich layer, just within the surface of the austenite.

At the austempering temperature, some of the austenite may transform gradually to bainite to leave a mixture of bainite and residual retained austenite. The latter could be in several positions, for example, between the laths of the bainite structure[137] or along ferrite grain boundaries.[138]

When the hold time and austempering temperature are relatively low, the carbon in the austenite remains at its initial relatively low value formed on cooling. On further cooling to room temperature, therefore, much of this austenite transforms to martensite because the M_s temperature can be above room temperature. As the transformation to bainite proceeds, carbon rejected from the bainite builds up in the austenite with the result that its M_s temperature eventually decreases to below room temperature. The austenite is then stable at room temperature and no martensite is formed on cooling. The volume fraction of austenite retained at room temperature, therefore, increases. On further holding, the volume fraction of retained austenite at room temperature decreases because more of it will have transformed to bainite before final cooling. Eventually, the tendency to form carbide in the austenite will increase and carbide may form, but it is well known that this tendency is inhibited by the presence of silicon. The result is that a silicon addition leads to the retention of a higher volume fraction of austenite with a higher carbon content.

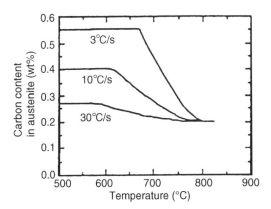

Figure 1.85 *Evaluation of carbon enrichment in untransformed γ during cooling from 825°C. After Minote et al.*[136]

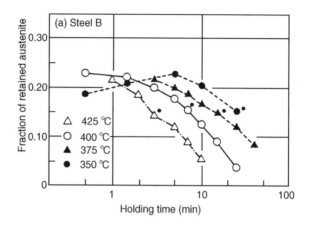

Figure 1.86 *Change in retained austenite volume fraction with isothermal holding time for a steel containing 1.2% silicon for different holding temperatures (After Matsumura[139])*

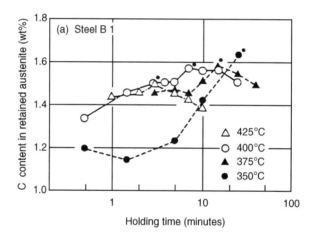

Figure 1.87 *Variation of carbon content in retained austenite for different isothermal holding times for a steel containing 1.2% silicon for different holding temperatures (After Matsumura[139])*

Figure 1.86 illustrates how the volume fraction of austenite retained at room temperature in a steel containing 1.2% manganese and 1.2% silicon varies with time and temperature of austempering, and the equivalent variation of the carbon content of the retained austenite is given in Figure 1.87. An example of the effect of holding time at 400°C on the nature of a stress-strain curve is given in Figure 1.88. It is seen that for the low holding time of 1 minute, the curve is typical of a classical dual-phase ferrite–martensite steel with continuous yielding. Higher hold times lead to an increasing yield point elongation and a higher elongation. Figure 1.89 shows that an increasing hold time leads to a progressive decrease in tensile strength and an increase in yield stress, but that the elongation values pass through a peak for an intermediate holding time of 6 minutes.

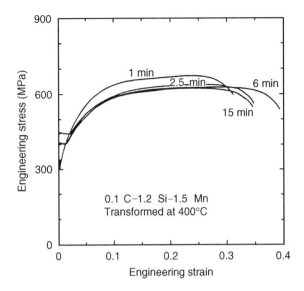

Figure 1.88 *Room temperature engineering stress-strain curves of specimens of a 0.1 C–1.2 Si–1.5 Mn steel transformed at 400°C for different times. Curves plotted to failure (After Sakuma* et al.[137]*)*

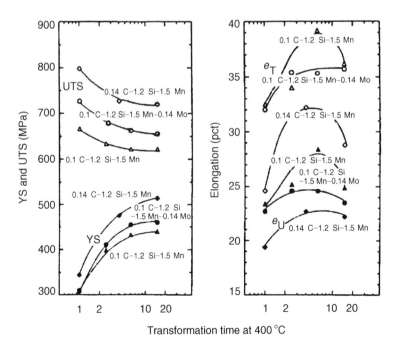

Figure 1.89 *Changes in room temperature mechanical properties as a function of transformation time for isothermally transformed specimen of the three steels (After Sakuma* et al.[137]*)*

When the austenite retained at room temperature is deformed, for example during forming, there is a tendency for it to transform to martensite, as indicated previously, and this is the process that leads to the high work-hardening rate and consequently high values of uniform elongation. It is found that austenite regions with different carbon contents have different stabilities in the presence of the deformation.[136] Austenite with a very high carbon content may be so stable that most of it remains untransformed even after high levels of strain. Alternatively, austenite containing lower levels of carbon could transform completely to martensite under the action of relatively small amounts of strain. Neither of these two conditions give an optimum effect on formability. Thus the mechanical stability is critical if the best ductility is to be obtained.[136] The austenite should have an intermediate stability in order that it transforms gradually over the complete strain range needed.[140]

An example of the effect of holding time in the bainitic region on the build-up of carbon in the retained austenite is illustrated for a steel containing 0.14% carbon, 1.94% silicon and 1.66% manganese in Figure 1.90. It is seen that holding at 400°C for 10 seconds, 1 minute or 8 minutes leads to a progressive build-up of carbon in the retained austenite. The effect on the stability of the austenite during deformation is illustrated in Figure 1.91. This figure shows that much of the austenite retained after the shorter holds transforms in a tensile test during the first few per cent of strain, whereas the austenite retained after the 8-minute hold transforms more gradually over 30% strain. The effect on the work-hardening coefficient n throughout the test is illustrated in Figure 1.92, and compared with the effect for a ferrite–martensite dual-phase steel. It is evident that the n value decreases sharply at low strains for the dual-phase steel. The n values for the two shorter holds are generally similar to those of the dual-phase steels, but the reduction with increasing strain is at a lower rate. For the longest hold time,

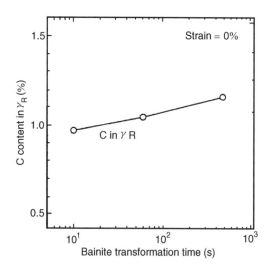

Figure 1.90 *Effect of bainite transformation time at 400°C on C content in retained austenite (After Itami et al.[141])*

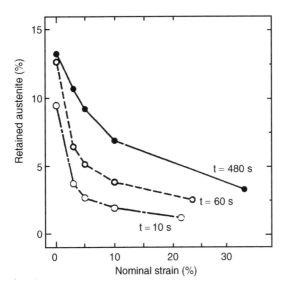

Figure 1.91 *Effect of bainite transformation time (t) on change in volume fraction of retained austenite with deformation (After Itami et al.[141])*

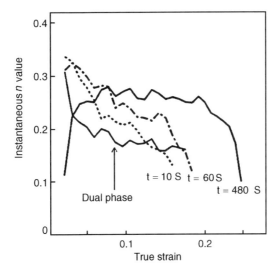

Figure 1.92 *Effect of bainite transformation time (t) on instantaneous n value (After Itami et al.[141])*

giving the highest carbon content in the retained austenite, the *n* value is retained at a high level up to 0.2 true strain corresponding to 22% elongation. The effect on properties is illustrated in Figure 1.93. This figure shows that the 8-minute sample with the highest carbon content in the retained austenite has the highest yield stress as well as the highest elongation. This is clearly a reflection of the gradual transformation of austenite to ferrite over the complete strain range.

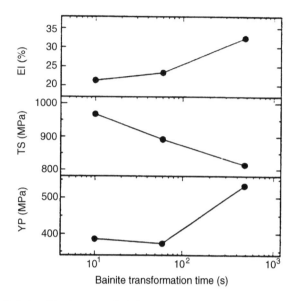

Figure 1.93 *Relationship between bainite transformation time and tensile properties (After Itami et al.*[141]*)*

Ultra-high-strength steels

These are steels with a tensile strength above about 1000 N/mm^2, though steels with a tensile strength above 800 or 900 N/mm^2 may sometimes be regarded as ultra-high-strength steels. They are of two basic types, those in which the microstructure contains a high proportion of martensite and those containing a high proportion of bainite. Each may, however, contain smaller proportions of the other phase, together with significant proportions of ferrite or retained austenite.

One approach to produce a steel with a tensile strength above 980 N/mm^2 was to produce a dual-phase, ferrite–martensite structure, comprising a high hardness martensite in a ductile ferrite containing a minimum of dissolved carbon.[142] The method involved intercritical, continuous annealing at the lowest possible temperature subject to dissolving the carbides, cooling and then quenching from just above the bainite start temperature to give a transformation product that is enriched as far as possible with carbon. The highest elongation was obtained with the highest silicon content used (1.9%) which led to a homogeneous distribution of fine martensite islands. Quenching from just above the bainite start temperature also led to the highest proportion of retained austenite.[143]

A substantially complete martensite structure may be obtained by quenching from the single-phase austenite region using, for example, steels containing niobium and a high-carbon equivalent expressed as $C_{eq} = C + Si/24 + Mn/6$.[144] It was found that the steels in the as-quenched condition softened during a bake-hardening treatment at 170°C. It was necessary, therefore, that the steel was tempered to precipitate most of the carbon before use. The work showed, as expected, that the tempering treatment reduced the strength, particularly for tempering temperatures above about 300°C, but also showed that there was a

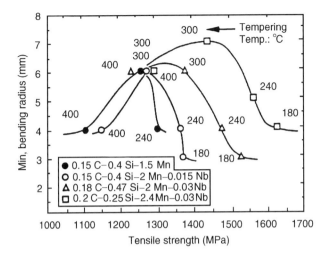

Figure 1.94 *Effect of tempering temperature on the balance of bendability and tensile strength of various steels (After Hosoya et al.[144])*

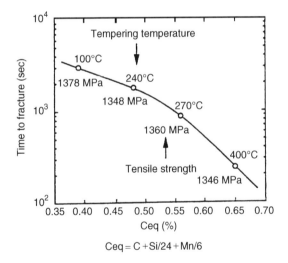

Figure 1.95 *Effects of carbon equivalent and tempering temperature on time to fracture of 4-point bent samples (After Hosoya et al.[144])*

minimum in bendability (maximum in minimum bend radius) for tempering temperatures close to 300°C. Figure 1.94 shows that for the highest carbon equivalent studied, values of tensile strength up to above 1600 N/mm^2 could be obtained, but that for each chemistry, the minimum bendability occurred at the intermediate values of tensile strength. For a given tensile strength close to 1350 N/mm^2, it was also found (Figure 1.95) that the time for delayed fracture decreased with increasing carbon equivalent and increasing temperature used. It was considered that controlling the carbide distribution within the grains was important to achieve good bendability whereas avoiding the carbides that would be formed at the grain boundaries at the higher temperatures was necessary to give good delayed fracture characteristics. Both factors were controlled by tempering at temperatures below 240°C.

Ultra-high-strength steels containing a high proportion of bainite may clearly be obtained by rapidly cooling a steel with a suitable composition to temperatures close to 400°C, as has been used for the TRIP steels. As before, optimization of the strength/elongation balance depends on retaining a sufficient proportion of retained austenite. An example is illustrated in Figure 1.96, which shows how the strength elongation and structure vary with annealing temperature for a steel containing 0.4% carbon, 1.55% silicon and 0.8% manganese. The highest values of elongation were obtained for low annealing temperatures for which the carbon in the austenite would have been high which led to the highest proportion of retained austenite. Quenching from the single-phase austenite region was,

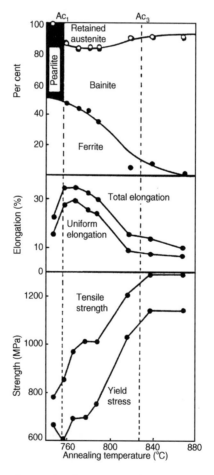

Figure 1.96   *Variation of strength, elongation and structure with annealing temperature, samples quenched to 400°C and held for 5 min (Matsumura et al.[145])*

however, able to give a strength above 1200 N/mm^2 but with a lower elongation. Other work,[146] as a further example, confirmed that values of tensile strength up to above 1400 N/mm^2 could be obtained using a similar type of annealing cycle but with a richer chemistry.

Zinc-coated steels

Coating with zinc (*galvanizing*) is one of the most widely used and cost-effective means of protecting mild steel against atmospheric corrosion. Zinc itself has good resistance to corrosion through the formation of protective surface films of oxides and carbonates but zinc coatings protect steel in two ways:

1. By forming a physical barrier between the steel substrate and the environment.
2. By providing *galvanic* or *sacrificial* protection by virtue of the fact that zinc is more electronegative than iron in the electrochemical series. This effect is illustrated schematically in Figure 1.97.

Production methods

Zinc coatings are applied continuously by the *hot dip galvanizing* and *electro-galvanizing* processes, and in order to appreciate the difference in properties between the products, the two processes will be briefly described.

Hot dip galvanizing (HDG)

The various stages of a modern HDG line are shown schematically in Figure 1.98. Cold-rolled strip is first welded to the trailing end of the previous coil and enters

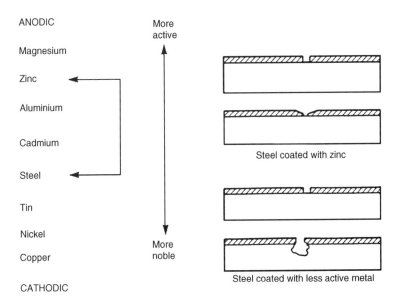

Figure 1.97 *Galvanic protection of steel by zinc*

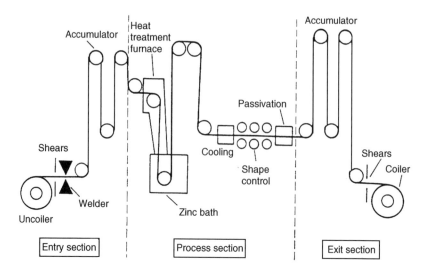

Figure 1.98 *Hot dip galvanizing line*

an accumulator or storage station. This enables the process section of the line to function whilst the cutting and joining operations are being carried out.

The strip then passes into a heat treatment furnace with an inert atmosphere. Initially, the strip is heated to a temperature of about 680°C but electrolytic precleaning is now often used before entry and the material then enters the annealing section of the furnace. In this section, the strip attains a temperature of about 730°C to 850°C and the cold-rolled strip recrystallizes. Still under a protective atmosphere, the strip is cooled rapidly by means of a gas stream, before entering the zinc bath which is maintained at a temperature of 465°C.

Small amounts of aluminium, typically 0.15%, are added to the bath to restrict the thickness of the Zn–Fe layer and thus provide an adherent coating that can accommodate major strains in subsequent forming operations. The aluminium reacts preferentially with the steel as it enters the galvanizing bath, forming a thin layer of Fe–Al compounds ($FeAl_3$, Fe_2Al_5), and thereby retards the reaction between iron and zinc. The lead content of the bath affects the zinc crystal size of *spangle* on the surface of the strip, and for applications requiring a minimum or spangle-free appearance, the lead level of the bath is held below 0.15%. However, in order to develop a spangle, additions of antimony are now preferred to high lead contents because they result in less intergranular corrosion.

After the coating operation, the strip passes through a series of *gas-knives* which use air, nitrogen or steam to control the amount of zinc on the surface of the steel. Depending on the end application, the strip may then be temper rolled or stretch levelled to provide shape control or to eliminate the formation of strain markings (*Lüders lines*) during cold forming. The final stage in the process is generally a passivation treatment in which the strip is sprayed with a solution of chromic acid. This inhibits the formation of a corrosion product known as *white rust* which will be discussed later.

The HDG process is also used to produce *galvanneal* strip in which the plain zinc coating of the traditional galvanized product is converted to a Zn–Fe alloy with an approximate composition of 90% Zn, 10% Fe. Galvanneal strip is produced by reheating the strip after it has left the coating bath and giving it a thermal cycle to develop a suitable alloy layer of Fe–Zn intermetallic compounds. The reheating was formerly carried out in a gas-fired chamber but is now more usually achieved by induction heating. The formation of this layer is critically dependent on the thermal cycle if the most suitable phase distribution is to be formed. The coating weights produced on galvanneal steel are at the bottom of those produced on plain zinc coatings.

Galvanneal coatings have a matt grey appearance of low reflectivity and have the following characteristics and have a major market in the motor industry:

1. Similar corrosion resistance as plain zinc coatings.
2. Good paint adherence.
3. Better resistance welding performance than plain zinc coatings.

Electro-galvanizing

Zinc coatings. In contrast to the HDG process, electro-galvanizing is carried out at or near ambient temperature, the feedstock having been annealed in the conventional manner. Prior to the coating operation, the strip is first passed through a series of chemical or electrolytic cleaning baths to remove dirt or oil. The plating solution is made up primarily of zinc sulphate and zinc was previously supplied to the electrolyte by means of zinc anodes. However, zinc is now more usually supplied by adding zinc compounds to the electrolyte in the form of zinc oxide or zinc carbonate and the anodes are made of lead. Oxygen is, therefore, released at the anodes during the plating process. After plating, the strip is passed through chromate baths for passivation.

87% Zn 13% Ni. Both forming and welding problems are introduced with the application of heavy zinc coatings and therefore there is a demand for coatings that will provide good corrosion resistance, even when applied in thin layers. One approach to this problem is the incorporation of nickel into zinc-based coatings, and a coating containing 87% Zn 13% Ni is now being used for automotive applications. The coating is applied by the electroplating route and the material is marketed in the UK under the name *Nizec*.

Fe–Zn coatings. An iron–zinc coating may also be produced by an electrolytic process and is used mainly in Japan and North America in the automotive industry. Care is needed, however, in controlling the plating conditions because changes in temperature and concentration of the electrolyte may cause relatively large changes in the composition of the coating. The usual coating weight is about 50 g/m^2 and the iron content of the coating is often in the range 8–20%.

According to Denner,[147] electrolytic iron–zinc coatings are particularly good for chipping and corrosion resistance and also for coating and paint adhesion.

Corrosion resistance

The life of a zinc coating on steel is roughly proportional to its thickness, and therefore, galvanized strip is produced in a range of coating thicknesses to satisfy different conditions and end uses. In the hot dip product, the amount of zinc on the steel is expressed in g/m^2, the figures after the letter Z (plain zinc) and ZF (iron–zinc alloy) in Table 1.4 indicating the minimum mass per unit area, including the coatings on both sides of the strip.

In electro-galvanized strip, the coatings are significantly lighter, as shown in Table 1.5. In this table, it should be noted that the coating mass relates to a single surface rather than the combined mass for the two surfaces, which is the convention used for the hot dip product. Electro-galvanized material can also be supplied with single-sided (e.g. ZE 25/00) or differential (e.g. ZE 75/25) coatings.

As with other steel products, the atmospheric corrosion resistance of zinc-coated steel is affected greatly by the nature of the environment and it is common to differentiate between the performance in rural, industrial and marine locations. The performance of hot dip galvanized steel, in terms of life to first maintenance, is illustrated in Table 1.6. This indicates the following features:

1. The detrimental effect of sulphur dioxide (industrial) and chloride ions (marine) in the atmosphere.
2. The beneficial effect of increasing the thickness of the zinc coatings.

Table 1.4   *Use of hot dip coatings*

Light coating	Z 100 Z 200	Where corrosion conditions are mild or where forming requirements preclude heavier coatings
Medium coating	Z 275	Standard coating
Heavy-duty coating	Z 350 Z 450 Z 600	For longer life requirements
Iron–zinc coating	ZF 100 ZF 180	Fe–Zn coatings for good painting and welding characteristics

Table 1.5   *Coating thickness/mass for electro-galvanized coatings*

Coating designation	Nominal zinc coating values for each surface		Minimum zinc coating values for each surface	
	Thickness μm	Mass g/m^2	Thickness μm	Mass g/m^2
ZE 25/25	2.5	18	1.7	12
ZE 50/50	5.0	36	4.1	29
ZE 75/75	7.5	54	6.6	47
ZE 100/100	10.0	72	9.1	65

After BS EN 10152: 1994.

Table 1.6 *Hot dip galvanized steel–typical period (years) to first maintenance*

Mass g/m^2 including both surfaces	Coastal	Industrial and Urban	Suburban and Rural
275	2–5	2–5	5–10
350	2–5	2–5	5–10
450	5–10	2–5	10–20
600	10–20	5–10	20–50

After *Galvatite Technical manual*
(British Steel Strip Products.)

3. The need for paint protection in internal applications if the conditions are wet and polluted.

When zinc-coated products are stored under wet conditions or where condensation can occur, *white rust* can form on the surface. This is due to the formation of zinc carbonate and detracts from the appearance of smooth zinc coatings. However, although it may form in large amounts, the appearance of white rust does not necessarily indicate severe degradation of the zinc coating and it will generally convert to a protective layer. As indicated earlier, the formation of white rust can be retarded by immersing galvanized steel in chromic acid solutions to passivate the surface.

Care must be taken when fixing zinc-coated steels to other metals in order to avoid galvanic corrosion. In particular, copper or brass should not be coupled directly to galvanized steel since the coating can fail rapidly in wet and polluted atmospheres.

In aqueous media, the performance of zinc-coated steels is affected by a number of factors:

1. *pH of solution* – the corrosion rate is generally low in the pH range 6–12 but can be rapid outside this range.
2. *Hardness of water* – hard water precipitates carbonates on zinc surfaces which reduce the rate of corrosion.
3. *Water temperature* – whereas zinc provides sacrificial protection to the steel substrate at ambient temperature, the reverse situation occurs at temperatures above 60–70°C. Therefore, zinc-coated steels should not be used in hot-water systems.
4. *Chlorides* – soluble chlorides can produce rapid attack, even within the pH range 6–12.

Cold-forming behaviour

The cold-forming behaviour of zinc-coated strip is governed primarily by the forming characteristics of the substrate. As indicated earlier, in hot dip galvanizing, cold-reduced strip is subjected to rapid heating and cooling cycles, similar to those employed in continuous annealing but generally without an over-ageing stage. Therefore, the formability of hot dip galvanized steel is poorer than that of uncoated, batch-annealed material. However, the formability of the hot dip product can be improved through the use of modified annealing cycles but the special IF (interstitial-free) steels can also be employed as a substrate for highly formable, hot dip galvanized strip. The IF steels were referred to earlier on p. 39 and owe their good formability to the elimination of carbon and nitrogen from solid solution and the development of a favourable crystallographic orientation texture.

As described earlier, electro-galvanizing takes place at near-ambient temperature and therefore the coating operation has little effect on the forming behaviour.

Standard specifications

Hot dip galvanized products are covered by BS EN 10142: 1991 *Continuously hot-dip zinc coated mild steel strip and sheet for cold forming* and the available grades together with the associated mechanical properties are given in Table 1.7. These grades are arranged in increasing suitability for cold forming, ranging from DX 51D (bending and profiling quality) to DX 54D (special deep-drawing quality). Differentiation is made between coating type, namely zinc (Z) or zinc-alloy (ZF) coatings. Further designations are also applied in order to denote coating finish, surface quality and surface treatment. As indicated in Table 1.8, the grades can be supplied with a range of coating mass, but it should be noted that the figures for coating mass (g/m^2) include both surfaces.

A possible area of confusion in BS EN 10142: 1991 is the change in designation of the basic grades. When this standard was first introduced, the grades were

Table 1.7 *BS EN 10142: 1991 Steel grades and mechanical properties*

Designation			Yield strength $R_e^{1)}$ N/mm^2 max.$^{2)}$	Tensile strength R_m N/mm^2 max.$^{2)}$	Elongation A_{80} % min.$^{3)}$
Steel grade		Symbol for the			
Steel name	Steel number	type of hot-dip coating			
DX51D	1.0226	+Z	–	500	22
DX51D	1.0226	+ZF			
DX52D	1.0350	+Z	300$^{4)}$	420	26
DX52D	1.0350	+ZF			
DX53D	1.0355	+Z	260	380	30
DX53D	1.0355	+ZF			
DX54D	1.0306	+Z	220	350	36
DX54D	1.0306	+ZF			

[1] The yield strength values apply to the 0.2% proof stress if the yield point is not pronounced, otherwise to the lower yield strength (R_{eL})

[2] For all steel grades a minimum value of 140 N/mm^2 for the yield strength (R_e) and of 270 N/mm^2 for tensile strength (R_m) may be expected.

[3] For product thicknesses ≤0.7 mm (including zinc coating) the minimum elongation values (A_{80}) shall be reduced by 2 units.

[4] This value applies to skin-passed products only (surface qualities B and C).

Table 1.8 *BS EN 10142: 1991 Coating masses*

Coating[1]	Coating mass in g/m^2, including both Surfaces[2] min	
	Triple spot test[3]	Single spot test[3]
100	100	85
140	140	120
200	200	170
225	225	195
275	275	235
350	350	300
450	450	385
600	600	510

[1] The coatings available for the individual steel grades are give in tables 2 and 3.
[2] The coating mass of 100 g/m^2 (including both surfaces) corresponds to a coating thickness of approximately 7.1 μ m per surface.
[3] See 7.4.4 and 7.5.3.

Table 1.9 *BS EN 10142: 1991 List of corresponding designations*

Designation according to EN 10142-A1: 1995			Designation according to EN 10142: 1990
Steel name	Steel number	Symbol for the type of hot-dip coating	Steel name
DX51D	1.0226	+Z	Fe P 02 G Z
DX51D	1.0226	+ZF	Fe P 02 G ZF
DX52D	1.0350	+Z	Fe P 03 G Z
DX52D	1.0350	+ZF	Fe P 03 G ZF
DX53D	1.0355	+Z	Fe P 05 G Z
DX53D	1.0355	+ZF	Fe P 05 G ZF
DX54D	1.0306	+Z	Fe P 06 G Z
DX54D	1.0306	+ZF	Fe P 06 G ZF

designated Fe P01 to FE P06, i.e. those adopted in BS EN 10130: 1991 (see Table 1.3). However, BS EN 10142: 1991 was amended in 1995 and the Fe P0 system of designation was changed to the DX 51D/DX 54D style shown in Table 1.9. The corresponding designations under both systems are identified in Table 1.9.

Electro-galvanized grades are specified in BS EN 10152: 1994.

Other hot dip coatings

Whereas galvanneal strip has a Zn–Fe coating which is produced by diffusion from a plain zinc layer, zinc alloy coatings are also deposited directly onto the steel.

55% Al–Zn

In 1972, the Bethlehem Steel Corporation introduced steel strip with a coating of 55% Al 43.5% Zn 1.5% Si under the name *Galvalume*. Since that time, the manufacture of the product has been licensed throughout the world and it is marketed in the UK under the name *Zalutite*. The 55% Al–Zn coating is applied by the hot dip process route in a similar manner to that used for the conventional, plain zinc coatings. On cooling from the coating bath, an aluminium-rich phase is the first to solidify and makes up about 80% of the volume of the coating. The remainder is made up of an interdendritic, zinc-rich phase and an Al/Fe/Zn/Si intermetallic compound bonds the coating to the steel substrate, providing further resistance to corrosion. Silicon is added to the Al–Zn alloy in order to restrict the growth of the brittle intermetallic layer.

The coating was developed specifically to provide an improved corrosion performance compared with plain zinc coatings and benefit is derived from the separate effects of zinc and aluminium. In the initial stages of corrosion, attack takes place preferentially on the zinc-rich phase until its corrosion products stifle further activity in these areas. However, as well as acting as a barrier to the transport of corrodents, the zinc also provides sacrificial protection at cut edges and areas of damage. As the zinc-rich phase is leached away, corrosion protection is provided by the aluminium-rich phase which forms protective films of oxides and hydroxides on the surface of the material. As indicated in Table 1.10, the 55% Al–Zn layer provides between two and four times the life of conventional zinc coatings, depending on the nature of the environment.

The material is produced with standard coating masses of 150 (AZ 150) and 185 (AZ 185) g/m^2, including both surfaces. These values equate to coating thicknesses of 20 and 25 μm respectively, on each surface. The forming properties of the material are generally similar to those of continuously annealed, hot dip zinc-coated steel. However, it has an increased tendency to spring back and lacks the self-lubricating properties of hot dip zinc coatings. The application of an effective lubricant is therefore essential.

Table 1.10 *Comparative corrosion losses as a decrease in thickness (micrometres) for 55% Al–Zn alloy coated and hot dip zinc-coated steel strip at Australian test sites*

Site	Years exposed	(A) Hot dip zinc-coated (μm)	(B) 55% Al–Zn alloy (μm)	Ratio A/B
Severe marine	2.5[a]	16.8	5.2	3.2
Industrial marine	7	10.5	4.7	2.2
Industrial	7	9.8	3.4	2.9
Marine	4	5.9	1.4	4.2
Rural	4	1.4	0.8	1.8

[a]Exposure was discontinued after $2\frac{1}{2}$ years because all the coating on the groundward surface of the hot dip zinc-coated sample had been lost by that time. No rust on remaining 55% Al–Zn alloy-coated samples still on exposure after seven years.
After *Zalutite Technical Manual*
(British Steel Strip Products).

The 55 Al–Zn coating has a smooth, silvery appearance with a very fine spangle and is said to be attractive in the unpainted condition. However, the material is also supplied in the factory-painted condition and painting is recommended for use in severe marine and very corrosive environments. The coating exhibits good resistance to heat and can withstand discoloration in air at temperatures up to 310°C, whereas a limit of 230°C is prescribed for hot dip zinc coatings.

95% Zn 5% Al

A coating of this composition was first introduced in 1982 and is marketed worldwide under the name *Galfan*. The coating is again deposited by the hot dip process and, in addition to good corrosion properties, it is claimed to have particularly good forming characteristics. The structure of the coating is dependent on the rate of cooling from the bath and cooling rates greater than 20°C/s result in the formation of a fine eutectic of zinc-rich and aluminium-rich phases. Slower cooling rates result in the separation of a primary zinc-rich phase.

Hot dip aluminium coatings

Aluminium-coated steel may be produced using a hot dip process and therefore combines the strength and formability of steel with the corrosion resistance of aluminium. The steel base may be an aluminium-killed steel, but the use of an IF steel provides high formability and improved higher temperature oxidation resistance. A product with the basic coating weight may be used for applications such as motor car silencers, but a product with about twice the coating weight at 250 g/m^2 may be used for building applications. The thicker coating also contains a silicon addition. This reduces the thickness of the alloy layer and imparts greater ductility than would be obtained from a pure aluminium coating.

Aluminium-coated steels withstand temperatures up to 500°C for long periods while retaining their original surface appearance. Above this temperature, the coating gradually converts to an iron–aluminium alloy layer, but under certain conditions, it may still retain good heat-resisting properties.

Organic-coated steels

During the 1980s there was a dramatic growth in the production of pre-finished strip, coated with various types of paint or plastic. These coatings are available in a wide range of colours and textures and are used to advantage where corrosion resistance and a decorative appearance are of major concern. The coatings are applied to a range of steel substrates, but generally to zinc-coated strip, and are formulated specifically for various manufacturing requirements or end uses. Coatings are available that provide long life in external applications, good deep-drawing characteristics and resistance to heat or chemical attack.

The organic coatings are applied continuously as a liquid film or as a laminate which is bonded to the substrate with an adhesive. In either case, the steel substrate is thoroughly cleaned in a multi-stage process to ensure uniform and optimum adherence of the coating. In the liquid film route, the substrate is first coated with a primer, which is cured, and the top coating is then applied by the reverse-roll method. Finally, the topcoat is cured in a finishing oven. Where embossing is required, typically on PVC coatings of 200 µm, a patterned steel roll is applied to the hot PVC as it emerges from the finishing oven. The material is then immediately quenched in water in order to 'freeze' in the embossed texture. Coil-coated strip is marketed in the UK under the name *Colorcoat*.

In the laminate coating, an adhesive and backing coat are applied by roller-coating and the strip is passed through an oven to activate the adhesive and cure the backing coat. The coating film, generally PVC, is bonded to the steel and cooled immediately by water quenching. The material is marketed in the UK under the name *Stelvetite* and is intended for internal applications.

Organic-coated strip is specified in Euronorm 169–85. The following information on the characteristics of the more important coatings has been derived from Annex A of Euronorm 169–85 and a British Steel publication:[148]

- *PVC Plastisol* (200 µm) – a plasticizer-bearing coating with very good flexibility. Can be drawn and formed easily. Suitable for embossing for decorative purposes and can be used in internal and external applications. *Typical applications* – roofing and cladding on buildings, curtain walling, furniture, vehicle fascia panels, garage doors.
- *PVC Organosol* (50 µm) – a coating with good flexibility and specially recommended for deep-drawn parts. Not recommended for exterior use. *Typical applications* – electric light fittings, cable trunking, deep-drawn parts.
- *Acrylic* (25 µm) – unplasticized coating with good flexibility. Suitable for continuous operation at temperatures up to 120°C. Good resistance to chemical attack.
- *Epoxy* (5–15 µm) – hard, chemical-resistant coating with good flexibility. *Typical applications* – used extensively as a primer for two-coat systems. Good adhesion to polyurethane foam. Not recommended for external use.
- *Polyesters* (25 µm) – widely applied coatings with good flexibility and suitable for continuous exposure at temperatures up to 120°C. Various formulations are available that offer good deep-drawing properties, resistance to chemical attack and that are suitable for exterior use. *Typical applications* – consumer durables, deep-drawn components, building components.
- *PVF_2* (27 µm) – a coating with good flexibility and highly resistant to chemicals and solvent attack. Highly resistant to weather and particularly suitable for exterior use. *Typical applications* – building components.

As indicated above, many of these coatings exhibit good formability and are amenable to forming operations such as press braking and folding, roll forming and deep drawing. However, the flexibility of the coatings varies with ambient temperature and, to minimize the risk of cracking during forming, it is recommended that the liquid film and laminate-coated materials should be allowed to attain minimum temperatures of 16 and 20°C respectively.[149]

The resistance welding of organic-coated steel is not possible by conventional methods because the coating prevents the flow of current across the electrodes. On the other hand, resistance welding is possible if the coating is removed locally, or if special capping pieces are inserted to melt the coating.[149] However, mechanical joining techniques have been developed specifically for coated steels and adhesives are also available that are suitable for particular types of coating.

Steel prices

The basis prices of standard hot-rolled and cold-reduced strip grades are shown in Figure 1.99. These prices were effective in February 1997 and it must be borne in mind that steel prices are adjusted periodically. This figure shows that a substantial differentiation is made between the hot-rolled and cold-reduced conditions, reflecting the additional processing costs associated with the latter. However, the cost differential within these two conditions with respect to the various formability criteria is relatively small.

Figure 1.99 also includes the basis prices for zinc-coated steels. This indicates that electro-galvanized strip costs about £28/tonne more than uncoated, cold-reduced steel. However, the cost differential is substantially more in the case of hot dip galvanized strip, which has a heavier zinc coating than the electro-galvanized product.

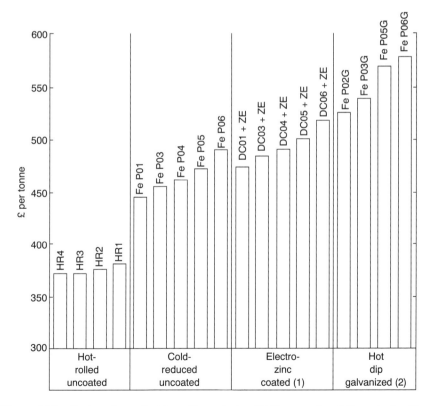

Figure 1.99 *Basic prices of strip mill products: (1) Zinc coating 2.5 μm both sides; (2) Zinc coating 275 g/m², including both sides: as of February 1997*

Tinplate

Tinplate enjoys a pre-eminent position in the packaging industry, particularly in relation to cans for food and beverages. Substantial amounts of tinplate are also used in aerosol containers and for the packaging of paints and oils. This section reviews briefly the manufacturing process for tinplate and the basic procedures employed in canmaking. These specific aspects, together with other important technical details relating to tinplate, are covered in a very definitive and comprehensive publication by Morgan.[150]

Method of manufacture

Virtually all tin mill products in the UK are produced from continuously cast steel to the chemical composition shown in Table 1.11. Where special corrosion resistance is required, a steel with 0.015% P max. and 0.06% Cu max. (type L) is supplied.

The starting point in the manufacture of tinplate is hot-rolled strip with a typical thickness of about 2 mm. This is pickled to remove the scale formed during hot rolling and the material is then cold rolled to the required thickness in either a single-stage or a two-stage operation. For *single-reduced* tinplate, the cold-rolling reduction is of the order of 90% and the material is available in the thickness range 0.16–0.6 mm. After cold rolling, the strip is softened by either batch or continuous annealing in order to restore ductility, but the annealed material is then given a light cold reduction, termed *temper rolling*, before the tinning operation. Temper rolling improves the surface finish and flatness of the strip and also provides the required mechanical properties for particular applications. Thus single-reduced tinplate is available in a range of tempers with *different strengths* and this is illustrated in Table 1.12.

In *double-reduced* tinplate, cold-rolled and annealed strip is subjected to a second cold reduction of 30–40%. No further annealing is undertaken and the material is substantially work hardened, exhibiting a marked directionality in properties. The bulk of double-reduced tinplate is produced to the thickness range 0.16–0.18 mm but is available in the range 0.13–0.27 mm. Typical mechanical properties are shown in Table 1.13. By virtue of its higher strength, double-reduced tinplate provides the facility to reduce the cost of a can or other components by decreasing the thickness of the material without loss of rigidity.

Following the cold-rolling sequences described above, the strip is coated with tin. A small amount of material is still coated by hot dip tinning, similar to that described earlier for galvanizing, but the *Ferrostan* process involving the

Table 1.11   *Tinplate composition*

				Weight % max. (Type MR)					
C	Mn	P	S	Si	Cu	Ni	Cr	Mo	Al
0.13	0.6	0.02	0.05	0.03	0.2	0.15	0.1	0.05	0.02–0.10

Table 1.12   *Tinplate–single-reduced tempers*

Temper number	Euronorm/ISO temper designation	Typical application
T1	T50	Mainly used for components which make the maximum demand on the formability of the steel base, e.g. deep-drawn containers, bakeware, puddings basins, oil filter bodies
T2	T52	Typically used for forming operations which are less severe than above, e.g. shallow-drawn bakeware, rectangular caps, cushion rings for paint cans
T3	T57	Used for a wide range of applications where moderate formability is required, e.g. shallow-drawn components, can bodies
T4	T61	Typically used for ends, bodies, stampings where a stiffer and stronger product is required
T5	T65	This is the strongest product available in the conventional temper range and is typically utilized for stiff ends and bodies

After British Steel Tinplate–Product Range

Table 1.13   *Tinplate – double-reduced tempers*

Designation		Target proof stress (0.02% non-proportional elongation) longitudinal MPa	Target hardness HR 30TM	Typical applications
New	Previous			
DR550	DR8	550	73	Round can bodies and can ends
DR620	DR9	620	76	Round can bodies and can ends
DR660	DR9M	660	77	Beer and carbonated beverage can ends

electro-deposition of tin now accounts for the bulk of tinplate production. Prior to plating, the strip is thoroughly cleaned in electrolytic pickling and decreasing units, followed by washing. The Ferrostan process uses a bath of acid stannous sulphate and tin with a purity of not less than 99.85%. After plating, the coating is flow-melted by resistance heating to a temperature above the melting point of tin (232°C), e.g. 260–270°C, followed by water quenching. This treatment produces a reflective surface and also results in the formation of an iron–tin compound ($FeSn_2$) which plays an important role in the corrosion resistance and soldering characteristics of the material. The product is then passivated by immersion in a dichromate solution which deposits a very thin film of chromium on the surface. After passivation, a thin film of oil is applied in order to preserve the surface from attack and also to enhance the lubrication properties in subsequent handling operations. The oil applied, dioctyl sebacate, is a synthetic organic oil and is acceptable for use in food packaging.

Electrolytic tinplate can be produced with equal or differential coatings on each surface, the former carrying the prefix E in the designation and the latter having the prefix D. The mass on each surface is expressed in g/m^2 and thus E 2.8/2.8 has 2.8 grams of tin per square metre on each surface, giving a total of 5.6 g/m^2. A tin coating of 5.6/5.6 g/m^2 is equivalent to a coating thickness of 0.75 μm per surface. The normal range of equally coated products is from E 1.4/1.4 to E 11.2/11.2. However, low-tin coatings have been developed down to 1.0/1.0 g/m^2. In differentially coated products, the normal range is from D 2.0/1.0 to D 11.2/5.6.

Tinplate manufacturers also produce other products such as uncoated and oiled sheet (*blackplate*) and material coated with metallic chromium. According to Morgan[150] blackplate has not achieved significant usage in canmaking operations due to problems in providing adequate resistance to corrosion by lacquering techniques. However, electrolytic chromium/chromium oxide-type coatings have enjoyed greater success and were developed primarily because of the extremely variable price of tin. The coating is duplex and consists of about 80% metallic chromium adjacent to the steel substrate and 20% hydrated chromic oxide/hydroxide in a layer above. It is recommended that this type of coating is lacquered to provide added surface protection and to enhance fabrication.

Canmaking processes

Three basic procedures are employed:

1. Three-piece can manufacture.
2. Drawing and wall-ironing.
3. Draw and redraw.

Three-piece cans consist of a welded cylindrical body and two ends, one of which (the base) will be attached to the body of the can by the canmaker and the other will be applied after filling. The cylinder is rolled into shape from flat, pre-lacquered, rectangular blanks and the two edges are joined by electrical welding. A further coat of lacquer is then applied to the weld seam. The can ends are pressed from circular blanks in an operation that requires a high degree of precision. The ends are contoured with a series of expansion rings so that they can support the internal pressure through tensile rather than bending stresses. The rims of the ends are also carefully stamped and curled so as to accept a sealing compound which forms an airtight seal with the body of the can. Cans of this type are used for most human foods and also for paints, oils and chemicals.

The *drawing and wall-ironing* (DWI) process is more modern than three-piece canmaking and is illustrated schematically in Figure 1.100. Thus DWI eliminates

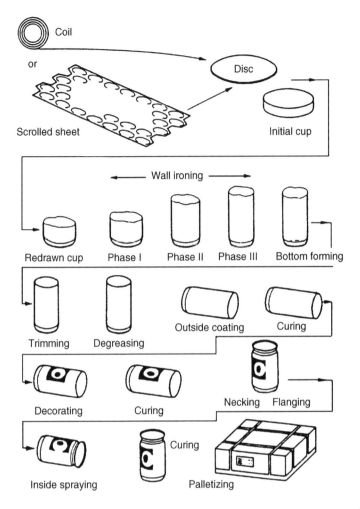

Figure 1.100 *Drawing and wall-ironing operation sequence (After Morgan[150])*

the need for a welding operation and relies solely on presswork. Morgan cites the following advantages for DWI over three-piece cans:

1. More effective double seaming of the top end to body, because of the absence of the sensitive side seam junction (particularly valuable for processed foods and carbonated beverages).
2. Significantly lower metal usage and cost.
3. More attractive appearance due to the absence of a side seam.

Following the initial drawing operation, the cup is redrawn to final can diameter. During *wall-ironing*, the can may pass through a series of dies which produce a substantial reduction in wall thickness and a complementary increase in body height. The principle of wall-ironing is shown in Figure 1.101. Following trimming, the can is degreased prior to applying decoration to the outside and lacquer

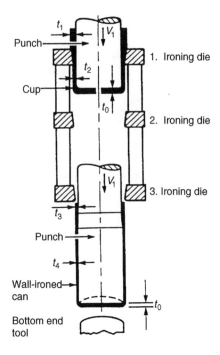

Figure 1.101 *Wall-ironing with three ironing dies (After Morgan[150])*

to the inside. DWI cans are used for beer and soft drinks, for pet foods and some human foods. Aluminium is a major competitor to tinplate in DWI cans for these products, although tinplate enjoys a larger share of the UK market at this time.

In the *draw and redraw* (DRD) process, the initially drawn cup is redrawn to one of smaller diameter and greater height in one or two redrawing operations. Like the DWI process, it eliminates the side seam but the maximum height/diameter ratio that can be produced by multiple drawing is less than that achievable by DWI. However, DRD has a major advantage over DWI in that pre-lacquered tinplate can be used, thus eliminating the costly cleaning and spray-lacquering operations at the end of the process[150]. DRD cans are used for pet foods and some human foods, including baby foods.

Canmaking via DWI and DRD takes place at high speed and involves severe plastic strain. The steel, therefore, needs to be of the highest quality and a very low level of non-metallic inclusions is essential to the efficient operation of these processes. Gauge control is also important. The DRD process requires higher $\bar{r}$ values so that extensive drawing can take place. However, care must be taken to avoid an excessively large ferrite grain size which can give rise to an orange peel effect and a poor surface for lacquering.

Joining methods

One of the advantages of strip steel, as mentioned previously, is that it has the ability to be formed into components economically with relatively complicated shapes. A further advantage is that these components may be satisfactorily joined together to form complete assemblies using a wide variety of methods. Some of these methods involve joining at discrete points or larger areas at intervals along the join depending on the load-bearing requirements of the component. Some methods involve continuous joining along edges between components whereas other techniques, including the use of adhesives, involve joining over an area. Some of the methods involve melting of the steel (welding) or the use of additional materials including the use of adhesives, brazing or soldering. Some fastening techniques involve deformation or cutting of the steel whereas others involve the use of special joining components such as screws or rivets.

Each joining method has its own particular advantages and disadvantages, including the ease of joining, whether fume is produced, the time required, whether the method may be adapted to mass-production methods and whether access to both sides of the strip is necessary. It is clear, however, that the most important consideration is that any joint must provide the rigidity, strength (including fatigue strength) and resilience (ability to absorb energy) necessary for the application. It is also essential that any joint should not downgrade too seriously as a result of contact with environmental conditions. Some joining methods may be used in combination to complement each other's characteristics and others may be chosen because of the need to be able to undo a join and reassemble it later.

In service, a joint could in principle be subjected to a wide range of different modes of stress. It has been considered,[151] however, that there are three modes that can normally occur. These are two modes of tensile shear, parallel to or across the joint, and cross-tension normal to the joint.

Welding

A wide variety of welding processes may be used for the joining of strip steels which may be categorized into resistance and fusion types. Some of the more commonly used processes are resistance spot welding, resistance seam welding, projection welding, laser welding and MIG/TIG welding.

Resistance spot welding

In this method, heat is generated by passing an electric current through two or more sheets to be welded together by means of two electrodes which clamp on opposite sides of the sheets.

Spot welding is generally used for joining steel strip up to 3 mm thick, though thicknesses up to 10 mm can be welded by this process. The size and shape of the welds are controlled primarily by the size and shape of the water-cooled electrodes. In general, the weld nugget should be oval in cross-section but should not extend to the outer surfaces. The electrodes exert a significant clamping force on the strip materials and the contacting surfaces are generally heated by the passage of a high current (5000–20 000 A) using a low voltage (5–20 V). When sufficient melting is achieved, the current is switched off but the clamping force of the electrodes is maintained until the weld pool has completely solidified. The cycle is completed in a fraction of a second. Spot welding is generally carried out using AC current, but DC is also used.

The integrity of a spot weld is judged by means of a peel test or a chisel test, which separate the sheet materials after welding. In some cases, separation occurs through the weld and along the original interface of the materials (*interface failure*), or else the material tears around the weld nugget (*plug failure*). Tensile tests, using tensile-shear and cross-tension configurations, are also employed for evaluating the static strength of spot welds.

Weldability lobes

The heat generated during resistance heating can be expressed by the following equation:

$$H = I^2RT$$

where $H =$ the heat in joules,
$\quad\quad\ I =$ the current in amps,
$\quad\quad R =$ the resistance in ohms, and
$\quad\quad T =$ the time in seconds.

For a fixed welding time T_1, therefore, the diameter of the weld nugget will increase with current according to the relationship shown in Figure 1.102(a). At a particular level of current a_1, the weld diameter will reach what is regarded as the minimum acceptable size, namely a diameter of $3.5\sqrt{t}$, where t is the sheet thickness. As the current is increased beyond this critical level, the size of the nugget will increase until the stage is reached when the electrodes can no longer contain the molten metal, giving a condition which is termed *expulsion* or *splash*. This is achieved at current a_2 and this is regarded as unacceptable, irrespective of weld diameter. For a given welding time, therefore, there exists a range of current that will produce acceptable welds, i.e. from that which just meets the minimum acceptable size criterion to that which just avoids the splash condition. This type of exercise can be repeated for different weld times and an acceptable range of current can be defined for each particular time. These combinations of current and time that produce acceptable welds are then expressed in the form of weldability *lobes*, as illustrated in Figure 1.102(b). Thus currents or times below the lower bound of the lobe produce welds that are below the minimum acceptable diameter and which generally exhibit interface failure. Conversely, the combination of currents and times above the lobe will lead to the splash condition.

The width of the lobe can be taken as a measure of the weldability of a material, wider lobes indicating a greater tolerance to changes in production conditions.

Resistance spot welding of high-strength steels

Experience has shown that the weldability lobes of higher-strength steels are only slightly narrower than those of low-strength steels.[152,153] The acceptable range of current (*available current*) can be improved by using larger electrode forces and the optimum results are obtained by increasing these forces by about

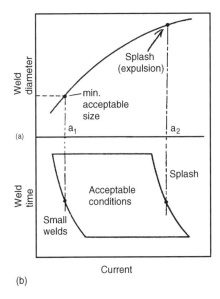

Figure 1.102 *Weldability lobes–resistance welding; (a) fixed time T_1; (b) for various times (After Jones and Williams[152])*

50% compared with comparable thicknesses of plain carbon steel. The available current can also be increased significantly by using larger electrode tip sizes. The width of the lobe, for example, can be increased by 700–2000 A by increasing the tip diameter from 4.8 to 6.4 mm.

Alloying elements increase the resistivity of low-carbon ferritic steels and consequently less current is require to produce a weld of a given size. The increased electrode forces, however, reduce the contact resistance. The overall effect, therefore, is generally a reduction in the current required.

The mode of failure can be different in higher-strength steels compared with low-strength steels. Both carbon and phosphorus can influence the fracture mode of spot welds and partial plug failures are the most common type of failure in peel tests on rephosphorized steels with a diameter of $5\sqrt{t}$. 100% plug failures, however, can be obtained in rephosphorized steel of 0.67 mm thickness by increasing the weld size to $7.5\sqrt{t}$.[152] The mode of fracture is, however, influenced by steel type, weld diameter, sheet thickness, geometrical factors and test type. Plug failures are promoted by large weld/high heat conditions. The optimum balance between strength and weldability is obtained by generally limiting the combined carbon plus phosphorus contents to a maximum of 0.18%.

It is useful to note that a considerable improvement to the fatigue life of welded joints may be obtained by increasing the weld size or by increasing the number of welds.[152] It was found, however, that the increase in strength was generally rather less than a simple multiplication of the strength of each weld by the number of welds due to the unequal loading of each weld during testing. It was considered that weld design, pitch and number need to be considered carefully if adequate benefit is to be derived from the use of higher-strength steel.

Resistance spot welding of zinc-coated steels

In relation to hot dip products, a distinction must be made between plain zinc and galvanneal coatings. Because of its lower contact resistance, higher currents are required for plain zinc-coated strip such as G275, compared with equivalent thicknesses of uncoated mild steel. On average, the currents have to be increased by about 20%. On the other hand, galvanneal-coated strip has a higher surface resistance than G275 and requires only a slight increase in current compared with uncoated steel. The weld times employed for zinc-coated steels may be 50% longer than those used for uncoated steel and the electrode pressure is generally increased.

The spot welding of zinc-coated steels is characterized by an electrode life which is shorter than that obtained with uncoated steels. This results from the alloying, mushrooming and erosion between the electrode material and the coating. Work by Williams[154] showed, however, that electrode life is affected very considerably by the type of coating. Thus for both the hot dip and electrolytically deposited coatings, the longest lives are obtained with iron–zinc and nickel–zinc coatings and values in excess of 500 welds can be achieved. Under non-optimum shop floor conditions, however, the number of welds which can be produced can be considerably less.

Resistance seam welding

The principle of resistance seam welding is similar to that of resistance spot welding except that the current is supplied to the work using electrodes in the form of wheels and the work moves continuously between the wheels to form a continuous weld. With coated steels, the electrodes can become contaminated and poor welds are produced. This problem is overcome in the Soudronic process by using continuously moving copper wires between the wheels and the work to act as intermediate electrodes, as illustrated in Figure 1.103. At each moment during the weld, therefore, the work is contacted by an effectively new and clean electrode.

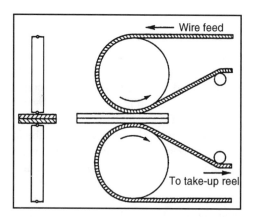

Figure 1.103   *Principle of the Soudronic welding process*[155]

Adequately wide ranges of current and welding speed are available for practical use. With some coated steels, however, the limiting factor is the tendency for cracking to occur in the weld or in the heat-affected zone. Interrupted current programmes, for example 2 cycles on and 2 cycles off, give a lower tendency for cracking and consequently a wider welding range.[155] The use of interrupted currents also reduces the tendency for surface brassing of the weld and for wire breaks.

An alternative method for coated steels is narrow seam welding, as illustrated in Figure 1.104. In this method, scraping devices and high pressure profile rollers are used continuously to clean and profile the wheel electrodes. Such machines are used for welding petrol tanks and exhaust systems. Welding speeds of up to 5 metres per minute may be achieved and narrow welds in the range 1.5–3.5 mm may be produced. These welds are stronger than the parent steel and are suitable for watertight and airtight applications.[155]

Laser seam welding

A laser provides a uniquely concentrated form of energy and is now being used for the manufacture of tailor-made welded blanks. Sheets of different strength, thickness or coating type are welded together prior to pressing the component. The method produces very narrow welds (~1 mm wide), free from distortion, with a minimal heat-affected zone. The welded sheets may be used to form complex pressings and when used with metallic-coated steels, there is only minimal damage to the coating and galvanic protection of the weld.

It has been found[156] that by using a 5 kW laser, satisfactory welds can be achieved at speeds of 10 metres per minute for 0.7 mm strip, dropping to 5 metres

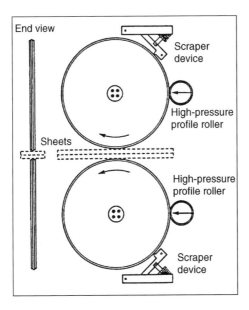

Figure 1.104 *Schematic detail of electrode wheel showing scraper and profiled friction drive wheel*[155]

per minute for 2.0 mm strip. A good fit-up between the pieces to be joined and good joint beam alignment is, however, necessary for these rates to be employed. Simulative forming tests showed that cup heights of 80–100% of those of the parent material may be obtained with the weld before failure. The situation is more complicated when the sheet components have a significantly different thickness.

Laser-welded blanks are used in automotive construction for components such as door inners, monosides and longitudinal rails, as previously illustrated in Figure 1.33. The benefits depend on the part and on the production volumes. Laser seam welding is also now used for vehicle construction to produce continuous joints and improve torsional stiffness. The number of applications is small at present but is likely to increase in the future. The narrow weld results in low distortion. Close tolerances are, however, required on joint fit-up.

MIG and TIG welding

These abbreviations refer to metal inert gas and tungsten inert gas respectively. Both processes are arc welding methods in which an inert gas is used to protect the weld. In the MIG process the weld is prepared using a consumable electrode which melts to form part of the weld metal. In the TIG process a non-consumable tungsten electrode is used and the weld metal comes from the pieces to be joined.

Mechanical joining

One of the simplest methods of joining along a straight edge is to use what is called *lock forming* or *clinching*, as illustrated in Figure 1.105. Modifications of this method are particularly suitable for the manufacture of box sections, as shown in Figure 1.106. Another technique involves the use of a pneumatic tool which cuts holes in the pieces to be joined and bends two tags through the holes to make a joint, as illustrated in Figure 1.107.

Press joining provides a further technique for joining sheet materials.[157] The tooling for this method is illustrated in Figure 1.108 and consists of a punch, an

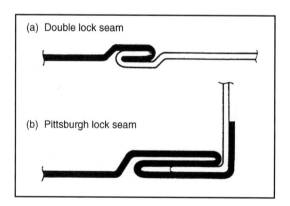

(a) Double lock seam

(b) Pittsburgh lock seam

Figure 1.105   *Two forms of lock seaming*[155]

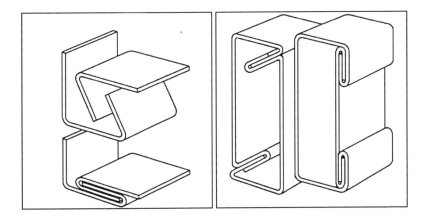

Figure 1.106   *Types of clinched sections*[155]

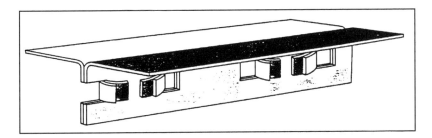

Figure 1.107   *A diagram illustrating the type of joining using a stitch folding gun*[155]

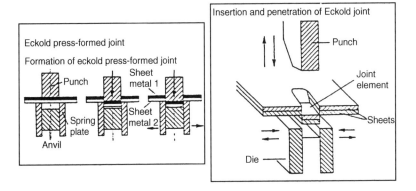

Figure 1.108   *Diagram illustrating the formation of an Eckold press-formed joint*[155]

anvil and an expanding die. When the punch has penetrated the double thickness of steel to be joined, the displaced element of steel is compressed between the punch and the anvil. This causes the element of the upper sheet to spread beneath the lower sheet thus forming the joint. It is found that the strength of press-formed joints depends on the thickness and strength of the steel being joined. The maximum load that can be generated in a shear test is higher in the direction perpendicular to the two cuts, but the elongation to failure is several times greater in the direction parallel to the cuts than in the perpendicular direction.

The method has been adapted by the use of multi-point tooling to make a large number of separate joins by means of one movement of the tool. The orientations of each joint may be selected to give the optimum directionality of the strength of the final composite joint.

Joining using fasteners

Fasteners include screws and rivets. Formerly, these required holes to be drilled prior to the insertion of the screw. Self-drilling and self-tapping screws, however, are now available which greatly simplify assembly since access to the reverse side of the steel is not needed. A similar benefit may be obtained, however, using compressive fasteners which have a nut which forces the legs apart to clamp two steel sheets together.

A recent advance is the development of self-piercing rivets which are now coming into common use. The basic form of this device is semi-tubular, as illustrated in Figure 1.109(a). The joining process itself occurs in two stages but these are combined into a single continuous operation. In the first stage, the rivet is pressed into the two sheets to be joined and its shank cuts a hole through the top sheet and partially through the bottom sheet. In the later stages of travel through the bottom sheet, the sheet material is displaced by the die causing the shank of the rivet to flare out and lock itself into the steel. A typical cross-section of a riveted joint is illustrated in Figure 1.109(b).

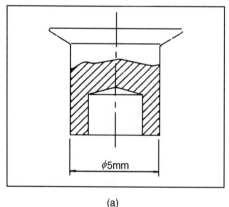
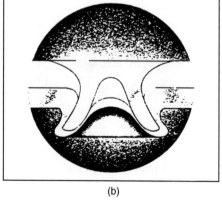

(a)   (b)

Figure 1.109 *Diagrams illustrating the cross-section of a self-piercing rivet (a) before and (b) after riveting*[158]

A large force (typically 40 kN) is required to make the joint and this is provided by means of a C-frame which is able to accommodate the rivet setter and the upsetting die. When used to join two sheets of unequal thickness, it is preferable for the thin material to be on the side that is fully pierced. It has been claimed that the fatigue performance and energy absorption of self-piercing rivet joints are equivalent to or better than spot welded joints.[159]

Use of adhesives in weldbonding, clinchbonding and rivbonding

Adhesives may be used to join various components and are widely used to join steel to dissimilar materials. Developments are still in progress, however, concerning joints that are required to carry a structural load which are likely to be subject to severe weather conditions in service. For potential applications in the automotive industry, for example, the adhesive is generally considered for use in combination with other joining methods such as spot welding, press joining or riveting. These hybrid methods are known as weldbonding, clinchbonding and rivbonding respectively.

Detailed studies[160] have shown that each method could be used under normal automotive production conditions and that they could be used to join two, three or four thickness combinations of steel together. It has been found that for simple joints and arrays, each hybrid outperforms its parent method (without the adhesive), but the extent of the improvement was markedly greater for shear- than for peel-loading conditions. This can be attributed to the fact that adhesives are significantly weaker in peel than in shear because of the reduced ability of the steel to dissipate stress along the bond line under this mode of loading. The use of the hybrid methods leads to an improvement in both the collapse resistance and torsional stiffness of box hat structures. The extent of the benefit was, however, greater for the former than for the latter loading regime. This was attributed to the limited ability of the adhesive to withstand large deformations associated with impact loading.

Strip steel in use

Fatigue

The load-bearing capability of any component manufactured from sheet steel depends on its size and shape and on the thickness and strength of the steel used to make the component. Most components are, however, joined together with other components to form complete structures and the strength of any complete structure clearly depends on the way the components are joined together as well as on the strength of the individual components. Examples of such structures would include the complete body shell of a motor car, the frame of a steel-framed building or a cylindrical steel drum with each end connected to the side wall in a suitable way (see below).

Many components and structures are subject to vibration in use. The components and joints may, therefore, be subject to fatigue damage, whereby crack formation and failure could occur at a lower stress than would cause failure by the single application of the stress. The design of any structure must, therefore, take the possibility of fatigue damage into account. This is done by estimating fatigue life at critical positions in any structure and by modifying the design, if necessary, to ensure that the life is adequate for the intended application. The critical positions are usually at discontinuities such as holes, joints or corners where the local stress is high. The resulting estimate will depend on an interaction between the local stresses developed in the structure and the properties of the material used in its formed or welded condition.

Total fatigue failure usually occurs in one of two stages. The first is crack initiation and the second is crack propagation. The total fatigue life or number of strain cycles to failure N_t may, therefore, be estimated by combining the crack initiation life N_i with the crack propagation life N_p. These two contributions may be estimated using data obtained from different types of specimen. The motor industry is, however, more concerned with crack initiation.[161]

Fatigue crack initiation

The crack initiation fatigue life may be estimated using data obtained from smooth specimens, often in the form of a small tensile specimen with a gauge length of only a few millimetres. For tests involving both tension and compression, the ratio of length to thickness must be small enough to enable sufficient strain to be given in compression without leading to buckling. Each specimen may be subjected to a cyclic strain of constant amplitude, alternating between fixed amounts of deformation in tension and compression and often using a triangular waveform. It is often found that the cyclic stress developed during each test varies gradually during the test, particularly for the higher strains, and then reaches a constant value. It finally falls rapidly as the first cracks are initiated at which point the test is terminated. For lower strains which are almost completely elastic, the stress developed remains substantially constant up to the stage of crack initiation.

The number of cycles to crack initiation always increases as the applied strain decreases and these parameters may be plotted against each other to give what is called a strain–life curve. Examples of this relationship are given in Figure 1.110. Plastic deformation occurs at the higher values of strain and the number of cycles to give crack initiation becomes relatively small, depending on the strength and other characteristics of the steel. At very low values of applied strain, sufficient cycles to give crack initiation can never be applied. For practical purposes, therefore, N_i becomes infinite.

The stresses and strains developed half way through the various tests and in the stable region may be plotted against each other to form what is known as the *cyclic stress–strain curve* for the steel. It may be used to calculate the local strain that would be developed as a result of any localized applied cyclic stress.

In service, any component may be subjected to a fairly random sequence of strains and it is clearly necessary to know the expected stress history of any critical part of a structure before any estimate of fatigue life can be made. The local strain history may then be calculated using the cyclic stress–strain curve. A method must also be available to be able to calculate the contribution to fatigue damage that would arise from each of the strains in the expected strain history.

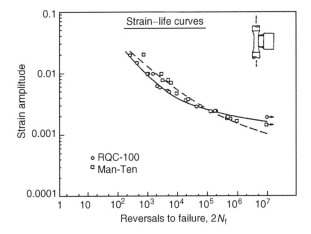

Figure 1.110   *Strain-life curves for two structural steels (After Landgraf[162])*

In this way the total fatigue damage and hence fatigue life N_i to crack initiation may be calculated. The simplest method to use is the linear damage rule put forward by Palmgren[163] and Miner.[164] According to this rule, if the proportions of the various strains m in the strain history are p_1, p_2, $p_3 \ldots p_m$, etc., and the numbers of cycles to crack initiation for these strains taken from the strain-life curve are N_1, N_2, $N_3 \ldots N_m$, etc., then the cumulative damage is given by:

$$\text{Damage} = p_1 N_i / N_1 + p_2 N_i / N_2 + p_3 N_i / N_3 \ldots, \text{ etc.}$$

and failure is expected to occur when $\Sigma p_m N_i / N_m = 1$.

A difficulty with this method (or any other method) is that the calculation of the proportion of each active strain from a highly irregular and random sequence of cyclic strains is not straightforward. Several methods have been proposed of which the rainflow method is probably the most accurate one. Further details of these methods are given in a complete conference proceedings.[165]

An alternative way of carrying out a fatigue test is to use a series of constant cyclic stresses to develop a stress–life curve. It is said[162] that this approach is more suited to long-life fatigue problems but less suited to low-cycle problems involving relatively large plastic strains. With this method, a stress may be identified below which the number of cycles to crack initiation is never reached. This stress is regarded as the fatigue strength of the steel. The relationship between the fatigue strength and the monotonic tensile strength depends on the yield to tensile strength ratio of the steel[166] since the yield or proof stress is the parameter which influences the onset of plastic deformation.

Fatigue crack growth

Fatigue crack growth rates may be measured using a standard testing method[167] and a type of specimen, as illustrated in Figure 1.111. Before testing, a crack is formed at the notch tip using a suitable load cycle and a stress intensity factor at the crack tip is defined which depends on the crack length, the specimen

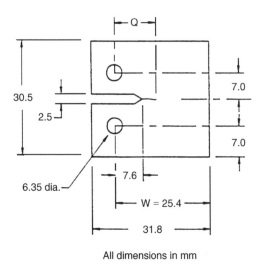

All dimensions in mm

Figure 1.111   *Compact tension specimen used for crack growth test*[167]

geometry and the difference between the minimum and maximum loads applied. During the test the crack length is monitored using a travelling microscope and a plot of crack length versus number of cycles enables the crack growth rate per cycle to be obtained. A fatigue threshold is defined as the maximum stress intensity factor below which an existing crack does not propagate. It represents a fatigue limit related to crack growth.

The cyclic stress–strain curve

The method of obtaining the cyclic curve mentioned above involves the use of several specimens. Alternative methods have been proposed for obtaining the cyclic curve from a single specimen. Some of these involve the waveforms illustrated in Figure 1.112.

It was indicated above that the cyclic stresses developed as a result of the application of a constant cyclic strain may often vary during the test. The implication is that the cyclic stress–strain curve is often different from the monotonic tensile curve obtained from a conventional tensile test. The monotonic curve would correspond with the first part of the first cycle in a fatigue test. Examples of cyclic softening, cyclic hardening and cyclic stability, or a mixture of all three, have been observed for various types of steel when plastic strain is involved in the test but the strength is stable when the deformation is substantially elastic. It is found that the effects tend to be associated with the strengthening mechanisms used in the steel.[168] Strengthening by solid solution effects, grain refinement and precipitation, for example, tend to be stable in the presence of plastic deformation whereas strengthening by temper-rolling or bake-hardening effects tend to be unstable. It must be remembered, however, that components made from bake-hardening steels are not subject to plastic deformation over their main area during use. The increase in dent resistance arising from the bake hardening will not, therefore, be affected by fatigue. Studies of the cyclic stability of steel cut

Method	Strain waveform
Constant strain amplitude (plastic or total strain) (several specimens)	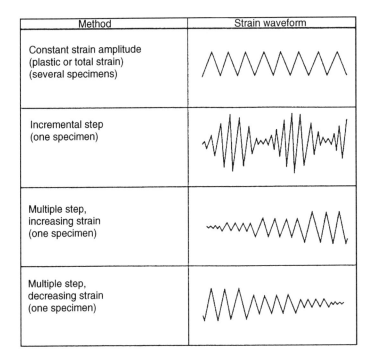
Incremental step (one specimen)	
Multiple step, increasing strain (one specimen)	
Multiple step, decreasing strain (one specimen)	

Figure 1.112 *Different strain waveforms used to measure a cyclic stress-strain curve*

from pressings involving different types of forming strain[169] have also shown that the increase in strength from the cold work during forming may be lost during cyclic deformation. It is a safe procedure, therefore, to disregard the increase in strength due to forming strains for design calculations.

Cold work embrittlement

Any structure must clearly have sufficient strength to survive normal use and sufficient ductility to withstand any unusual impacts. It has been found, however, that certain IF steels may be subject to a condition known as *secondary cold work embrittlement* (SCWE). This condition may lead to brittle fracture after the material has previously received severe strains such as those encountered during deep drawing. The cracking is accompanied by very little plastic deformation and consequently very little energy absorption and the initial fracture surfaces have been reported to be mainly intergranular.[170,171] There is little evidence that SCWE has caused any serious problem in service,[172] but nevertheless it has been the subject of extensive research to show how the condition may be avoided.

The condition has been mainly studied using three types of laboratory test, but the most commonly used test involves a cylindrical cup. A circular blank of steel is first deep drawn to form the cup and a truncated conical plunger is used to open the cup end to determine whether the rim of the cup is brittle. The plunger may be pushed slowly into the cup as illustrated in Figure 1.113, using a

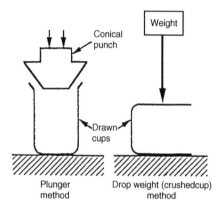

Figure 1.113   *Schematic diagrams illustrating two forms of an SCWE test (After Henning[173])*

tensile machine, or it may be dropped from a suitable height, usually 1 metre. In each case, the test is carried out at various temperatures below room temperature and a temperature is established at which a brittle crack is first formed. This temperature, known as the ductile to brittle transition temperature (DBTT), is a measure of the brittle tendency of the steel and decreases as the tendency for brittle behaviour decreases. It is found, however, that the transition temperature is influenced by the drawing ratio used to produce the cup and increases with increasing drawing ratio.[172]

In a variant of the above test, a cylindrical cup is also used but the weight is dropped onto the side of the cup, as also illustrated in Figure 1.113. Again the DBTT is defined as that at which brittle cracks first form. The third method involves bending.[173] Suitable strips of steel are bent through 180° and then opened at various temperatures below room temperature using five samples at each temperature. At a sufficiently low temperature, the samples crack open in a brittle manner and the transition temperature is defined as the temperature at which all five specimens crack in this brittle manner. The three methods do not give the same transition temperature for the same steel but there is a correlation between the results. Figure 1.114 shows, for example, a correlation between the transition temperature from the bend test and that from the side crush test. It is useful to note that the transition temperature from the bend test is higher than that from the side impact test. This implies that the bend test is more severe. It has recently been suggested that the bend test could become more widely used.[174]

2244225824742460 white space

As indicated earlier, the concern about SCWE covers IF steels and particularly those containing phosphorus.[175,176] In the absence of interstitial carbon, the phosphorus may segregate to the grain boundaries during annealing and promote intergranular fracture by causing grain boundary embrittlement. The effect is greater for batch-annealed steel than for continuously annealed steel[177] since slow cooling gives more time for segregation. Carbon in solution reduces the effect,[178] but the main way of controlling the condition is by the addition of boron.[176] A very small quantity, such as 0.0005%, is sufficient to reduce the

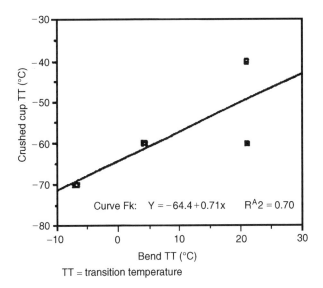

Curve Fk: Y = -64.4 + 0.71x $R^2 = 0.70$

TT = transition temperature

Figure 1.114 *Relation between brittle transition temperature measured using the crushed cup and the bend test methods (After Henning[173])*

tendency to SCWE,[179] but not sufficient to have a noticeable detrimental effect on the *r* value.

Strip steel in automotive applications

The automotive industry is the most important single market for steel strip and has provided the greatest stimulus and challenge for the development of new products, new ways of using them and greater product consistency. In spite of the competition from aluminium alloys and plastics, steel has maintained its position as the predominant material for the car body and structural components due to its good formability, high modulus of elasticity, ease of welding and relatively low cost. Various grades of steel now comprise 50–60% of the vehicles weight. Since the 1970s, however, significant changes have taken place in the selection of strip steels for automotive construction:

1. The need to reduce fuel consumption and fume emissions while maintaining structural performance.
2. The need to improve corrosion performance to be able to provide customers with better warranties against structural and cosmetic deterioration.
3. The need to improve passenger safety.

Reduced fuel consumption may clearly be achieved by improved engine efficiency, but it has also been achieved by reducing vehicle weight. Figure 1.115 shows, for the top 30 registered European cars, how the kilometres travelled per litre increases with decreasing weight, but also shows that, for a given weight, there was a general increase from 1982 to 1992 due to other reasons. Clearly, a

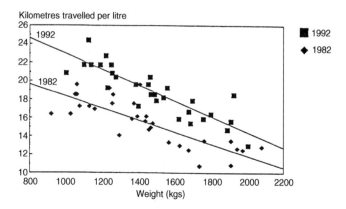

Figure 1.115 *Plots illustrating the variation in kilometres travelled per litre of fuel with car weight for 1982 and 1992 (After[180])*

car with a low weight and hence low fuel consumption, will be attractive from an economic point of view but it will also lead to a reduction in the formation of the 'so-called' *greenhouse gas*, CO_2, that can lead to global warming.[175]

Weight reduction and the use of high-strength steels

During the early 1970s, higher-strength steels were introduced in the United States for safety-related or structural members such as bumper reinforcements, side door intrusion beams and seat belt anchors. These components were manufactured mainly from hot-rolled, niobium-treated, micro-alloyed steels which provided a favourable cost/strength/weight ratio compared with lower-strength unalloyed steel or any other strip steel available at that time. In addition, their use necessitated only minor changes in manufacturing methods and facilities. It was unfortunate that since these were new components, their use inevitably led to an increase in vehicle weight. More recently, higher-strength thinner gauge steels have been used for some of these components to take out some of the additional weight that had been introduced. These have included the ultra-high-strength steels, with a tensile strength above 1000 MPa, strengthened by a high proportion of transformation product, and steels which develop their ultra-high strength by heat treatment and quenching after forming.

The oil crises of 1973 and 1979 provided the initial stimuli for weight reduction itself. The car body is assembled from large body panels and constitutes about 25–30% of the total weight of a medium-size car. It is, therefore, the heaviest vehicle component. It was clear, therefore, that reducing the weight of the body could have a significant impact on the total weight of the complete car. It was realized that this could be done by the substitution of steel by aluminium or plastics, but the penalty would have been increased cost as illustrated in Figure 1.116. It was generally accepted that the use of high-strength steel, which would enable component performance to be maintained at a reduced thickness, was the only way to achieve both weight and cost savings.

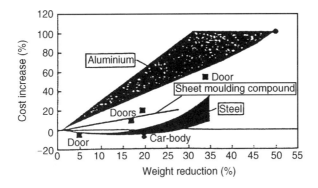

Figure 1.116   *Costs of different lightweight construction methods*[181]

The micro-alloyed steels mentioned earlier were suitable for use for relatively simple structural parts, but their reduced formability and increased springback compared with lower-strength mild steel prevented their use for the large body panels, which in general involved difficult forming operations. In addition, the formability of any steel decreases with decreasing thickness as indicated by the lower plane strain intercept on the forming limit diagram. Any decrease in thickness, therefore, only exacerbates the loss in formability that arises from the increase in strength.

The reaction of the automobile industry world-wide was to request steels with a relatively small increase in strength over mild steel, with the loss in formability reduced to a minimum while still retaining high r values. The first of these steels were the rephosphorized steels which, as indicated previously, retained higher r values and higher elongation values than would have been obtained from any micro-alloyed steel of similar strength. Steels with minimum proof stresses of 180, 220, 260 and 300 N/mm^2 have been available in Europe, though the grades with the lower strength are the ones which have been more commonly used. The early rephosphorized steels were batch annealed and based on an aluminium-killed composition, but a more recent trend has been to anneal different types of steel continuously, either on a continuous annealing line or on a hot dip galvanizing line. The hot dip-coated steels, usually with a galvanneal coating for the motor industry, tend to be based on an IF-type composition to be more compatible with a continuous annealing cycle. These steels, as mentioned previously, may also be solid solution strengthened with manganese, boron or silicon as well as a small phosphorus addition to reduce problems that would otherwise be associated with high phosphorus levels.

The solid solution-strengthened steels made by alloying either an aluminium-killed or an IF base composition are sufficiently formable to enable difficult body panels to be produced without local necking or splitting, but the pressing has still been subject to greater springback than lower-strength mild steel. The trend has, therefore, been to use bake-hardening steels. These steels usually have a minimum bake-hardening index of 40 N/mm^2. Very approximately, therefore, it is possible with a bake-hardening steel to achieve the same formability and springback as the equivalent grade of non-bake-hardening steel but with the performance in service of one grade higher.

Many applications of bake-hardening steels aim to provide improved plastic dent resistance mainly in lightly deformed parts of panels with low curvature such as bonnets and boot lids. Drewes and Engl[183] presented a relationship between yield stress, sheet thickness and load required to give the same residual dent depth of 0.2 mm as illustrated in Figure 1.117. This may be used to calculate what reduction in gauge could be obtained for the same dent resistance by using a higher yield stress. The curves for rephosphorized, micro-alloyed and bake-hardening steel produce similar effects for the same sheet thickness but the dual-phase steel gives a different effect due to the high rate of work hardening.

Figure 1.118[184] gives an example of the various types of high-strength steel that may typically be used for the various parts of an existing motor vehicle with an indication of change for the future. Figure 1.119 gives a summary of the historical development of the use of high-strength steel in cars from a European point of view.

Hitherto, weight reduction has mainly been achieved by substitution of thinner steels for thicker ones without any major change in design concept. Recently, a major project[185] has been initiated, with phase 1 sponsored by 32 large steel companies across the world, to maximize weight reduction using any existing manufacturing processes and steel grades that are available. The project known as the ULSAB (ultra light steel auto body) project, aims to design a body-in-white (BIW) with the same or higher-performance features as existing designs but with a reduced weight. Phase 1 of this project has been completed[185] and progress has been reviewed.[186,187]

A bench mark of existing designs was obtained by taking mean values over a number of suitable, mid-size, existing designs and normalizing the figures to take

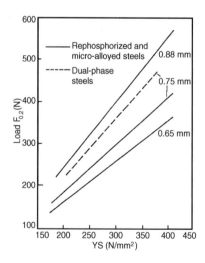

Figure 1.117   *Effect of yield strength and sheet thickness on dent resistance under quasistatic localized loading ($F_{0.2}$ = load producing 0.2 mm remaining dent depth) (After Drewes and Engl.[183])*

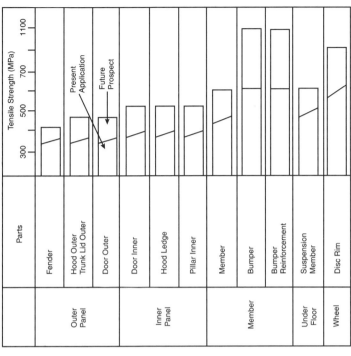

Figure 1.118 *Examples of applications of high-strength steels*

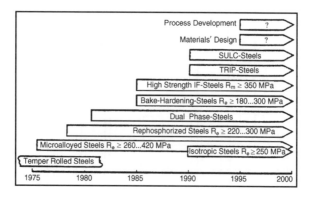

Figure 1.119 *The development of high-strength, cold-rolled steels for auto-body applications. The time axis reflects the approximate first industrial usage, from a European perspective, in the automotive industry (After Bleck[94])*

Table 1.14 *Progress of the ULSAB project at the end of phase 1*

	Representative figure post-2000	Performance target	Achieved figure
Mass	250 kg	200 kg	205 kg
Static torsional rigidity	13 000 Nm/deg.	>13 000 Nm/deg.	19 056 Nm/deg.
Static bending rigidity	12 200 N/mm	>12 200 N/mm	12 529 N/mm
First BIW mode	40 Hz	>40 Hz	51 Hz

account of different vehicle size. The main performance features being considered are the static torsional rigidity, the static bending rigidity and the first BIW mode. The latter is the resonant frequency for vibration of the body and should be as high as possible for passenger comfort. Table 1.14 gives the properties expected to be representative of a typical mid-size body in the post-2000 time period, after reasonable further improvements to existing types of design have been made. It compares them with the targets for the ULSAB design and the figures that have been achieved. It is seen that the main objective was to give a 20% reduction in weight from 250 to 200 kg while maintaining or improving the body performance. It is evident that the performance objectives have been achieved comfortably but that the weight reduction target has not been quite achieved.

The approach to the design has been a holistic one. Thus, parts of the body have been increased in weight to be able to make greater reductions elsewhere. Hydroforming is to be used to make the roof-rail sections, which could not easily be made using conventional forming techniques. Some parts are made by roll forming and some are made from *sound deadened steel* as described below. The torsional and bending rigidities of the structure have been improved by weld bonding, but a number of continuous joints have also been achieved using laser welding. The complexity of the structure has been reduced by the use of laser-welded blanks thus obviating the need for additional strengthening members. The estimated cost of production has been given as $154, less than for a typical mid-size BIW and representing a saving of 14%. It has been said, therefore, that there would be a case for manufacturing a ULSAB body even if no savings in weight had been predicted.

It is useful to note that steels with a yield stress above 210 MPa have been specified for two-thirds of the parts in the ULSAB design and a high proportion of the steel used is also coated with one or other of the corrosion protective coatings available. It is clear, therefore, that corrosion protection, as well as mechanical performance, has been an important consideration of the design. Several steel bodies are being built to confirm and demonstrate the phase 1 design and performance. These are to be exhibited around the world during 1998.

Improved corrosion resistance

In recent years, the use of zinc-coated steels in car body construction has increased dramatically as manufacturers have attempted to improve their warranties against corrosion. Two types of corrosion have to be considered, namely:

1. Perforation corrosion which takes place from the inside of the vehicle to the outside, causing a hole to appear.
2. Cosmetic corrosion which takes place on the external surfaces of the vehicle.

Most manufacturers have incorporated about 70% coated steel in the body-in-white and it is reported that Audi has adopted a 100% coated steel construction. Both hot dip and electro-galvanized steels are being used in place of uncoated, cold-reduced steel and their application to different parts of the car body has been described by Dasarathy and Goodwin.[45] This is illustrated in Figure 1.120 which refers to the materials listed below. It is useful to note, however, that there is a general trend away from the use of electro-coated material towards the use of galvanneal and that hot dip aluminium coated steels are used for car silencers.

- CRIFF: uncoated steel
- Galvatite: hot dip zinc
- Galvanneal: hot dip iron–zinc alloy

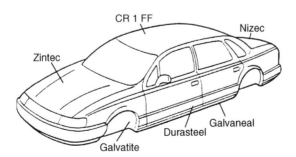

Figure 1.120 *Typical automotive body panels fabricated from metallic-coated steels (After Dasarathy and Goodwin*[45]*)*

- Nizec: electro-deposited 87% Zn 13Ni
- Durasteel: Nizec + 1.0 μm organic coating

Durasteel is one of a number of proprietary steels with a thin organic coating which is intended primarily as a lubricant for complex forming operations but which also provides corrosion resistance. After forming, the organic coating is on the inside surface of the car where it is subsequently electroprimed but not finished with a surfacer or topcoat.

The substitution of uncoated steel by coated products in automobiles introduces a number of production implications and the experience in the Rover Group has been described by Thompson.[188] One very important consideration is that the zinc surface is significantly softer than an uncoated steel surface and is therefore more susceptible to damage during decoiling and handling. However, these problems have been overcome at Rover by training programmes rather than facility changes. The soft nature of the zinc coating can also cause zinc to be wiped off the steel by the tools during deep-drawing operations. However, it is reported that the wiping action has little effect on the corrosion behaviour as only a small amount of zinc is actually removed and the coating maintains sacrificial protection. On the other hand, the pick-up of zinc on the tool necessitates more frequent tool cleaning and a reduction in the output of the press.

Thompson reports that the blanking process can leave small particles of zinc around cropped edges which could deposit on the tools and produce surface imperfections during pressing. However, this problem has been overcome by the installation of blank-washing equipment. Although there was concern initially that the introduction of coated steels might require major press tool modifications because of the different frictional characteristics of the surface, no tool modifications were necessary and the problems have been overcome by improved housekeeping and routine maintenance. In relation to the welding implications, the author comments on the need for changes in machine parameters to cope with the increased current (10–20%) requirements of coated steel and also on the greater frequency of weld tip dressing due to the build-up of zinc on the tip surface.

Reference was made earlier to *Durasteel*, which has a 1 μm organic coating over a zinc-rich substrate. However, duplex coatings with organic components of 5–7 μm thick are also finding application in car body construction in order to improve the corrosion performance and resistance to stone-chip damage. Bowen[189] has stated that the incidence of stone-chip damage has increased due to the greater use of road salt and also because of changes in vehicle design. The latter is related to the practice of body streamlining in order to reduce air resistance and the drag coefficient. Thus the paintwork on a sloping bonnet has become subjected to greater attack than was the case on unpainted radiator grilles. This problem is also exacerbated to some extent by the fact that zinc coatings are inferior substrates to bare steel from the point of view of achieving good paint adhesion and therefore stone-chip damage is more evident. Bowen states that this problem can be overcome by using duplex coatings and, as an example, describes the use of *Zincal Duplex*, an electro-galvanized steel coated with *Bonnazine 2000*, an epoxy-based primer, filled with metallic zinc and aluminium and containing molybdenum sulphide as a slip agent during forming. It is reported that Ford are applying duplex coatings to the bonnets of all Granadas/Scorpios and Fiestas, achieving a 20% reduction in stone-chip damage and the complete elimination of red-rust corrosion.

Dasarathy and Goodwin[45] comment on the need for preprimed steel for automotive construction which would replace the electropriming and surfacer primer stages of finishing at the manufacturers. Such a product would be hot dip or electro-deposited zinc-coated steel with a 25 μm organic coating. However, the authors state that this type of material is still very much at the development stage and one of the major problems that must be overcome is that current organic coatings are not weldable.

Before leaving automotive applications, reference should also be made to the use of terne-coated steel products, i.e. with tin–lead coatings. *Ternex*[190] has a hot dip coating of 92% Pb 8% Sn, applied over a nickel flash, electrolytically deposited over the steel surface. Such materials are resistant to attack by automotive fuels and are used extensively for the manufacture of petrol tanks.

Strip steels in buildings

The construction industry is a major consumer of strip steel but still represents a major growth sector for this product.

Steel-framed houses

Following earlier problems, interest has now been revived in steel-framed housing since it offers many advantages over traditional construction. In particular, the frame can be erected very quickly and made weatherproof so as to allow internal work to proceed at an early date. Labour costs and construction times are also reduced which leads to lower interest costs and faster financial returns. Benefits also accrue to the householder since the steel frame does not absorb moisture and is not subject to shrinkage or warping. The risk of cracking in wall linings is therefore reduced and the steel frame is also fire resistant.

Earlier attempts to introduce steel-framed houses were unsuccessful because the designs were geared to rapid construction rather than aesthetic appeal. The initial steel frames were also manufactured in painted mild steels which were prone to corrosion problems.[191] In current construction, these problems have been overcome by:

1. Preserving a traditional exterior of bricks and mortar around a load-bearing steel frame.
2. Employing hot dip galvanized steel in place of painted steel for corrosion resistance.

An illustration of the construction of a steel-framed house with the preservation of a traditional exterior is shown in Figures 1.121(a) and (b).

The structural members that make up the steel frame are roll-formed U channels in Z28 (280 N/mm^2 min. YS) grade steel with a zinc coating of 275 g/m^2

Figure 1.121   *Steel-framed housing (Courtesy of Precision Metal Forming Ltd)*

(including both sides). According to Haberfield,[191] modular frames, typically 5 m × 2.4 m high, are fabricated at the factory and bolted together on site on traditional foundations. The internal partitioning walls are constructed and assembled in the same way, followed by the second storey and roof. Steel joists are used to support the floor loadings from the second floor.

The above author states that up to 2 tonnes of hot dip galvanized steel sections may be used in a three-bedroomed house and the thickness of the sections ranges from under 1 mm to 2.5 mm or greater, depending on the function. The basic frame is designed to withstand all forces and no strengthening or stiffening contribution is assumed from the external brick or internal plasterboard linings.

Steel cladding

From the early days of 'corrugated iron', strip products have featured prominently as cladding materials, particularly in the construction of industrial buildings. A considerable amount of effort has been devoted to the development of structurally efficient profiles for steel cladding and some typical profiles are shown in Figure 1.122.

Although a significant part of the market for cladding and roofing is still satisfied by unpainted, or zinc- or zinc–aluminium-coated strip, this area has been revolutionized by the introduction of pre-finished, organic-coated strip. These

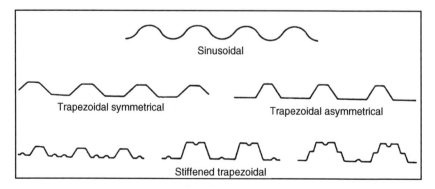

Figure 1.122   *Typical profiles of organic-coated steel cladding (After Lewis et al.[192])*

products were described in an earlier section and involve a variety of paint or plastic formulations which are applied to zinc-coated steels. These coatings provide enhanced corrosion resistance and are available in a wide range of colours and surface textures.

The types of coating that are recommended for external use include PVC Plastisol (200 μm), PVF 2 (27 μm), Silicone Polyester (25 μm) and Architectural Polyester (25 μm).[148,149]

Steel lintels

Lightweight steel lintels have now virtually replaced reinforced concrete lintels in domestic housing. These components are specified in BS 5977: Part 2: 1983 *Specification for prefabricated lintels*, which provides guidance on the design and testing of all types of lintel for use in domestic buildings up to three storeys high. This specification deals with post-galvanized lintels with heavy zinc coatings and also those fabricated from pre-galvanized strip with minimum zinc coatings of up to 600 g/m^2 (including both sides).

Other applications for strip steels

Domestic appliances

On page 94, information was provided on the production and properties of prefinished strip, coated with various types of paint or plastic. During the late 1960s, work was initiated on the use of pre-painted strip for the high-gloss, white domestic appliance market and this proved to be a major commercial success. Thus the majority of wrap-arounds for washing machines, refrigerators, dish washers, domestic boilers and freezers are now produced from pre-painted strip, providing the following benefits to the manufacturers:

1. Higher productivity.
2. Savings in floor space.
3. Reduced capital expenditure.
4. Elimination of the effluent problems associated with paint work.

The organic coating used for this type of work is a high-gloss polyester, 25 μm thick over a galvanized steel base, which provides a high flexibility/hardness ratio and good resistance to elevated temperatures and detergents. Although the degree of cold forming is not very severe for the majority of the components involved, areas such as the ports for front-loading washing machines are subjected to a limited drawing operation.

Steel drums

Steel drums, with capacities ranging from 25 to 210 litres, constitute a major market for cold-reduced strip but one that is now facing major competition from plastics. The formability requirements are not high and steel drums are normally made from uncoated, Fe PO1 strip. However, limited use is made of galvanized steel to provide greater corrosion resistance against particular products.

For the future, there is the prospect of using higher strength steels in lighter gauges in an attempt to stem the greater penetration of the market by plastics. However, attention must also be given to improved designs in order to preserve adequate rigidity in down-gauged drums.

Sound deadened steel

Since 1991 there has been in Europe a regulation requiring that noise levels must be below certain specified values that would otherwise cause damage to hearing. An indication of the noise level that could damage hearing is given in Figure 1.123. In addition, a low noise level has become an important selling point for cars and other products. This may be achieved by appropriate design but noise levels may also be reduced by the application of various sound absorbing materials, such as mastic, onto hidden surfaces.

A special product is now available from at least five companies in Europe to provide sound absorption without the need for the application of any additional material. It is known as *sound deadened steel* and consists of two sheets of steel enclosing a thin layer of a viscoelastic polymer.[194] In use, vibratory shear strains are developed in the polymer by relative motion of the two sheets and this leads to the damping of the vibration and the development of a small amount of heat. The polymer may be formulated to give maximum damping at various temperatures up to 150°C as required and is stable up to 200°C. It does not, therefore, lose its damping capability as a result of paint baking. The degree of damping achieved depends on the shape of the component and the method of support.[195] It also varies with the frequency of the vibration, as illustrated in Figure 1.124, which gives an example of the effect of sound deadened steel on the airborne noise from a garage door.[193] It is seen that the noise is reduced by an average of about 10% within the frequency range 200–10 000 Hz. Another possible advantage is

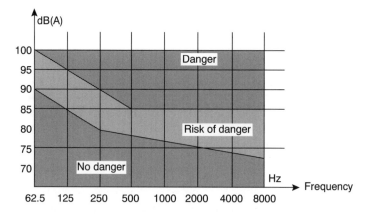

Figure 1.123 *Noise levels expected to have an effect on hearing ability (After Mathieu[193])*

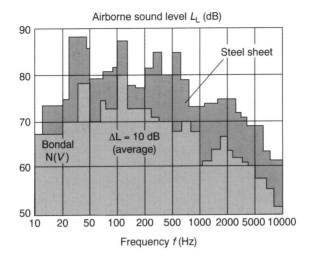

Figure 1.124   *A comparison of the sound level from a garage door made from sheet steel and from sound-deadened steel. The average difference was 10 dB (After Mathieu[193])*

that the noise reduction may be accompanied by a weight reduction as a result of the elimination of the conventional materials that would otherwise be needed to reduce noise, such as felt.

The formability of the composite depends on the formability of the steel and the adhesion of the polymer since the latter influences the tendency for wrinkling in areas of a pressing subject to in-plane compression. During forming the two sheets may move relative to each other to give an offset of the edge. Holes for fixing must, therefore, be made after forming. The polymer acts as an insulator. Spot welding may, therefore, be achieved by the use of shunts which lead to local heating and softening of the core material. This enables it to be compressed and allows the weld current to pass and make a satisfactory weld.

The material already has many applications in the manufacture of cars, including engine enclosures, oil sumps and rocker covers and its use is spreading to other industries where noise control is becoming increasingly important.

Vitreous enamelled products

Vitreous enamelling is a long-established finishing process and one that is being applied to an increasing range of steel products. Cold-reduced steel is the most widely used substrate and the grade is selected according to the strength and formability required and the enamelling process that is used. After fabrication, the product is degreased, pickled, rinsed and dried and in the most common process, a ground coat enamel slip or slurry is applied to the component, either by dipping or spraying. After drying, the enamel is fused by firing at temperatures above 800°C. A second coat, the cover coat enamel, is applied, dried and fired in a similar way. Some enamel is applied by means of a single-coat process and also by means of a two-coat one-fire process. The ability to produce a satisfactory product depends on the nature of the steel as well as on the nature of the enamel and the firing conditions. Special grades of steel specifically developed for enamelling are, therefore, often used.

The main potential problem that may affect an enamelled component is called *fish scaling*, whereby small slivers of enamel split off from the surface to leave a fish-scale appearance. This may occur soon after enamelling or many weeks later. Fish scaling arises because the solubility of atomic hydrogen in steel is as much as 1000 times greater at the firing temperature than at room temperature. During cooling the solubility limit is exceeded with the result that gaseous molecular hydrogen is formed at high pressure. If this gaseous hydrogen forms at the enamel steel interface and if there is not sufficient enamel adherence and toughness, the pressure developed is able to crack the enamel and remove a fragment to form a fish scale.

The tendency to form fish scales may be minimized by keeping the dew point of the furnace atmosphere low since water vapour is the main source of atomic hydrogen by a reaction with iron to form iron oxide. It is also beneficial to avoid unnecessarily high firing temperatures since such temperatures lead to even higher solute hydrogen contents. The most important method of avoiding fish scaling, however, is to use a steel which contains sites that can accommodate high-pressure gaseous hydrogen within the steel and which, therefore, prevent the build-up of gaseous hydrogen at the steel enamel interface.

The sites that are used may be physical or chemical. Physical sites include voids, cracks, vacancies and dislocations. Chemical sites include precipitates such as TiN, TiC, AlN, Fe_3C or BN and inclusions such as Al_2O_3, MnS and MnO. Certain steels are hot rolled with a high coiling temperature to produce coarse carbides that crack on cold reduction to provide physical sites. Titanium-bearing IF steels are commonly used since they provide chemical sites as well as good formability.

The enamelling process provides a hard, durable finish in a wide range of attractive colours, which is resistant to heat, abrasion and chemical attack. Applications for vitreous enamelled steels include domestic cookers, baths, hot-water tanks and cookware, architectural facing panels, flue pipes, silos, heat exchanger panels and tanks for bulk storage.

References

1. Hewitt, B.J., *Review of Annealing Technology*, IISI, Technical Exchange Session, p. 51 (1996).
2. Imose, M., *Trans. Iron Steel Inst. Japan*, **25**, 911 (1985).
3. Lubensky, P.J., Wigman, S.L. and Johnson, D.J., *Microalloying '95, Conf. Proc.*, Pittsburgh, The Iron and Steel Society, p. 225 (1995).
4. Meyer, P. and Frommann, K., *ABM Int. Congress*, Sao Paulo, p. 249 (1994).
5. *The Making, Shaping and Treating of Steel* (Ninth Edition), Ed. Harold E. McGannon, United States Steel (1971).
6. Swartz, J.C. *Trans. Met. Soc. AIME*, **239**, 68 (1967).
7. Hall, E.O. *Proc. Phys. Soc.*, Series B 64, 747 (1951).
8. Petch, N.J. *J. Iron and Steel Inst.* **174**, 25 (1953).

9. Pickering, F.B. In *Materials Science and Technology Vol. 7: Constitution and Properties of Steel*, VCH, p. 41 (1992).

10. Turkdogan, E.T. *Iron and Steelmaker*, **16**, 61 (1989).

11. Copreaux, J., Gaye, H., Henry, J. and Lanteri, S. *ECSC Report 7210*. EC/303 (1994).

12. Hudd, R.C., Jones, A. and Kale. *JISI*, **209**, 121 (1971).

13. Liu, W.J. and Jonas, J.J. *Met. Trans.*, **20A**, 1361 (1989).

14. Kozasu, I. In *Materials Science and Technology Vol. 7: Constitution and Properties of Steel*, VCH, p. 183 (1992).

15. Kozasu, I., Ouchi, C., Sampei, T. and Okita, T. In *Microalloying 75*, Union Carbide Corp., p. 120 (1977).

16. Ouchi, C. In *High Strength Low Alloy Steel*, Proc. of Wollongong Int. Conf., August 1984, p. 17 (1985).

17. Steven and Haynes. In *An Atlas of Continuous Cooling Diagrams for Engineering Steels*, British Steel Corporation.

18. Krauss, G. In *Materials Science and Technology Vol. 7: Constitution and Properties of Steel*, VCH, p. 1 (1992).

19. Andrews, K.W.*J. Iron and Steel Inst.*, **203**, 721 (1965).

20. Hutchinson, W.B. *Int. Mat. Rev.*, **29**, 25 (1984).

21. Roe, K.J. *J. Appl. Phys.*, **36**, 2024 (1965).

22. Bunge, H.J. *Z. Metallkd.*, **56**, 872 (1965).

23. Hu, H. *Textures of Crystalline Solids*, **4**, 13 (1980).

24. Emren, F., Schlippenbach, U. and Lucke, K. *Mat. Sci. and Tech.*, **5**, 238 (1989).

25. Randle, V. *Microtexture Determination and its Applications*, The Institute of Materials, London (1992).

26. Held, J.F. In *Mech. Working and Steel Processing IV: Proc. 8th Conf.*, Edgecombe, D.A. (ed.), Warrendale: AIME, pp. 3–38 (1965).

27. Schlippenbach, U. and Lucke, K. In *Proc. 7th Riso Int. Symp. on Met. and Mat. Sci.*, Roskilde, Riso Nat. Lab., pp. 541–546 (1986).

28. Leslie, W.C., Michalak, J.T. and Aul, F.W. In *Iron and its Dilute Solid Solutions*, Interscience Publishers, New York, p. 119 (1961).

29. Abe, M. In *Materials Science and Technology Vol. 7: Constitution and Properties of Steel*, VCH, p. 285 (1992).

30. Dillamore, I., Smith, C.J.E. and Watson, T.W. *Met. Sci. J.*, **1**, 49 (1967).

31. Every, R.L. and Hatherly, M. *Texture*, **1**, 183 (1976).

32. Takechi, H., Kata, H. and Nagashima, S., *Trans. AIME*, **242**, 56 (1968).

33. Decker, B.F. and Harker, D. *J. Appl. Phys*, **22**, 900 (1951).

34. Ray, R.K., Jonas, J.J. and Hook, R.E. *Int. Mat. Rev.*, **39**, 129 (1994).

35. Emren, F., Schlippenbach, U. and Lucke, K. *Acta Metall.*, **34**, 2105 (1986).

36. Ono, S. and Nishimoto, A. *Trans. Iron and Steels Inst. Japan*, **26**, B33 (1986).

37. Ushioda, K., Hutchinson, W.B., Agren, J. and Schlippenbach, U., *Mat. Sci. and Tech.*, **2**, 807 (1986).

38. Lankford, W.T., Snyder, S.D. and Bauscher, J.A. *Trans. ASM*, **42**, 1197 (1950).

39. Held, J.F. In *Mechanical Working and Steel Processing IV: Proc. 8th Conf.*, Edgecombe, D.A. (ed), Warrendale, AIME, p. 3 (1965).
40. Atkinson, M. and Maclean, I.M. *Sheet Met. Ind.*, **42**, 120 (1965).
41. Hudd, R.C. In *Proc. Low carbon Structural Steel for the Eighties*, Spring Residential Course, Series 3, No. 6, The Institution of Metallurgists, 111A-1, March (1977).
42. Horta, R.M.S.B., Roberts, W.T. and Wilson, D.V. *Int. J. Mech. Sci.*, **12**, 231 (1970).
43. Wilson, D.V. and Butler, R.D. *J. Inst. Met.*, **90**, 473 (1961).
44. Hudd, R.C. and Lyons, L.K. *Metals Technology*, **2**, 428 (1975).
45. Dasarathy, C. and Goodwin, T.J. *Metals and Materials*, **21**, January. (1990).
46. Wang, N.M. and Shammamy, M.R. *J. Mech. Phys. Solids*, **17**, 43 (1969).
47. Pearce, R. *Sheet Metal Forming*, Adam Hilger (1991).
48. Backhoven, S.P. and Keeler, W.A. *Trans. ASM*, **56**, 25 (1963).
49. Haberfield, A.B. and Boyles, M. *The Metallurgist*, **7**, 453 (1975).
50. Lee, A.P. and Hiam, J.R. ASM-AIME Metals Congress (1973).
51. Iwaya, J., Tanaka, Y., Shirasaws, H. and Miyahara, M. *Journal of the JSTP*, **35**, No. 404, 1123 (1994).
52. Yamazaki, K., Mizuyama, Y., Oka, M. and Tokunaga, Y. *Journal of the JSTP*, **36**, No. 416, 1973 (1995).
53. Gerber, W.W. *Hydrogen in Metals*, ASM, pp. 115–147 (1974).
54. Hosoya, Y., Tsuyama, S., Nagataki, Y., Kanetoh, S., Izuishi, T. and Takada, Y. *NKK Technical Review*, **72**, 20 (1995).
55. Deutscher, *Stahl u Eisen*, **115**, 39 (1995).
56. *High Strength Steel Bulletin*, p. 1, Winter. (1995).
57. *Metallurgia*, p. 148, April. (1992).
58. Winter, D. *Wards Auto World*, p. 54, June (1993).
59. Ferry, J.W. In 'High Production Roll Forming', Society of Manufacturing Engineers, Dearborn, p. 31 (1983).
60. Cain, E. In 'High Production Roll Forming', Society of Manufacturing Engineers, Dearborn, p. 7 (1983).
61. Miles, J. *J. Metals and Materials*, p. 398, July (1989).
62. Comstock, G.F., Urban, S.F. and Cohen, M. *Titanium in Steel*, New York, Pitman (1949).
63. Leslie, W.C. and Sober, R.J. *ASM Trans. Quarterly*, **60**, 99 (1967).
64. Elias, J.A. and Hook, R.E. In Mech. Working and Steel Processing IV, Proc. 13th Conf. Pittsburgh, *Met. Soc. AIME*, p. 348 (1971).
65. Krupic, V., Gladman T. and Mitchell P. 37th MWSP Conf. Proc., *ISS*, **33**, 907 (1996).
66. Takahashi, N., Shibata, M., Furuno, Y., Hayakawa, H., Asai, T. and Yamashita, Y. *Tetsu-to-Hagane*, **68**, S588 (1982).
67. Tokunaga, Y. and Kato, H. *Metallurgy of Vacuum-Degassed Steel Products*, Pradhan, R. (ed.), The Minerals, Metals and Materials Soc., p. 91 (1990).
68. Tokunaga, Y., Yamada, M. and Ito, K. *Tetsu-to-Hagane*, **73**(2), 109 (1987).

69. Kino, N., Matsumura, Y., Tsuchiya, H., Furukawa, Y., Akagi, H. and Sanagi, S. *CAMP-JISI*, **3**, 785 (1990).
70. Satoh, S., Obara, T., Takasaki, J., Yasuda, A. and Nishida, M. *Kawasaki Steel Technical Report*, **12**, 36 (1985).
71. Goodenow, R.H. and Held, J.F. *Met. Trans.*, **1**, 2507 (1970).
72. Satoh, S., Obara, T., Nishida, M., Matsuno, N., Takasaki, J., and Satoh, H. *Kawasaki Steel Technical Report*, **8**, 1 (1983).
73. Tither, G., Garcia, C.I., Hua, M. and DeArdo, A.J. Int. Forum, Physical Metallurgy of IF Steels, *ISIJ*, p. 293 (1994).
74. Ruiz-Aparicio, L.J., Hua, M., Garcia, C.I. and DeArdo, A.J. Proc. Conf., Thermo-Mechanical Processing, Theory, Modelling and Practice, Stockholm, September (1996).
75. Gawne, D.T., *Steel Research*, London, British Steel Corporation (1976).
76. Meyzaud, Y., Michaut, B. and Parniere, P. In *Texture and the Properties of Materials*, 4th Int. Texture Conf., Cambridge, The Metals Society, London, p. 255 (1975).
77. Leslie, W.C., Rickett, R.L., Dotson, C.L. and Walton, C.S. *Trans. ASM*, **46**, 1470 (1954).
78. Whiteley, R.L. and Wise, D.E. In *Flat Rolled Products III*, Earhart, E.W. (ed.), New York, Interscience, p. 47 (1962).
79. Parayil, T.R. and Gupta, I. In *Mechanical Working and Steel Processing XXVII, Proc. 31st Conf.* Kuhn, L.G. and Hultgren F.A. (eds), Warrendale AIME, p. 314 (1989).
80. Osawa, K., Matsudo, K., Kurihara, K. and Suzuki, T. *Tetsu-to-Hagane*, **70**, S552 (1984).
81. Shimisu, M., Matsukura, K., Takahashi, N. and Shinagawa, H. *Tetsu-to-Hagane*, **50**, 2094 (1964).
82. Takahashi, M. and Okamoto, A. *Sumitomo Met.*, **27**, 40 (1974).
83. Henin, C. and Brun, C. *Rev. de Metall.*, CIT, 599 (1981).
84. Takahashi, N., Abe, M., Akisue, O. and Katoh, H. In *Proc. Symp. on Metallurgy of Continuous Annealed Sheet Steel*, Bramfitt, B.L. and Mangonon, P.L. (eds), Dallas, AIME, p. 51 (1982).
85. Toda, K., Gondoh, H., Takechi, H., Abe, M., Uehara, N. and Komita, K. *Trans. Iron and Steel Inst. Japan*, **15**, 305 (1975).
86. Osawa, K., Matsudo, K., Kurihara, K. and Suzuki, T. *Tetsu-to-Hagane*, **70**, S552 (1984).
87. Ono, S., Nozoe, O., Shimomura, T. and Matsudo, K. In *Proc. Symp. on Metallurgy of Continuous Annealed Sheet Steel*, Bramfitt, B.L. and Mangonon, P.L. (eds), Dallas, AIME, p. 99 (1982).
88. Matsudo, K., Osawa, K. and Kurihara, K. In *Proc. Symp. on Metallurgy of Continuously Annealed Cold Rolled Sheet Steel*, Pradhan, R. (ed.), Detroit, AIME, p. 3 (1984).
89. Pradhan, R. In *Proc. Symp. on Metallurgy of Continuously Annealed Cold Rolled Sheet Steel*, Pradhan, R, (ed.), Detroit, AIME, p. 185 (1984).

90. Prum, N., Meers, U., Mathy, H., Messien, P. and Leroy, V. In *Proc. Symp. on Hot and Cold Rolled Sheet Steels*, Pradhan, R. and Ludkovsky, G. (eds), Cincinnati, The Metallurgical Society, p. 3 (1988).

91. Obara, T., Sakata, K. and Irie, T. In *Proc. Symp. on Metallurgy of Continuous Annealed Sheet Steel*, Bramfitt, B.L. and Mangonon, P.L. (eds), Dallas, AIME, p. 83 (1982).

92. Katoh, H., Koyama, K. and Kawasaki, K. In *Proc. Symp. on Metallurgy of Continuously Annealed Cold Rolled Sheet Steel*, Pradhan, R. (ed.), Detroit, AIME, p. 79 (1984).

93. Herman, J.C. and Leroy, V. *AISE, Future of Flat Rolled Steel*, Prod. Conf., Chicago, V.2, paper 6 (1995).

94. Bleck, W., *JOM*, p. 26, July (1996).

95. Hayami, S. and Furukawa, T. In *Microalloying 75*, Korchynsky, M. (ed.), New York, Union Carbide Corporation, p. 311 (1977).

96. Takechi, H. In *Vehicle Design and Components – Materials Innovation and Automotive Technology*, 5th IAVD Conf., Dorgham, M.A. (ed.), Geneva, Interscience (1990).

97. Paules, J.R. In *Microalloyed Vanadium Steels*, Korchynsky, M., Gorczyca, S., and Blicharski, M. (eds), Krakow, Association of Polish Metallurgical Engineers, p. 19 (1990).

98. Cuddy, L.J. and Raley, J.C. *Metall. Trans. A.*, **14A**, 1989 (1983).

99. Cuddy, L.J. *Plastic Deformation of Metals*, Academic Press, New York, 129 (1975).

100. Jonas, J.J. Proceedings of Conference, Mathematical Modelling of the Hot Rolling of Steels, Hamilton, Canada, p. 99, August (1990).

101. Bai, D.Q., Yue, S., Sun, W.P. and Jonas, J.J., *Met. Trans. A.* **24A**, 2151 (1993).

102. Kasper, R., Lotter, U. and Biegus, C. *Steel Research*, **65**, No. 6, p. 242 (1994).

103. Sekine, H. *et al.*, *Seitetsu Kenkyu*, **287**, 11920 (1976).

104. Pickering, F.B. In *Materials Science and Technology Vol. 7: Constitution and Properties of Steel*, VCH, p. 335 (1992).

105. Bordignon, P.J.P., Hulka, K. and Jones, B.L. *High Strength Steels for Automotive Applications*, Niobium Technical Report Nb TR-06/84, December (1984).

106. Pradhan, R. In *Proc. Symp. on Metallurgy of Continuous Annealed Sheet Steel*, Bramfitt, B.L. and Mangonon, P.L. (eds), Dallas, AIME, p. 203 (1982).

107. Pradhan, R., *Scand. J. Met.*, **13**, 298 (1984).

108. Toda, K., Gondoh, H., Takechi, H. and Masui, H., *Met. Trans.* **7A**, 1629 (1976).

109. Goodman, S.R. In *Proc. Symp. on Metallurgy of Continuously Annealed Cold Rolled Sheet Steel*, Pradhan, R. (ed.), Detroit, AIME, p. 239 (1984).

110. Gawne, D.T. and Lewis, G.M.H. *Mat. Sci. and Tech.*, **1**, 183 (1985).

111. Katoh, H., Takechi, H., Takahashi, N. and Abe, M. In *Proc. Symp. on Metallurgy of Continuously Annealed Cold Rolled Sheet Steel*, Pradhan, R. (ed.), Detroit, AIME, p. 37 (1984).

112. Ohashi, N., Irie, T., Satoh, S., Hashimoto, O. and Takahashi, I. SAE Paper 810027, (1981).

113. Takechi, H. In *Proc. Symp. Hot and Cold Rolled Sheet Steel, Cincinnati*, Pradhan, R. and Ludkovsky, G. (eds), The Metallurgical Society, p. 117 (1988).

114. Shimada, M., Nashiwa, M., Osaki, T., Furukawa, Y., Shimatani, Y., Fujita, Y. and Aihara, H. *Review of Annealing Technology*, Technical Exchange Session, IISI, p. 17 (1996).

115. Hanai, N., Takemoto, N., Takunaga, Y., Mizuyama, Y. *Trans. Iron Steel Institute Japan*, **24**, 17 (1984).

116. Mizui, N., Okamoto, A., Amagasaki and Tanioku, T., *Proc. Int. Conf. Steel in Motor Vehicle Manufacture*, Würzburg, 24–26 September, p. 85 (1990).

117. Yamazaki, K., Horita, T., Umehara, Y., Morishita, T. In *Proc. Micro Alloying '88*, Materials Park, Ohio, ASM, p. 327 (1988).

118. Rana, F., Skolly, R. and Gupta, I. In *Proc. Symp. High-Strength Sheet Steels for the Motor Industry*, Baltimore, Pradhan, R. (ed.), The Iron and Steel Society, p. 89 (1994).

119. Hayashida, T., Oda, M., Yamada, T., Matsukawa, Y. and Tanaka, J. In *Proc. Symp. High-Strength Sheet Steels for the Motor Industry*, Baltimore, Pradhan, R. (ed.), The Iron and Steel Society, p. 135 (1994).

120. Sakata, K., Satoh, S., Kato, T. and Hashimoto, O. *Int. Forum for Physical Metallurgy of IF Steels, Tokyo*, Iron and Steel Institute of Japan, p. 279, May (1994).

121. Zackay, V.F., Parker, E.R., Fahr, D. and Busk, R. *Trans. Am. Soc. Met.*, **60**, 252 (1967).

122. Williams, E.W. and Davies, L.K. In *Recent Developments in Annealing*, ISI Special Report 79 (1963).

123. Irie, T., Satoh, S., Hashiguchi, K., Takahashi, I. and Hashimoto, O. *Trans. Iron and Steel Inst. Japan*, **21**, 793 (1981).

124. Lawson, P.D., Matlock, D.K. and Krauss, G. In *Proc. Conf. on Fundamentals of Dual Phase Steel*, Kot, R.A. and Bramfitt, B.L. (eds), Warrendale, AIME, p. 347 (1981).

125. Gladman, T. In *Advances in the Physical Metallurgy and Applications of Steel*, Eyre, B.L. (ed.), The Metals Society, London, p. 55 (1981).

126. Maid, O., Dahl, W., Strassburger, C. and Muschenborn, W. *Stahl u. Eisen*, **108**, p. 355 and 365 (1988).

127. Speich, G.R. and Miller, R.L. In *Proc. Conf. on Fundamentals of Dual Phase Steel*, Kot, R.A. and Bramfitt, B.L. (eds), Warrendale, AIME, p. 279 (1981).

128. Messien, P., Herman, J.C. and Greday, T. In *Proc. Symp. on Metallurgy of Continuous Annealed Sheet Steel*, Bramfitt, B.L. and Mangonon, P.L. (eds), Dallas, AIME, p. 271 (1982).

129. Speich, G.R., Demarest, V.A. and Miller, R.L. *Metall. Trans.*, **12A**, 1419 (1981).

130. Bleck, W., Drewes, E.J., Engl, B., Litzke, H. and Muschenborn, W. *Stahl u. Eisen*, **106**, 21 (1986).

131. Hehemann, R.F. *Phase Transformations*, Aaronson, H.I. (ed.), Metals Park, OH, Amer. Soc. Met., p. 397 (1970).
132. Shirasawa, H. and Thomson, J.G. *Trans. Iron Steel Inst. Japan*, **27**, 360 (1987).
133. Sakuma, O., Matsumura, O. and Akisue, O. *ISIJ International*, p. 1348, November (1991).
134. Okamoto, A., Nagao, N., Takahashi, M. and Saiki, K. In *Proc. Symp. on Metallurgy of Continuous Annealed Sheet Steel*, Bramfitt, B.L. and Mangonon, P.L. (eds), Dallas, AIME, p. 287 (1982).
135. Sudo, M. and Kokubo, I. *Scand. J. Met.*, **13**, 329 (1984).
136. Minote, T., Torizuka, S., Ogawa, A. and Niikura, M. ISIJ International, **36**, 201 (1996).
137. Sakuma, Y., Matlock, D.K. and Krauss, G. *Developments in the Annealing of Sheet Steels*, Pradhan, R. (ed.), The Minerals, Metals and Materials Society, p. 321 (1992).
138. Sugimoto, K., Usui, N., Kobayashi, M. and Hashimoto, S. *ISIJ International*, **32**, p. 1311, No. 12 (1992).
139. Matsumura, O., Sakuma, Y. and Takechi, H. *ISIJ International*, **32**, 1014 (1992).
140. Jones, A. and Evans, P.J. *Int. Symp. Low Carbon Steel for the 90's*, Asfahani, R. and Tither, G. (eds), Pittsburgh, TMS, (1993).
141. Itami, A., Takahashi, M. and Ushioda, K. *ISIJ International*, **35**, 1121 (1995).
142. Tanaka, Y., Shirasawa, H. and Miyahara, M. SAE Paper 850117 (1985).
143. Rigsbee, J.M. and VanderArend, P.J. In *Formable HSLA and Dual-Phase Steel*, Davenport, A.T. (ed.), Warrendale, AIME, p. 56 (1979).
144. Hosoya, Y., Tsuyama, S., Nagataki, S., Kanetoh, S., Izuishi, T. and Takada, Y. *NKK Technical Review*, No. 72, p. 20 (1995).
145. Matsumura, O., Sakuma, Y. and Takechi, H. *Trans. Iron Steel Inst. Japan*, **27**, 570 (1987).
146. Evans, P.J. Report, Project 7210.MB/815, Commission of the European Communities (1995).
147. Denner, S.G. Proc. Int. *Conf. on Zinc and Zinc Alloy Coated Sheet Steel (GALVATECH '89)*, ISIJ, Tokyo, p. 101 (1989).
148. British Steel, Colourcoat and Stelvetite Brochure, British Steel.
149. British Steel, Colourcoat and Stelvetite Pre-finished Steels Technical Manual, British Steel.
150. Morgan, E. *Tinplate and Modern Canmaking Technology*, Pergamon Press (1985).
151. Singh, S., Kotting, G. and Schmidt, G. DVS 165, SMD 1156, 26 (1996).
152. Jones, T.B. and Williams, N.T. *Resistance Spot Welding of High Strength Steels*, C287/81, I. Mech. E. (1981).
153. Jones, T.B. and Williams, N.T. *Resistance Spot Welding of Rephosphorised Steel, A Review of Welding in the World*, **23**, No. 11/12, 248.
154. Williams, N.T. In *Proc. Vehicle Design and Components – Materials Innovation and Automotive Technology*, 5th IAVD Conf., Dorgham, M.A. (ed.), Geneva, Interscience (1990).

155. Galvatite Technical Manual, British Steel Strip Products, May (1993).
156. Lallay, M. and Waddell, W. *La Revue de Metallurgie*, ATS-JS 94, p. 24, (1994).
157. Leibig, H.P. *British Engineer*, Winter (1987).
158. Bokhari, N. *Welding and Metal Fabrication*, p. 187, May (1995).
159. Budde, L. and Lappe, W., Bander Bleche Rohre, May (1991).
160. Jones, T.B. Report, Project 7210.MB/817, Commission of the European Communities (1996).
161. Yan, B. 37th Mechanical Working and Steel Processing Conf. Proc., ISS-AIME, XXXIII 101 (1996).
162. Landgraf, R.W. In Proceedings of the SAE Fatigue Conference, P-109, Dearborn, Michigan, Paper 820677, p. 11 (1982).
163. Palmgren, A. *Z. Vereins Deutscher Ingenicure*, **68** (1924).
164. Miner, M.A. *Journal of Applied Mechanics*, **12**, p. A-159 (1945).
165. Proceedings of the SAE Fatigue Conference, P-109, Dearborn, Michigan, April, (1982).
166. Sperle, J.O. and Trogen, H. *Scand. J. Met.*, **18**, 147 (1989).
167. *Annual Book of ASTM Standards*, Vol. 03.01, E647-93.
168. Jones, A. and Hudd, R.C. Report EUR 11220 EN, Commission of the European Communities (1988).
169. Godwin, M.J. Report EUR 12749 EN, Commission of the European Communities (1990)
170. Irie, T., Satoh, S., Yasuda, A. and Hashimoto, O. In *Metallurgy of Continuous Annealed Sheet Steel*, Bramfitt, B.L. and Mangonon, P.L. (eds), AIME, Dallas, PA, p. 155 (1982).
171. Kino, N., Yamada, M., Tokunaga, T. and Tsuchiya, H. In *Metallurgy of Vacuum Degassed Steel Products*, Pradhan, R. (ed.), The Minerals, Metals and Materials Society, Indianapolis, p. 197 (1990).
172. Bhat, S.P., Yan, B., Chintamani, J.S. and Bloom, T.A. In *Proc. Symp. High-Strength Steels for the Motor Industry*, Pradhan, R. (ed.), Baltimore, p. 209 (1994).
173. Henning, L.A. 33rd Mechanical Working and Steel Processing Conf. Proc., ISS-AIME, XXIX, p. 9 (1992).
174. Yan, B. and Gupta, I. 38th Mechanical Working and Steel Processing Conf. Proc., ISS-AIME, XXXIV, p. 417 (1997).
175. Yaluhara, E., Sahata, K., Kato, T. and Hashimoto, O., *ISIJ Int.*, **34**, No. 1, 99 (1994).
176. Machara, Y., Misui, N. and Arai, M. In *Interstitial Free Steel Sheet: Processing, Fabrication and Properties Conf. Proc.*, Collins, L.E. and Baragor, D.L. (eds), CIM/ICM, 19–20 August, Ottawa, Ontario, p. 135 (1991).
177. Pradhan, R. In *Int. Forum Phys. Met. of IF Steel*, ISIJ, Tokyo, p. 165 (1994).
178. Konishi, M., Ohashi, N. and Yoshida, H. Kawasaki Steel Technical Report, No. 6, p. 305 (1974)
179. Takahashi, N., Shibata, M., Furuno, Y., Hayakawa, H., Kabuta, K. and Yamamoto, K. In *Metallurgy of Continuous Annealed Sheet Steel*, Bramfitt, B.L. and Mangonon, P.L. (eds), AIME, Dallas, PA, p. 133 (1982).

180. *Car 2000, A Mira Report*, published by The Motor Research Association, October (1993).
181. Takechi, H. *Steel Today and Tomorrow*, January–March., p. 5 (1996).
182. Innovations in Steel. *Steel Times*, April, p. 141 (1996).
183. Drewes, E.J. and Engl, B. In Proc. Coated and High Strength Steels – Auto Tech 89, I. Mech. E., C399/37, (1989).
184. Lowe, K. Private Communication (1997).
185. *ULSAB Report Phase 1*, September (1995).
186. Hughes, R.L. *IBEC '95, Body Design and Engineering*, p. 59 (1995).
187. Crooks, M.J. and Miner, R.E. *JOM*, p. 13, July (1996).
188. Thompson, S.J.A. In Proc. Coated and High Strength Steels – Auto Tech 89, I. Mech. E., C399/37 (1989).
189. Bowen, M. In Proc. Coated and High Strength Steels, – Auto Tech 89, I. Mech. E., C399/37 (1989).
190. British Steel Ternex Brochure, British Steel.
191. Haberfield, A.B. Steelresearch 1987–88, British Steel, p. 42 (1988).
192. Lewis, K.G., Jones, D. and Godwin, M.J. *Metals and Materials*, **4**, 357 June (1988).
193. Mathieu, S. Conf. Trans. ECCA General Meeting, DT378, May (1996).
194. *Design Engineering*, p. 17, March (1996).
195. British Steel Distribution, Auto Update.

2 *Low-carbon structural steels*

Overview

Although open to wide interpretation, the term *structural steels* is commonly used to identify the predominantly C–Mn steels, with ferrite–pearlite microstructures, which are used in large quantities in civil and chemical engineering. The steels are produced in plates and sections, sometimes up to several inches thick, and generally with yield strength values up to about 500 N/mm^2. However, structural steels also include low-alloy grades which are quenched and tempered in order to provide yield strengths up to about 700 N/mm^2.

As illustrated in this chapter, these steels are used in a wide and diverse range of applications, including buildings, bridges, pressure vessels, ships and off-highway vehicles. More recently, structural steels have been used extensively in very demanding applications such as offshore oil and gas platforms and the associated pipelines that often operate in extremely cold and chemically aggressive environments.

A major feature of most forms of construction in structural steels is the high level of welding employed and the requirement for high-integrity welds. Welding began to replace riveting as the principal joining process in the 1940s but, at that time, structural steels were characterized by high carbon contents and were therefore prone to cold cracking. The requirement for lower carbon grades with improved weldability was illustrated very dramatically in the construction of the first all-welded merchant ships (*Liberty ships*) during World War II. However, the break-up of these vessels on the high seas also led to the recognition of a further major property requirement in structural steels, namely toughness as opposed to ductility.

In the early 1950s, the work of Hall and Petch revolutionized the design of structural steels with the concept that refinement of the ferrite grains led to an increase in both the yield strength and toughness of ferrite–pearlite steels. Thus steels with yield strength values up to about 300 N/mm^2 could be produced in aluminium-grain-refined compositions, with good impact properties and with good welding characteristics. Ferrite grain refinement remains the single most important metallurgical parameter in the make-up of modern structural grades but the demand for higher strength steels required a further strengthening mechanism, namely precipitation strengthening. Thus small additions of niobium, vanadium and titanium were added to structural steels to raise the yield strength up to a level of about 500 N/mm^2. Since they were added in levels of up to only 0.15%, these additions became known as *micro-alloying* elements and the compositions were designated *High-strength Low-alloy* (HSLA) steels.

The late 1950s and 1960s represented a period of major research on the structure–property relationships and fracture behaviour of structural steels. However, it also heralded the introduction of an important new technique in the production of structural steels, namely *controlled rolling*. In essence, this enabled fine-grained steels to be produced in the as-rolled condition, thereby eliminating the need for costly normalizing heat treatments. More importantly, controlled rolling led to the generation of steels with properties far superior to those that could be obtained in the normalized condition.

In the 1970s and 1980s, controlled rolling was augmented with controlled cooling and the combination is now referred to as *thermomechanical processing*. In its more severe form (*direct quenching*), controlled cooling is now used as an alternative to reheat quenching for the production of quenched and tempered grades.

Structural steels have therefore undergone very significant changes, each change producing a substantial improvement in an important property such as strength, toughness or weldability. Aspects such as improved cleanness and inclusion shape control have also been adopted, leading to improvements in fabrication and service performance. These factors, coupled with very favourable cost comparisons, have meant that structural steels have remained virtually unchallenged by competitive materials, other than reinforced concrete, in most of their traditional applications.

Underlying metallurgical principles

Structural steels, in the form of plates and sections, are produced in the hot-rolled condition and, in many respects, the underlying metallurgy is essentially similar to that involved in the strip grades. However, structural steels are generally used in thicker sizes and higher strengths, and toughness rather than ductility/formability is the more important requirement. The higher levels of elements such as carbon and manganese in structural steels also impose more detailed consideration of the control of microstructure and maintenance of properties in the welded condition.

A vast amount of research has been carried out on factors affecting the strength and toughness of structural steels, particularly those involving ferrite–pearlite microstructures which account for the bulk production of these steels. However, the most important issue is the control of ferrite grain size since refinement of the ferrite grains leads to an increase in both yield strength and toughness. This effect contrasts sharply with other strengthening mechanisms, such as solid solution strengthening and precipitation strengthening, which are accompanied by a reduction in toughness. In conventional hot rolling, structural steels would be reheated to a temperature of around 1250°C and rolling would be completed at temperatures of the order of 1000°C. This will result in a fully recrystallized, coarse austenitic structure which transforms to a coarse ferrite–pearlite structure on cooling to ambient temperature. In turn, this will yield material with a low level of toughness and the additional process of *normalizing* is required to refine the microstructure and improve the impact strength. This involves reheating the hot-rolled product to a temperature of about 910°C and, during the heating cycle, AlN is precipitated from solid solution in the ferrite which restricts the growth of the austenite grains at the normalizing temperature and leads to the formation of fine ferrite grains on air cooling to ambient temperature. However, as indicated in the *Overview*, the costly process of normalizing has now been largely superseded by controlled rolling, whereby lower temperatures are employed in the finishing stages of hot rolling in order to produce a fine austenite grain size which transforms subsequently to a fine-grained ferrite microstructure.

In general, alloy additions are not employed specifically in structural grades in order to produce solid solution strengthening. However, solid solution strengthening effects will arise from the presence of carbon and nitrogen in solution and also from silicon and manganese, which are added primarily for deoxidation and sulphide control purposes. On the other hand, structural steels can contain manganese contents up to 1.5% which also result in substantial strengthening due to the depression of the austenite to ferrite transformation and the consequent refinement of the ferritic grains.

As indicated later in this chapter, precipitation strengthening reactions are of major importance in the production of high-strength structural steels, particularly those involving the carbides or nitrides of elements such as niobium, vanadium and titanium. In this context, these elements are called *micro-alloying elements* and are taken into solution in the austenite phase during the reheating stage, but form compounds such as $Nb(CN)$, V_4C_3 and TiC on transformation to ferrite. However, the metallurgy of these high-strength low-alloy (HSLA) steels is complex, requiring detailed consideration of solubility/temperature effects at the reheating stage and precipitate size/cooling rate effects on transformation to ferrite. A further consideration is that these micro-alloying elements are also employed to retard the recrystallization kinetics of steels which are subjected to controlled rolling. This involves the strain-induced precipitation of $Nb(CN)$ or TiC in the austenitic condition at a temperature below 950°C which retards recrystallization, producing an elongated, pancake morphology. On cooling to ambient temperature, the deformation substructure in the austenite grains produces a fine ferritic structure with higher strength and toughness than that achieved by normalizing.

There is obviously a limit to the strength that can be developed in ferrite–pearlite microstructures through grain refinement, solid solution and precipitation strengthening and for yield strength levels greater than about 500 N/mm^2, transformation strengthening is employed. Thus the alloy content and cooling rate must be sufficient to produce a martensitic structure on quenching which in turn must be tempered in order to provide an adequate balance of strength and ductility/toughness. This invokes the concepts of hardenability and tempering resistance which are essentially those involved in *Engineering steels* which are discussed in Chapter 3.

Strengthening mechanisms in structural steels

Major research effort has been devoted to the detailed understanding of factors affecting the properties of low-carbon structural steels. Whereas considerable cost savings accrued from the use of lighter sections in higher strength steels, there was also the need to maintain, or indeed improve upon, other important properties such as toughness and weldability. Therefore detailed attention was given to identifying the strengthening mechanisms which were most cost-effective or that provided the best combination of properties.

The practical options for increasing the strength of steels are:

1. Refining the ferrite grain size.
2. Solid solution strengthening.
3. Precipitation strengthening.
4. Transformation strengthening.
5. Dislocation strengthening.

Whereas work hardening or dislocation strengthening can result in very high levels of strength, these are achieved at the expense of toughness and ductility. For this reason, little use is made of this method of strengthening but, as illustrated later in this chapter, work hardening is used in the production of high-strength reinforcing bars.

Ferrite grain refinement

In the early 1950s, work published by Hall[1] and Petch[2] laid the foundation for the development of modern, high-strength structural steels. The Hall–Petch equation, perhaps the most celebrated in ferrous metallurgy, is as follows:

$$\sigma_y = \sigma_i + k_y d^{-\frac{1}{2}}$$

where σ_y = yield strength
σ_i = friction stress which opposes dislocation movement
k_y = a constant (often called the dislocation locking term)
d = ferrite grain size

Thus refinement of the ferrite grain size will result in an increase in yield strength and the relationship is shown in Figure 2.1.

Whereas a strengthening effect usually leads to a decrease in toughness, it was shown that refinement of the ferrite grain size also produced a simultaneous improvement in toughness. The Petch equation linking toughness to grain size is given below:

$$\beta T = \ln \beta - \ln C - \ln d^{-\frac{1}{2}}$$

where β and C are constants, T is the ductile–brittle transition temperature and d is the ferrite grain size. Therefore, as illustrated in Figure 2.1, the impact transition temperature decreases as the ferrite grain size is reduced.

Refinement of the ferrite grain size can be achieved in a number of ways. Traditionally, fine-grained steels contain about 0.03% Al which is soluble at normal slab or bloom reheating temperatures of around 1250°C and which remains in solution during rolling and after cooling to ambient temperature. However, on subsequent reheating through the ferrite range to the normalizing or solution treatment temperature, the aluminium combines with nitrogen in the steel to

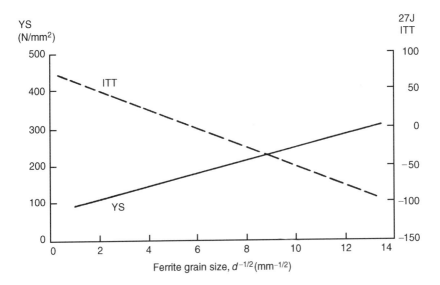

Figure 2.1 *Effect of ferrite grain size on yield strength and impact properties*

form a fine dispersion of AlN. These particles pin the austenite grain boundaries at the normal heat treatment temperatures just above Ac$_3$ (typically 850–920°C, depending upon carbon content) and therefore result in the formation of a fine austenite grain size. In turn, a fine austenite grain size results in the formation of a fine ferrite grain size on cooling to room temperature.

Although the austenite grain size is of major importance, other factors also play a part in developing a fine ferrite grain size. Thus the addition of elements such as carbon and manganese or an increase in the cooling rate from the austenite temperature range will lead to a refinement of the ferrite grains. In either case, this is achieved by depressing the temperature of transformation of austenite to ferrite. However, there is obviously a limit to the amount of strengthening that can be obtained by this mechanism before transformation is depressed to such an extent that it leads to the formation of bainite or martensite and the introduction of transformation strengthening.

Solid solution strengthening

The solid solution strengthening effects of the common alloying elements are illustrated in Figure 2.2 and work by Pickering and Gladman[4] has provided the strengthening coefficients shown in Table 2.1 for ferrite–pearlite steels containing up to 0.25% C and 1.5% Mn. These data illustrate the very powerful strengthening effects of the interstitial elements, carbon and nitrogen, but it must be borne in mind that these elements have only a very limited solid solubility in ferrite. However, both carbon and nitrogen also have a very adverse effect on toughness. Of the substitutional elements, phosphorus is the most potent and, as indicated in Chapter 1, additions of up to about 0.1% P are incorporated in the higher strength rephosphorized grades that are used in automotive body panels. However,

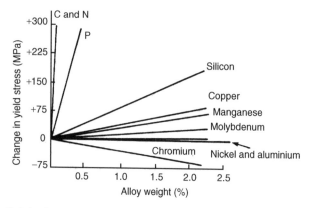

Figure 2.2   *Solid solution strengthening effects in ferrite-pearlite high-strength low-alloy steels (After Pickering[3])*

Table 2.1

Element	N/mm² per 1 Wt %
C and N	5544
P	678
Si	83
Cu	39
Mn	32
Mo	11
Ni	~0
Cr	−31

like carbon and nitrogen, phosphorus has a detrimental effect on toughness and therefore it is not used as a strengthening agent *per se* in structural steels. On the other hand, phosphorus is added to the so-called *weathering grades* because of its beneficial effect on atmospheric corrosion resistance. These steels will be discussed later in this chapter. Of the remaining elements, only silicon and manganese are cost-effective as solid solution strengtheners but silicon is added to steels primarily as a deoxidizing agent.

Precipitation strengthening

Precipitation strengthening can be induced by a variety of elements but, in the context of ferrite–pearlite steels, the systems of commercial significance are those involving niobium, vanadium and titanium. These elements have a strong affinity for carbon and nitrogen and, consequently, they have only a limited solid solubility in steel. Therefore they are added to steels in small amounts, e.g. up to about 0.06% Nb or 0.15% V, and as stated earlier are often referred to as micro-alloying elements.

As illustrated in Figure 2.3(a), a substantial amount of niobium will be taken into solution at a slab or bloom reheating temperature of 1250°C. On cooling,

Nb(CN) will precipitate at the austenite–ferrite interface during transformation (*interphase precipitation*) which leads to substantial strengthening. On the other hand, on reheating to a typical normalizing temperature of 920°C, very little Nb(CN) will dissolve and therefore virtually no precipitation strengthening can take place. However, the undissolved particles will act as pinning agents, restricting austenite grain growth and leading to the formation of a fine ferrite grain size. Therefore the reheating temperature controls the potential for precipitation strengthening and the strength increases progressively as the temperature is raised from 920 to 1250°C.

As indicated in Figure 2.3(b), vanadium dissolves more easily than niobium and complete solution of V_4C_3 would be expected to occur in commercial grades of structural steel at typical normalizing temperatures, e.g. 920°C. Slightly higher temperatures are required for the solution of VN which can act as a grain-refining

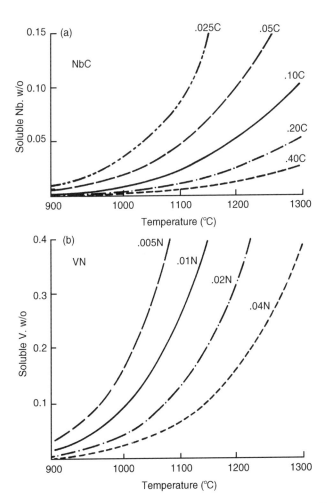

Figure 2.3 *Solubility of NbC and VN in austenite at various temperatures (After Irvine et al.[5])*

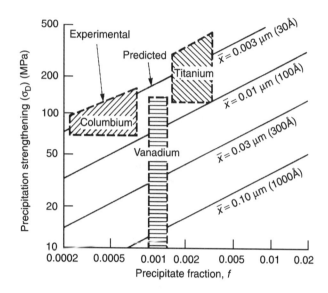

Figure 2.4   *The dependence of precipitation strengthening on precipitate size ($\bar{x}$) and fraction according to the Ashby–Orowan Model, compared with experimental observations for given micro-alloying additions (After Gladman et al.[6])*

agent at a temperature of 920°C. However, in Al–V steels, aluminium is the more powerful nitride former, and in the presence of 0.04% Al, significant levels of vanadium will go into solution at 920°C and be available for the precipitation of V_4C_3 on transformation to ferrite. Vanadium steels therefore provide significant precipitation-strengthening effects, i.e. up to 150 N/mm^2 per 0.10% V.

The strengthening effect of precipitated particles is dependent on both the volume fraction and particle size of the precipitates. This is illustrated in Figure 2.4, which was derived by Gladman *et al.* using the Ashby–Orowan model for precipitation strengthening. Whereas the volume fraction of precipitate is controlled by aspects such as solute concentration and solution treatment temperature, the particle size will be influenced primarily by the temperature of transformation, which is controlled by the alloy content and cooling rate effects.

Transformation strengthening

As stated earlier, both alloying elements and faster cooling rates depress the temperature of transformation of austenite to ferrite and, ultimately, the effect will be sufficient to cause transformation to bainite or martensite. The consequence of this progression is illustrated in Figure 2.5, which relates to steels containing 0.05–0.20% C. Thus the strength is increased progressively with the introduction of lower temperature transformation products but, of course, with some sacrifice to toughness and ductility. However, in the context of structural steels, there is a demand for quenched and tempered low-alloy grades with yield strengths up to 700 N/mm^2. Such steels are normally alloyed with molybdenum and boron to

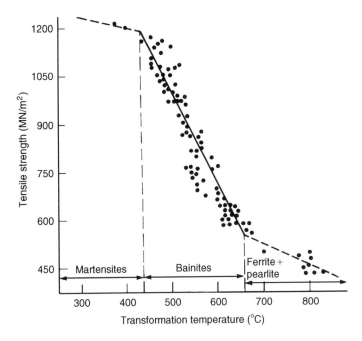

Figure 2.5 *Relationship between 50% transformation temperature and tensile strength (After Pickering[3])*

promote hardenability but there may also be a need to include elements such as vanadium to improve tempering resistance.

Structure–property relationships in ferrite–pearlite steels

Following the derivation of the Hall–Petch relationship:

$$\sigma_y = \sigma_i + k_y d^{-\frac{1}{2}}$$

it was proposed that this basic equation for yield strength could be extended to take account of the strengthening effects of alloying elements. Thus:

$$\sigma_y = \sigma_i + k' \, (\% \text{ alloy}) + k_y d^{-\frac{1}{2}}$$

During the late 1950s and early 1960s, Gladman and Pickering[4] pursued this line of development and provided the following quantitative relationship for yield strength, tensile strength and impact transition temperature:

$$\text{YS (N/mm}^2) = 53.9 + 32.3\% \text{ Mn} + 83.2\% \text{ Si} + 354\% \text{ N}_f + 17.4d^{-\frac{1}{2}}$$

$$\text{TS (N/mm}^2) = 294 + 27.7\% \text{ Mn} + 83.2\% \text{ Si} + 3.85\% \text{ pearlite} + 7.7d^{-\frac{1}{2}}$$

$$\text{ITT (}^\circ\text{C)} = -19 + 44\% \text{ Si} + 700\sqrt{(\%\text{N}_f)} + 2.2\% \text{ pearlite} - 11.5d^{-\frac{1}{2}}$$

where d is the mean ferrite grain size in mm and N_f the free (soluble) nitrogen.

Each of these equations illustrates very clearly the beneficial effects of a fine ferrite grain size in increasing the yield and tensile strength and depressing the impact transition temperature. It is also interesting to note that the pearlite content has no significant effect on the yield strength of these low-carbon, predominantly ferritic steels. On the other hand, pearlite increases the tensile strength and has a detrimental effect on toughness. The solid solution strengthening effects of manganese, silicon and free nitrogen are also highlighted in the above equations and it will be noted that free nitrogen is particularly detrimental to the impact properties.

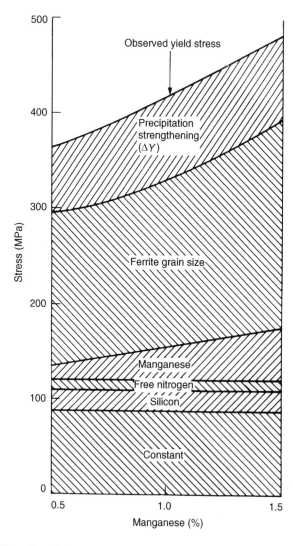

Figure 2.6   *The effect of increasing manganese content on the components of the yield stress of steels containing 0.2% carbon, 0.2% silicon, 0.15% vanadium and 0.15% nitrogen, normalized from 900°C (After Gladman et al.[6])*

Whereas the above equations clearly identify the solid solution strengthening effects of an element such as manganese, it must be borne in mind that manganese also contributes to strength by other means. Thus by depressing the temperature of transformation of austenite to ferrite, manganese causes further strengthening by:

1. Refining the ferrite grain size.
2. Refining the size of precipitation-strengthening particles, e.g. Nb(CN) and V_4C_3.

These effects are illustrated in Figure 2.6, which shows the effect of manganese on the yield strength of a V–N steel, normalized from 900°C. This figure also indicates that free nitrogen contributes very little to the overall strength of this particular steel in spite of having a very large strengthening coefficient. This is due to the fact that most of the nitrogen in this particular alloy remains out of solution as VN at a temperature of 900°C and therefore very little free nitrogen is available for solid solution strengthening. Whereas the VN particles refine the austenite grain size and produce a fine ferrite grain size, vanadium in solution leads subsequently to a dispersion-strengthening effect of the order of 75 N/mm^2.

Controlled rolling/thermomechanical processing

As stated earlier, the traditional route to a fine grain size in ferrite–pearlite structural steels has been to incorporate grain-refining elements, such as aluminium, and to normalize the materials from about 920°C after rolling. However, prior to the introduction of continuous casting, basic carbon steel plate was made from semi-killed (*balanced*) ingots and the additional costs associated with aluminium grain refinement were very considerable, as illustrated by the cost data from the early 1960s shown in Table 2.2.

In the late 1950s, steel users were also gaining experience with Nb-treated, micro-alloy steels which provided substantially higher strengths than plain carbon steel in the as-rolled condition but with a significant reduction in toughness compared with aluminium-grain-refined steels. However, when the micro-alloy steels were normalized to improve their impact properties, their strength advantage was forfeited. There was therefore the need for an alternative route to a fine grain size in structural steel plate which would overcome both the cost and strength penalties associated with traditional normalizing. In fact, the first

Table 2.2

Requirement	Extras (£/tonne)	Accumulative price Price (£/tonne)
nil (mild steel)	–	42.62
Si killing	7.50	50.12
Grain refinement	3.00	53.12
Normalizing	3.00	56.12
Impact testing < 0°C	3.50	59.62

)f a viable alternative to normalizing was published in 1958 when
[7] reported that European steel producers were adopting lower than
;hing temperatures during rolling, in order to refine the structure and
:chanical properties. This practice became known as *controlled rolling*
but in more recent years, the term *thermomechanical processing* has been used
increasingly to embrace both modified hot-rolling and in-line accelerated-cooling
operations.

Outline of process

The traditional hot-rolling operation for plates is shown schematically in
Figure 2.7(a). Typically, slabs are soaked at temperatures of about 1200–1250°C
and these are rolled progressively to lower plate thicknesses, often finishing at
temperatures above 1000°C. In plain carbon steels, soaking at 1200–1250°C
produces a coarse austenite grain size and rolling just below that range
results in rapid recrystallization. Even at a finishing temperature of 1000°C,
recrystallization and subsequent grain growth will be relatively rapid, resulting in
the generation of a coarse austenite grain size. On cooling to room temperature,
this results in the formation of a coarse ferrite grain size and the material must
be normalized in order to refine the microstructure.

In controlled rolling, the operation is a two-stage process and, as illustrated in
Figure 2.7(b), a time delay is introduced between roughing and finishing. This
allows the finishing operations to be carried out at temperatures below the recrystallization temperature, which results in the formation of fine *pancaked* austenite
grains and transformation to a fine-grained ferrite structure. In the 40 or so years
since its introduction, a considerable amount of research work has been carried
out world-wide on controlled rolling which has led to the development of materials with vastly improved mechanical properties compared with those obtained
in conventional, heat-treated steels. Whereas the early experiments in controlled
rolling were carried out on plain carbon steels, it soon became evident that the
process was greatly facilitated by the addition of carbide-forming elements. In
particular, it was shown that the addition of about 0.05% Nb caused a marked
retardation in recrystallization which allowed controlled rolling to be carried out
at significantly higher temperatures.

Slab reheating

The slab-reheating stage is important in that it controls the amount of micro-
alloying elements taken into solution and also the starting grain size. The
solubility curves for NbC and VN in steels of different carbon and nitrogen
contents are shown in Figure 2.3 and were referred to earlier. For most
commercial grades of steel, Sellars[8] states that complete solution of VC is
expected at the standard normalizing temperature of 920°C, and VN at somewhat
higher temperatures, whereas Nb(CN), AlN and TiC require temperatures in the
range 1150–1300°C. TiN is the most stable compound and little dissolution is
expected to take place at normal reheating temperatures. Whereas the presence of
fine, undissolved carbonitride particles will serve to maintain a fine austenite grain

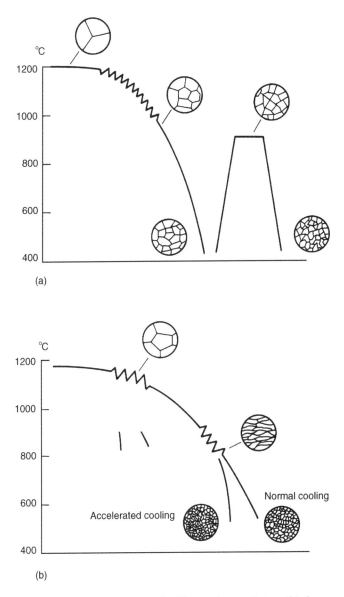

Figure 2.7 *Rolling schedules; (a) normal rolling and normalizing; (b) thermomechanical processing*

size at the reheating stage, it is equally important that micro-alloying elements are taken into solution so as to be available for the control of recrystallization and precipitation strengthening at later stages in the process. This dual requirement is often achieved by making multiple micro-alloy additions, incorporating the less soluble elements such as niobium and titanium for grain size control during reheating together with vanadium, which dissolves more readily and which provides substantial precipitation strengthening.

Rolling

Tamura *et al.*[9] recognize three distinct stages during controlled rolling:

1. Deformation in the recrystallization temperature range just below the reheating temperature.
2. Deformation in the temperature range between the recrystallization temperature and Ar₃.
3. Deformation in the two-phase $\gamma + \alpha$ temperature range between Ar₃ and Ar₁.

At temperatures just below the reheating temperature, the rate of recrystallization is rapid, increasing with both temperature and degree of deformation. However, refinement of the austenite structure is produced by successive recrystallization between passes, provided the strain per pass exceeds a minimum critical level. Recrystallization is retarded to some extent by the presence of solute atoms, such as aluminium, niobium, vanadium and titanium, and the process is known as *solute drag*. However, the major effect of niobium and other micro-alloying elements in retarding recrystallization and grain growth arises from the strain-induced precipitation of fine carbonitrides during the rolling process.

As the rolling temperature decreases, recrystallization becomes more difficult and reaches a stage where it effectively ceases. Cuddy[10] has defined the *recrystallization stop temperature* as the temperature at which recrystallization is incomplete after 15 seconds, after a particular rolling sequence. Using this criterion, the effect of the common micro-alloying elements on recrystallization behaviour is shown in Figure 2.8 and this illustrates the very powerful effect of niobium. The retardation effects of the various elements are dependent on their relative solubilities in austenite, the least soluble (niobium) having the largest driving force for precipitation at a given temperature and creating a proportionally greater effect in raising the recrystallization temperature than the more soluble elements such as aluminium and vanadium.

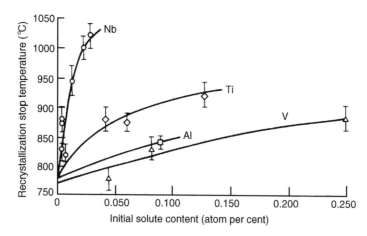

Figure 2.8   *Effect on* recrystallization stop temperature *of increase in micro-alloy content in 0.07% C 1.4% Mn 0.25% Si steel (After Cuddy[10])*

By introducing a delay between roughing and finishing, rolling can be made to take place at a temperature below 950°C, where the strain-induced precipitation of Nb(CN) or TiC is sufficiently rapid to prevent recrystallization before the next pass. According to Cohen and Hansen,[11] austenite recrystallization and carbonitride precipitation are interlinked during this process, substructural features in the deformed austenite providing nucleation sites for carbonitride precipitation which in turn pins the substructure and inhibits recrystallization. This results in an elongated pancake morphology in the austenite structure and the austenite is said to be *conditioned*. The deformation substructure that is introduced within the austenite grains has a particularly beneficial effect in developing a finer grain size. This arises from the fact that the substructure provides intragranular sites for ferrite nucleation in addition to those at the austenite grain boundaries.

The controlled-rolling operation can be intensified by depressing the deformation process below Ar_3 and into the two-phase $\gamma + \alpha$ region. In addition to further grain refinement, rolling in this region also produces a significant change in the microstructure. Thus a mixed structure is produced, consisting of polygonal ferrite grains which have transformed from deformed austenite and deformed ferrite grains which were produced during the rolling operation.

Transformation to ferrite

Although the mean ferrite grain size is related to the thickness of the pancaked austenite grains, other factors also play an important part in the control of the finished microstructure and properties. As discussed previously, alloying elements depress the austenite to ferrite transformation temperature and thereby decrease the ferrite grain size. A further important effect is the rate of cooling from the austenite (or $\gamma + \alpha$) range and a combination of controlled rolling and accelerated cooling is now being used to produce further improvement in properties.

The benefits of accelerated cooling can be used in two ways:

1. To increase the strength compared with air-cooled, controlled-rolled material.

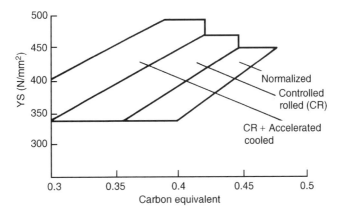

Figure 2.9 *Exploitation of controlled rolling and accelerated cooling for improved properties in plate steels*

2. To achieve the strength levels of controlled-rolled materials in steels of lower alloy content.

The latter approach is particularly attractive in that it utilizes steels of lower carbon equivalent and therefore provides improved weldability. This effect is shown schematically in Figure 2.9.

An extension of normal accelerated cooling after rolling is that employing the faster cooling rates of *direct quenching*. Whereas the former is concerned with refinement of the ferrite grains, the latter is concerned with the formation of lower temperature transformation products such as bainite and martensite. Direct quenching avoids the reheating costs associated with conventional off-line hardening treatments but still requires a subsequent tempering treatment.

Standard specifications

For many years, the relevant UK standard for structural steels was BS 4360 *Weldable structural steels* and provided the comprehensive yield strength–toughness matrix shown in Figure 2.10. Thus there were four levels of strength (40, 43, 50 and 55), the numbers referring to the minimum tensile strengths in kgf/mm². Within each strength grade, there were various sub-grades which represented increasing levels of toughness, designated by the following nomenclature:

Minimum Charpy V value of 27J at:

A	no requirement	DD	−30°C
B	20°C	E	−40°C
C	0°C	EE	−50°C
D	−20°C	F	−60°C

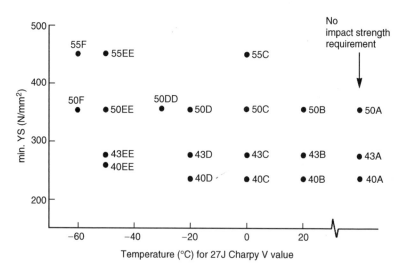

Figure 2.10 *Yield strength–impact strength requirements in BS 4360: 1986 Weldable structural steels (product forms–plates, strip and wide flats)*

However, BS 4360 has now been superseded by various European standards, of which the following are the more important:

BS EN 10025: 1993 *Hot rolled products of non-alloy structural steels.*
BS EN 10133: 1993 *Hot rolled products in weldable fine grain structural steels.*

The designation and property requirements of these European specifications are shown in Tables 2.3 (a) and 2.3 (b) which also provide a useful comparison with the grades specified in the former BS 4360 standard. Thus the former BS 4360 grades, 40, 43 and 50 A, B, C and D, are now covered by BS EN 10025: 1993 and the higher toughness grades DD, E and EE, together with the high-strength grades (55C and 55EE), are superseded by grades in BS EN 10113: 1993. To complete the puzzle, the former BS 4360 grades, 50F and 55F, have now been incorporated in the following, recently published European standard:

BS EN 10137: 1996 *High yield strength structural steels in the quenched and tempered or precipitation hardened condition.*

A European standard has also been introduced for *weathering steels* (BS EN 10155: 1993) and this will be discussed later in this chapter.

A possible source for confusion is that the current UK specifications dealing with the design of steel bridges and buildings have not yet been updated to deal with the changes from BS to European standards and they still refer to steels specified in BS 4360. However, this aspect will also be discussed in the appropriate parts of this chapter.

Steel prices

The basis prices of some of the plates, specified in BS EN 10025: 1993 and BS EN 10113: 1993, are shown in Figure 2.11. It should be stressed that the prices shown in this figure were those in force in March 1997 and it must be borne in mind that steel prices are adjusted from time to time. As indicated in Figure 2.11, the prices increase progressively at a given strength level as the specified impact properties are improved and also as the minimum yield strength values are increased. However, as illustrated in Figure 2.12, the ratio of cost: yield strength (at a given level of toughness) falls very significantly as the strength of these steels is increased. Therefore there is a major cost incentive to utilize higher-strength grades in structures where design is based primarily on yield strength.

Weathering steels

The term *weathering steels* is given to structural grades in which the resistance to atmospheric corrosion has been improved by the addition of small amounts of elements such as copper, phosphorus, silicon and chromium. These steels rust at a

Table 2.3(a) *Structural steels: comparison between grades in BS EN 10025: 1993 and BS 4360: 1986*

| | | | BS EN 10025: 1993 | | | | | | BS 4360: 1986 | | | | | |
| | | | | | | Impact Energy (J°C) Nominal thickness | | | | | | | Impact Energy (J°C) Nominal thickness | |
Grade		Former grade	Tensile strength ≥3 mm ≤100 mm N/mm²	Min. yield strength at 16 mm N/mm²	Max. Thk for specified yield N/mm²	Temp. °C	≤150 mm (1)	>150 mm ≤250 mm (1)	Grade	Tensile strength ≤100 mm N/mm²	Min. yield strength at 16 mm N/mm²	Max. Thk for specified yield N/mm² (2)	Temp °C	≤100 mm (3)
S185	(4)	Fe 310-0 (4)	290/510	185	25	–	–	–	–	–	–	–	–	–
S235	(5)	Fe 360A (5)	340/470	235	250	–	–	–	40A	340–500	235	150	–	–
S235JR	(4)	Fe 360B (4)	340/470	235	25	+20 (6)	27	–	–	–	–	–	–	–
S235JRG1		Fe 360B(FU)(4)	340/470	235	25	+20 (6)	27	–	–	–	–	–	–	–
S235JRG2		Fe 360B(FN)	340/470	235	250	+20 (6)	27	23	40B	340/500	235	150	+20 (6)	27
S235JO		Fe 360C	340/470	235	250	0	27	23	40C	340/500	235	150	0	27
S235J2G3		Fe 360D1	340/470	235	250	–20	27	23	40D	340/500	235	150	–20	27
S235J2G4		Fe 360D2	340/470	235	250	–20	27	23	40D	340/500	235	150	–20	27
S275	(5)	Fe 430A (5)	410/560	275	250	–	–	–	43A	430/580	275	150	–	–
S275JR		Fe 430B	410/560	275	250	+20 (6)	27	23	43B	430/580	275	150	+20 (6)	27
S275JO		Fe 430C	410/560	275	250	0	27	23	43C	430/580	275	150	0	27
S275J2G3		Fe 430D1	410/560	275	250	–20	27	23	43D	430/580	275	150	–20	27
S275J2G4		Fe 430D2	410/560	275	250	–20	27	23	43D	430/580	275	150	–20	27
S355	(5)	Fe 510A (5)	490/630	355	250	–	–	–	50A	490/640	355	150	–	–
S355JR		Fe 510B	490/630	355	250	+20 (6)	27	23	50B	490/640	355	150	+20 (6)	27
S355JO		Fe 510C	490/630	355	250	0	27	23	50C	450/640	355	150	0	27
S355J2G3		Fe 510D1	490/630	355	250	–20	27	23	50D	490/640	355	150	–20	27
S355J2G4		Fe 510D2	490/630	355	250	–20	27	23	50D	490/640	355	150	–20	27
S355K2G3		Fe 510DD1	490/630	355	250	–20	40	33	50DD	490/640	355	150	–30	27
S355K2G4		Fe 510DD2	490/630	355	250	–20	40	33	50DD	490/640	355	150	–30	27

E295	Fe 490–2	470/610	295	250	–	–	–	–	–	–	–	–
E335	Fe 590–2	570/710	335	250	–	–	–	–	–	–	–	–
E360	Fe 690–2	670/830	360	250	–	–	–	–	–	–	–	–

1 For sections up to and including 100 mm only.
2 For wide flats and sections up to and including 63 mm and 100 mm respectively.
3 For wide flats up to and including 50 mm and for sections no limit is stated.
4 Only available up to and including 25 mm thick.
5 The steel grades S235 (Fe 360A), S275 (Fe 430A) and S355 (Fe 510A) appear only in the English language version (BS EN 10025) as non-conflicting additions and do not appear in other European versions.
6 Verification of the specified impact value is only carried out when agreed at time of enquiry and order.

Symbols used in BS EN 10025
S = Structural Steel.
E = Engineering Steel.
'235' '275' '355' = min YS (N/mm^2) @ t $\leq$16 mm
JR = Longitudinal Charpy V-notch impacts 27J @ room temperature
JO = Longitudinal Charpy V-notch impacts 27J @ 0°C
J2 = Longitudinal Charpy V-notch impacts 27J @ −20°C
K2 = Longitudinal Charpy V-notch impacts 40J @ −20°C
G1 = Rimming steel (FU)
G2 = Rimming steel *not* permitted (FN)
G3 = Supply Condition 'N', i.e. normalized or normalized rolled
G4 = Supply Condition at the manufacturer's discretion.
Examples S235JRG1, S355K2G4

After BS EN 10025: 1993.

Table 2.3(b) Structural Steels: Comparison between grades in BS EN 10113: 1993 and BS 4360: 1990

	BS EN 10113									BS 4360: 1990							
			Max. Thk (mm)		Charpy (long)		Max. Thk (mm)					Max. Thk (mm)		Charpy (long)		Max. Thk (mm)	
Grade	UTS	Min.YS at t = 16 mm (N/mm²)	(1)	(2)	Temp. (°C)	Energy (J)	(1)	(2)	Grade	UTS	Min.YS at t = 16 mm (N/mm²)	(1)	(2)	Temp. (°C)	Energy (J)	(1)	(2)
S275N	370 to 510	275	150	150	−20	40	150	150	43DD	430 to 580	275	–	100	−30	27	–	(7)
S275NL			150	150	−50	27	150	150	43EE			150 (3)	–	−50	27	75 (5)	–
S355N	470 to 630	355	150	150	−20	40	150	150	50DD	490 to 640	355	150 (3)	100	−30	27	100 (5)	(7)
S355NL			150	150	−50	27	150	150	50E			–	100	−40	27	–	(7)
S420N	520 to 680	420	150	150	−20	40	150	150	50EE			150 (3)	–	−50	27	75 (6)	–
S420NL			150	150	−50	27	150	150	–								
S460N	550 to 720	460	100	100	−20	40	100	100	–								
S460NL			100	100	−50	27	100	100	55C	550 to 700	450	25	40	0	27	25	19
S275M	360 to 510	275	63	150	−20	40	63	150	55EE			63 (4)	–	−50	27	63 (4)	–
S275ML					−50	27											

Grade								
S355M	450 to 610	355	63	150	−20	40	63	150
S355ML					−50	27		
S420M	500 to 660	420	63	150	−20	40	63	150
S420ML					−50	27		
S460M	530 to 720	460	63	150	−20	40	63	150
S460ML					−50	27		

Longitudinal Charpy V-notch impacts

Grade	Min. ave. energy (J) at test temp (°C)						
	+20	0	−10	−20	−30	−40	−50
S_N/M	55	47	43	40			
S_NL/ML	63	55	51	47	40	31	27

After BS EN 10113: 1993.

Notes:
1 Applies to plates and wide flats.
2 Applies to Sections.
3 For wide flats max. thickness is 63 mm.
4 Not available as wide flats.
5 For wide flats max. thickness is 50 mm.
6 For wide flats max. thickness is 30 mm.
7 For sections no limit is given.

Symbols used in BS EN 10113
S = Structural Steel
'275' '355' '420' '460' = min YS (N/mm^2) @ t $\leq$ 16 mm.
N = Normalized or normalized rolled
M = Thermomechanical rolled.
L = Low temperature (−50°C) impacts.
Examples S275N, S355ML

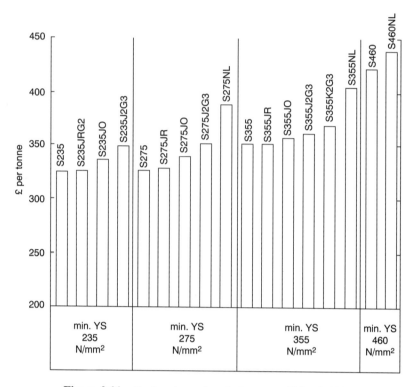

Figure 2.11 *Basis prices of steel plates as of March 1997*

lower rate than plain carbon steels and, under favourable climatic conditions, they can develop a relatively stable layer of hydrated iron oxide which retards further attack. Very often, reference is made to the brown 'patina' that develops on the surface of these steels and the attractive appearance that it presents in buildings and bridges. However, regardless of their aesthetic qualities, weathering steels can provide cost savings by eliminating the initial painting operation and subsequent maintenance work.

These steels were first introduced in 1933 by the United States Steel Corporation under the brand name *Cor-Ten* and have been licensed for manufacture throughout the world. However, they have not been used extensively in the UK because frequent rain inhibits the formation of the stable oxide layer and rusting continues, albeit at a relatively slow rate. Ideally, long dry summer periods are required to develop the adherent oxide layer.

Corrosion resistance

The mid-specifications of the chemical composition in the Cor-Ten series are given in Table 2.4. Cor-Ten A is the original high-phosphorus grade whereas Cor-Ten B and Cor-Ten C were introduced in the 1960s. These later steels have normal levels of phosphorus and are micro-alloyed with vanadium to provide high strength.

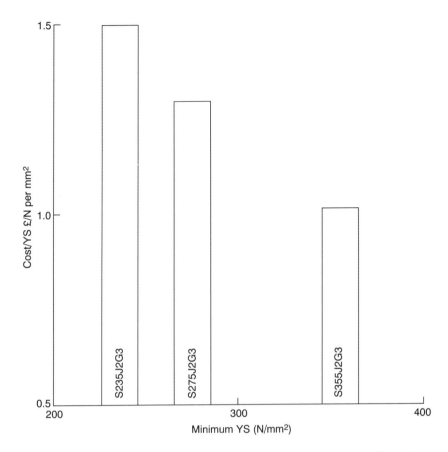

Figure 2.12 *Cost–yield strength relationships for plates (grades providing minimum impact energy of 27 J at −20° C*

Table 2.4

%	Cor-Ten A	Cor-Ten B	Cor-Ten C
C	0.08	0.14	0.16
Si	0.50	0.20	0.20
Mn	0.25	1.10	1.20
P	0.11	0.04 max.	0.04 max.
Cr	0.75	0.50	0.50
Ni	0.35	–	–
Cu	0.40	0.35	0.35
V	–	0.06	0.07

The corrosion behaviour of these steels is shown in Figure 2.13[12] where they can be compared with plain carbon steel. Very clearly, the weathering grades are superior to carbon steel in each of the atmospheres investigated, Cor-Ten A providing a better performance than Cor-Ten B. However, in keeping with the corrosion behaviour of other types of steel, the performance is worst in marine

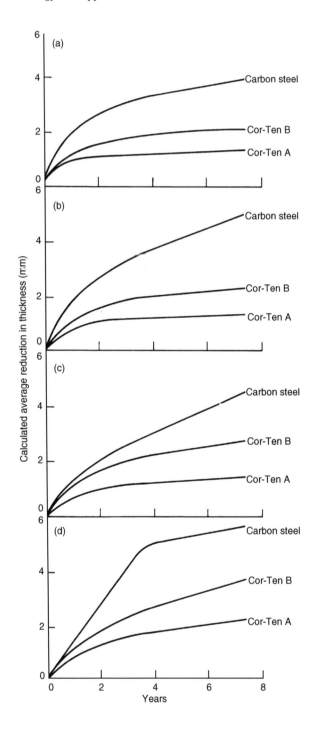

Figure 2.13 *Corrosion behaviour of Cor-Ten steels: (a) industrial; (b) semi-industrial; (c) semi-rural; (d) moderate marine*

atmospheres and weathering steels are not generally recommended for use in such environments.

In spite of the large amount of corrosion data that has been gathered on weathering steels, the mechanism of their superior performance is still relatively obscure. Studies by the United States Steel Corporation[13] showed that Cor-Ten steels rusted faster than carbon steel in the initial stages and it was only after a period of eight days that carbon steel showed a greater gain in weight. After that period, both materials had developed a continuous covering of rust mounds but those on the carbon steel grew to a larger size and eventually spalled from the surface. With Cor-Ten A, splitting was less frequent and no spalling was observed. X-ray diffraction work showed that the rust on both types of steel consisted essentially of $\gamma - Fe_2O_3.H_2O$ in the initial stages, but after about 30 days $\alpha - Fe_2O_3.H_2O$ was detected. However, in addition to iron oxides, iron sulphates ($FeSO_4.3H_2O$, $FeSO_4.7H_2O$ and $Fe_2(SO_4)_3$) have been detected in the rust layers formed on steel in polluted atmospheres. It is suggested tentatively that the beneficial alloying elements render these sulphates less soluble and thereby retard the penetration of air and moisture through the oxide layer to the steel interface.

Horton[14] has examined the effect of individual alloying elements on the corrosion resistance of Mayari R steel (Bethlehem Steel Corporation). The base composition used in this work was as shown in Table 2.5.

The results are summarized in Figure 2.14(a), which shows the effect of variations in a single alloying element in the above base. The corrosion penetration of the base steel was 2.9 mm and is represented as a horizontal line. Copper was not examined in this investigation but Horton lists the following order of effectiveness for other elements:

Most beneficial	P
	Cr
	Si
	Ni
No effect	Mn
Detrimental	S

Horton also analysed data on Cor-Ten A steel which involved the base composition shown in Table 2.6. These results are shown in Figure 2.14(b) and (c) which deal with industrial and marine (Kure Beach, North Carolina) sites respectively.

Table 2.5

C%	Si%	Mn%	P%	S%	Cr%	Ni%	Cu%
0.08	0.28	0.70	0.10	0.03	0.60	0.40	0.60

Table 2.6

C%	Si%	Mn%	P%	S%	Cr%	Ni%	Cu%
<0.10	0.22	0.25–0.40	0.10	<0.02	0.63	~0.5	0.42

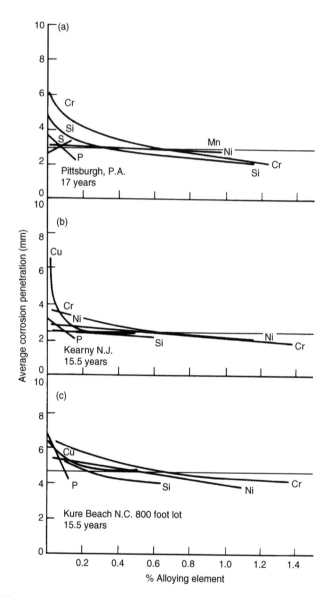

Figure 2.14   *Corrosion behaviour of weathering steels: (a) Bethlemen data on Mayari R; (b) USS Cor-Ten A in industrial atmosphere; (c) USS Cor-Ten A in marine atmosphere*

At the latter site, the order of effectiveness is as follows:

Most beneficial	P	
	Si	
	Cu	(up to 0.3%)
	Cr	
	Ni	
	Cu	(>3%)

However, at the industrial site (Kearny, New Jersey), copper exerts a more powerful effect and would be promoted to a higher ranking than that shown above.

Very similar results have also been obtained by Hudson and Stanners[15] and Larrabee and Coburn[16] and it is generally acknowledged that phosphorus produces a marked and progressive benefit up to at least 0.1%. However, phosphorus is very detrimental to both toughness and weldability and therefore it is not included in some grades of weathering steel, with some loss in corrosion resistance. Copper is regarded as an essential constituent in weathering steels but little benefit is gained by increasing the copper content above about 0.3%. Elements such as silicon and chromium are mildly beneficial and the greatest benefit is obtained at levels up to about 0.25% and 0.6% respectively. As indicated earlier, manganese can be regarded as being neutral in its effect on corrosion resistance, whereas sulphur is detrimental.

Steel specifications

Following the withdrawal of BS4360: 1990, a European standard was introduced for weathering grades, namely BS EN 10155: 1993 *Structural steels with improved atmospheric corrosion resistance*. The chemical compositions of these grades are shown in Table 2.7 which includes two high phosphorus compositions and specific ranges for chromium and copper.

The mechanical properties of these grades are shown in Table 2.8. BS EN 10155: 1993 offers a wider range of grades than the former British standard, extending the requirements for both tensile and impact properties.

Clean steels and inclusion shape control

In parallel with developments in micro-alloy steels and thermomechanical processing, major attention has also been given to methods of producing steels which are isotropic with regard to tensile ductility and impact strength. Initially, the need for these steels was precipitated by the incidence of *lamellar tearing* in which plate material separates along planar arrays of non-metallic inclusions under the forces generated in highly restrained welds. However, the need for higher levels of impact strength in structural steels and the requirement for better cold formability in strip products have also focused attention on the development of cleaner steels and inclusion shape control.

The practice of adding calcium to steels for the reduction of sulphide and oxide inclusions is now used world-wide and has the added benefit of modifying the shape and size of these inclusions. One of the pioneers in this field was Thyssen Niederrhein in Germany and the practice is described by Pircher and Klapdar.[17] After tapping, calcium in the form of calcium silicide or calcium carbide is injected deep into the ladle by means of a refractory lined lance, using argon as a carrier. The calcium vaporizes and as it bubbles through the molten bath it combines with sulphur and oxygen in the steel. In either case, the reaction products are carried into the slag. The steel is generally deoxidized

Table 2.7 BS EN 10155: 1993 Structural Steels with improved atmospheric corrosion resistance

| Designation | | Method of deoxidation | Chemical composition of the ladle analysis | | | | | | | | | |
According EN 10027–1 and ECISS IC 10	According EN 10027–2		C max. %	Si max. %	Mn %	P %	S max. %	N max. %	Addition of nitrogen binding elements[1]	Cr %	Cu %	Others
S235J0W	1.8958	FN	0.13	0.40	0.20–0.60	max. 0.040	0.040	0.009[2)5)]	–	0.40–0.80	0.25–0.55	3)
S235J2W	1.8961	FF	0.13	0.40	0.20–0.60	max. 0.040	0.035	–	yes	0.40–0.80	0.25–0.55	
S355J0WP	1.8945	FN	0.12	0.75	max. 1.0	0.06–0.15	0.040	0.009[5)]	–	0.30–1.25	0.25–0.55	3)
S355J2WP	1.8946	FF	0.12	0.75	max. 1.0	0.06–0.15	0.035	–	yes	0.30–1.25	0.25–0.55	
S355J0W	1.8959	FN	0.16	0.50	0.50–1.50	max. 0.040	0.040	0.009[2)5)]	–	0.40–0.80	0.25–0.55	3)4)
S355J2G1W	1.8963	FF	0.16	0.50	0.50–1.50	max. 0.035	0.035	–	yes	0.40–0.80	0.25–0.55	
S355J2G2W	1.8965	FF	0.16	0.50	0.50–1.50	max. 0.035	0.035	–	yes	0.40–0.80	0.25–0.55	
S355K2G1W	1.8966	FF	0.16	0.50	0.50–1.50	max. 0.035	0.035	–	yes	0.40–0.80	0.25–0.55	
S355K2G2W	1.8967	FF	0.16	0.50	0.50–1.50	max. 0.035	0.035	–	yes	0.40–0.80	0.25–0.55	

[1] The steels shall contain at least one of the following elements: Al total $\geq$0.020%, Nb: 0.015 to 0.060%, V: 0.02 to 0.12%, Ti: 0.02 to 0.10%. If these elements are used in combination, at least one of them shall be present with the minimum of content indicated.

[2] It is permissible to exceed the specified values provided that each increase of 0.001% N the P max. content will be reduced by 0.005%; the N content of the ladle analysis, however, shall not be more than 0.012%.

[3] The steels may show a Ni content of max. 0.65%.

[4] The steels may contain max. 0.30% Mo and max. 0.15% Zr.

[5] The max. value for nitrogen does not apply if the chemical composition shows a minimum total Al content of 0.020% or if sufficient other N binding elements are present. The N binding elements shall be mentioned in the inspection document.

Table 2.8 *Mechanical properties of steels to BS EN 10155: 1993*

Mechanical properties for flat and long products[1]

Designation		Minimum yield strength R_{eH}[1] N/mm² Nominal thickness mm					Tensile strength R_m[1] N/mm² Nominal thickness mm		Position of test pieces[1]	Minimum percentage elongation at fracture[1] %					
										$L_0 = 80$ mm Nominal thickness mm			$L_0 = 5.65\sqrt{S_0}$ Nominal thickness mm		
According EN 10027-1 and ECISS IC 10	*According EN 10027-2*	≤16	>16 ≤40	>40 ≤63	>63 ≤80	>80 ≤100	<3	≥3 ≤100		>1.5 ≤2	>2 ≤2.5	>2.5 <3	≥3 ≤40	>40 ≤63	>63 ≤100
S235J0W	1.8958	235	225	215	215	215	360–510	340–470	1	19	20	21	26	25	24
S235J2W	1.8961								t	17	18	19	24	23	22
S355J0WP	1.8945	355	345[2]	–	–	–	510–680	490–630	1	16	17	18	22	–	–
S355J2WP	1.8946								t	14	15	16	20	–	–
S355J0W	1.8959	355	345	335	325	315	510–680	490–630	1	16	17	18	22	21	20
S355J2G1W	1.8963														
S355J2G2W	1.8965														
S355K2G1W	1.8966														
S355K2G2W	1.8967								t	14	15	16	20	19	18

[1] The values in the table apply to longitudinal test pieces (1) for the tensile test. For plate, strip and wide flats with widths ≥600 mm transverse test pieces (t) are applicable.

[2] This value applies only in respect to shapes, sections and bars (see table 2).

with aluminium prior to calcium treatment with an initial oxygen content in the range 20–100 ppm. After calcium treatment, the oxygen content is reduced to 10–20 ppm. The authors report that the desulphurizing effect of calcium is determined largely by the amount of calcium added but, as illustrated in Figure 2.15, the effect is influenced greatly by the type of ladle refractories employed. With dolomite ladles, little reaction takes place between the molten steel and refractories and the low level of oxygen in the steel enhances the effect of calcium as a desulphurizing agent. With dolomite ladles, the addition of 1 kg of calcium per ton reduces the sulphur content from an initial level of 0.02% to a final level of 0.003%. However, the authors state that sulphur contents less than 0.001% can be produced by this technique. As illustrated in Figure 2.15, magnesium is also very effective as a desulphurizing agent. However, calcium is generally preferred because it is cheaper and more controllable.

As indicated earlier, anisotropy of toughness and ductility is caused by the elongation of inclusions into planar arrays and both manganese sulphides and stringers of oxide are damaging in this respect. However, the problem is largely eliminated if the inclusions are present as small, isolated, non-deformed particles. Therefore major attention has been devoted to inclusion shape control in addition to the reduction in the volume fraction of inclusions. The elements used as inclusion modifiers are calcium, zirconium, tellurium and the rare earth metals.

In aluminium deoxidized steels, the inclusion population will generally include elongated Type II manganese sulphides, alumina and some silicates. However, after calcium treatment, the inclusions are restricted to calcium aluminate of the type $CaO.Al_2O_3$. The sulphur in the steel is also associated with these inclusions, either as calcium sulphide or as sulphur in solution. The calcium aluminate particles are globular in nature and tend to retain their shape on hot rolling. The

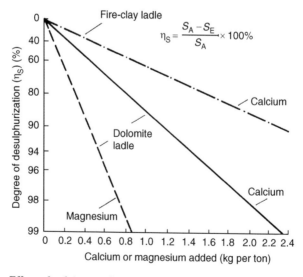

Figure 2.15 *Effect of calcium and magnesium on desulphurization. The desulphurizing additions were blown to a depth of 2.7 m in a steel bath having a temperature of 1580° C (After Pircher and Klapdar[17]*

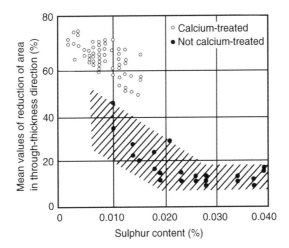

Figure 2.16 *Relationship between the sulphur content and the mean values of reduction of area of tensile test specimens in the through-thickness direction of steel grade FG 36 (After Pircher and Klapdar[17])*

beneficial effect of a reduction in sulphur content and calcium treatment on the reduction in area in the through-thickness direction is shown in Figure 2.16.

Ships

Although there has been a significant decline in the world's output of new ships since the mid-1970s, the production of large merchant vessels remains a major application for structural steels. Despite marked changes in material selection for other forms of transport, steel remains virtually unchallenged for hull construction in large tankers and carriers. However, like other forms of steel construction, shipbuilding has witnessed the following changes since the early 1940s:

1. The change from riveting to welding as the principal method of joining.
2. An appreciation of the need for high levels of toughness.
3. The adoption of higher strength steels for reduced construction costs or higher operating efficiency.

Standard-strength steels

The main specifications for shipbuilding materials are issued by organizations known as *Classification Societies*, namely:

- American Bureau of Shipping
- Bureau Veritas
- Det Norske Veritas
- Germanischer Lloyd

- Lloyd's Register of Shipping
- Nippon Kaiji Kyokai
- Registro Italiano Navale

Whereas each of these societies publishes its own design rules and steel specifications, they collaborate closely through the International Association of Classification Societies (IACS). Therefore there is a high degree of uniformity in steel specifications in terms of composition, tensile properties and impact resistance. This collaboration stems from the mid-1950s when there was an urgent need to harmonize the approaches that had been taken individually in formulating steel specifications with improved resistance to brittle fracture.

Until the early 1940s, shipbuilding had been based on riveted construction and only one grade of steel was in common use, namely 'shipbuilding quality'. The steel was specified very simply in terms of tensile and bend tests and no limitations were placed on chemical composition. However, during World War II, the emergency shipbuilding programme in the United States demanded higher production rates for the so-called Liberty Ships and brought about the change from riveted to welded construction. This necessitated consideration of steel composition in relation to weldability but of far greater significance was the emergence of the problem of brittle fracture. In some instances, brittle fracture led to the catastrophic break-up of cargo vessels at sea and also produced the spectacular failure of the SS *Schenectady* in January 1943, whilst lying alongside the outfitting berth of a shipyard. Whereas it was shown that structural performance could be enhanced by means of improved design in critical elements, it was also evident that there was an urgent need to improve the toughness characteristics of ship plate steel.

In the years following World War II, each of the classification societies took independent action in formulating steel specifications, but by 1952 pressure from shipbuilders, owners and steelmakers brought about the initial discussions for the harmonization of specifications. A detailed account of these discussions was published by Boyd and Bushell[18] and at the outset, the seven classification societies collectively had a total of 22 grades of steel. However, these steels could be classified into three main types:

1. Ordinary ship steel, which was used in modest thicknesses and in lightly stressed areas.
2. An intermediate grade for areas where there was a need for some control over notch toughness and for intermediate thicknesses.
3. A high-grade steel with good notch ductility and for heavy plate thicknesses.

Although the seven societies recognized these three broad categories of steel, they were not able to rationalize their individual grades into three commonly acceptable specifications. On the one hand, the American Bureau of Shipping (ABS) favoured specifications based on deoxidation practice, composition and heat treatment whereas the European societies preferred specifications based primarily on mechanical properties. It was agreed finally to adopt unified grades based on both approaches and this resulted in five specifications for the three basic types of steel:

- Grade A – ordinary shipbuilding steel
- Grade B – intermediate grade based on the ABS approach
- Grade C – highest grade based on the ABS approach
- Grade D – intermediate grade based on specified impact strength at 0°C (European approach)
- Grade E – highest grade based on specified impact strength at −10°C (European approach)

Since that time, further rationalization has taken place and Lloyd's[19] now specifies four grades of steel with increasing impact strength requirement at the standard minimum yield strength of 235 N/mm². Details of the composition and mechanical properties of these steels are given in Tables 2.9 and 2.10. Whereas normalizing was mandatory at one time for the higher toughness grades, these

Table 2.9   *Chemical composition and deoxidation practice for Lloyd's standard strength shipbuilding grades*

Grade	*A*	*B*	*D*	*E*
Deoxidation	Any method (for rimmed steel, see Note 1)	Any method except rimmed steel	Killed and fine grain treated with aluminium (see Note 2)	Killed and fine grain treated with aluminium
Chemical composition %				
Carbon	0.21 max. (*see* Note 3)	0.21 max.	0.21 max.	0.18 max.
Manganese	2.5×C% min.	0.80 min. (*see* Note 4)	0.60 min.	0.70 min.
Silicon	0.50 max.	0.35 max.	0.10–0.35	0.10–0.35
Sulphur	0.040 max.	0.040 max.	0.040 max.	0.040 max.
Phosphorus	0.040 max.	0.040 max.	0.040 max.	0.040 max.
Aluminium (acid soluble)	–	–	0.015 min. (*see* Note 5)	0.015 min. (*see* Note 5)

Carbon + $\frac{1}{6}$ of the manganese content is not to exceed 0.40%

Notes
1. For Grade A, rimmed steel may be accepted up to 12.5 mm thick inclusive, provided that it is stated on the test certificates or shipping statements to be rimmed steel and is not excluded by the purchaser's order.
2. Up to 25 mm in thickness, Grade D steel may be supplied in the semi-killed condition and without grain refinement, in which case the requirements for minimum silicon and aluminium contents do not apply.
3. The maximum carbon content for Grade A steel may be increased to 0.23% for sections.
4. For Grade B, when the silicon content is 0.10% or more (killed steel), the minimum manganese content may be reduced to 0.60%.
5. The total aluminium content may be determined instead of the acid soluble content. In such cases the total aluminium content is to be not less than, 0.020%.

After Lloyd's Register of Shipping, *Rules and Regulations for the Classification of Ships* (1995).[19]

Table 2.10   *Mechanical properties of Lloyd's standard strength shipbuilding grades*

Grade	Yield stress N/mm² minimum	Tensile strength N/mm²	Elongation on 5.65√S₀ % minimum	Charpy V-notch impact test (longitudinal)		
				Test temperature °C	Thickness mm	Average energy J minimum
A				–	–	–
B	235	400–490	22 (*see* Note 2)	0	≤50	27 (*see* Note 3)
D				–20	≤50	27
E				–40	≤50	27
					50 ≤ 70	34
					70 ≤ 100	41

Notes

1. Requirements for products over 50 mm thick in Grades A, B and D and 100 mm thick in Grade E are subject to agreement. *See* 2.1.1.
2. For full thickness tensile test specimens with a width of 25 mm and gauge length of 200 mm (*see* Fig. 2.2.4 in Chapter 2), the minimum elongation is to be:

Thickness mm	≤5	>5 ≤10	>10 ≤15	>15 ≤20	>20 ≤25	>25 ≤30	>30 ≤35	>35 ≤50
Elongation %	14	16	17	18	19	20	21	22

3. Impact tests are generally not required for Grade B steel of 25 mm or less in thickness provided that satisfactory results are obtained from occasional check tests selected by the Surveyor.

After Lloyd's Register of Shipping, *Rules and Regulations for the Classification of Ships* (1995).[19]

can now be supplied in the thermo-mechanically controlled processed condition, provided the specified mechanical properties can be achieved.

Higher-strength steels

By the mid-1960s, the higher-strength steels, based on micro-alloy additions, had become established and each of the classification societies introduced specifications with yield stress values in the range 300–400 N/mm². The mechanical properties of high-strength steels currently specified by Lloyd's are shown in Table 2.11. The higher strengths are achieved by grain refinement and precipitation strengthening and the steels can be supplied in the as-rolled, controlled-rolled or normalized condition. The steels are normally made to a restricted carbon equivalent (CE) of 0.41% max., based on the formula:

$$CE = C + \frac{Mn}{6} + \frac{Cr + Mo + V}{5} + \frac{Ni + Cu}{15} Wt\%$$

Lloyd's also specifies a range of quenched and tempered grades (grades 42–69) with minimum yield strengths in the range 420–690 N/mm² which are used in offshore construction. However, quenched and tempered grades are also used

Table 2.11 *Mechanical properties of Lloyd's higher strength shipbuilding grades*

Grade	Yield Stress N/mm² min.	Tensile Strength N/mm²	Elongation on 5.65√S₀ % min. (see Note 2)	Charpy V-notch impact tests			
				Thickness mm (see Note 1)	Test temperature °C	Average energy J minimum	
						Longitudinal	Transverse
AH 32				≤100	0		
DH 32	315	440–590	22	≤100	−20	31	22
EH 32				≤100	−40		
FH 32				≤50	−60		
AH 36				≤100	0	34	24
DH 36	355	490–620	21	≤100	−20	34	24
EH 36				≤50	−40	34	24
				> 50 ≤ 70	−40	41	27
				> 70 ≤ 100	−40	50	34
FH 36				≤50	−60	34	24
AH 40					0		
DH 40	390	510–650	20	≤50	−20	41	27
EH 40					−40		
FH 40					−60		

Notes

1. The requirements for products thicker than those detailed in the table are subject to agreement (*see* 3.1.4).
2. For full thickness tensile test specimens with a width of 25 mm and a gauge length of 200 mm (*see* Fig. 2.2.4 in Chapter 2), the minimum elongation is to be:

Thickness mm	≤5	>5 ≤10	>10 ≤15	>15 ≤20	>20 ≤25	>25 ≤35	>35 ≤50	>50
Elongation % — Strength level 32	14	16	17	18	19	20	21	To be specially agreed
Strength level 36	13	15	16	17	18	19	20	
Strength level 40	12	14	15	16	17	18	19	

After Lloyd's Register of Shipping, *Rules and Regulations for the Classification of Ships* (1995).[19]

for the construction of submarines. These can be considered as pressure vessels which have to withstand high hydrostatic pressures and the steels concerned have minimum yield strength values of 550 N/mm² (Navy Q1) and 690 N/mm² (Navy Q2), coupled with high levels of toughness.

Design considerations

For design purposes, naval architects regard the hull of a ship as a beam or girder in which the deck and bottom form the flanges and the sides constitute the web. Lloyd's and the other classification societies specify the minimum thicknesses of plate or *scantlings* that shall be used in various parts of a ship which, when acting together, give the structure the required stiffness or *section modulus*. In general, the plate thicknesses are related to the length of a ship, assuming that the depth and breadth conform to a reasonably fixed ratio of its length. For example, the ratio of length to depth is not expected to exceed 16:1 and the length to breadth ratio is generally greater than 5:1.

Naval architects have to legislate for the worst situation that a ship is likely to encounter due to wave action, namely a wave with a length equal to that of the ship. As illustrated in Figure 2.17, this can give rise to two extreme conditions of stress:

1. Suspension at the middle position putting the deck in tension *(hogging)*.
2. A wave at either end putting the bottom in tension *(sagging)*.

Thus the area of the ship that is given greatest attention in design is the middle section or *midships*, since this is the area which is subjected to greatest stress and deflection. For this reason, the plate thicknesses in this region, designated *0.4L amidships*, are heavier than those required towards the ends of the ship. Between the two extremes, classification societies quote a taper in terms of percentage decrease in thickness per metre so as to avoid abrupt changes in section. The longitudinal strength of an I-beam is located in the flanges and the thickness of the deck and bottom plating are greater than those in the sides of a ship. However, the longitudinal strength is not derived solely from the deck and bottom plating and

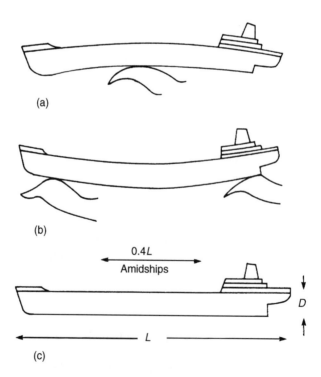

(a)

(b)

0.4L

Amidships

D

L

(c)

Figure 2.17   *Schematic illustration – hull deflection*

stiffeners in the form of rolled sections (*bulb or plain flats*) are used to reinforce these areas and also the hull sides. Plating is also used in the longitudinal and transverse partitioning walls or *bulkheads*, which also contribute to the strength of a ship with respect to buckling.

For the most part, a ship's hull is constructed from Grade A steel but Lloyd's Rules distinguish between the material grade requirements for different parts of the hull. As shown in Table 2.12(a), five classes of material are identified in ascending order of fracture toughness which are translated into steel grades, according to thickness requirements, in Table 2.12(b). A schematic illustration of the location of Grades A, D and E in the midship section of a large tanker is shown in Figure 2.18. Thus the use of the tougher Grades D and E is confined mainly to the more highly stressed deck and bottom regions whereas Grade A is adequate for the sides of a vessel.

The use of higher-strength steels in shipbuilding is attractive from two aspects:

1. Lower construction costs – from reduced steel weight and lower fabrication costs.
2. Lower operating costs – from reduced weight/lower fuel costs or higher carrying capacity for the same constructed weight.

Given the depressed state of the shipbuilding market, the former is probably the more important but the reduction in thickness that can be tolerated is governed by modulus and deflection considerations. From the analogy with a simple beam, the deflection of a ship is a function of the ratio of length to depth (*L/D*) and also the thicknesses of the plates used in the girder construction. Caution has therefore had to be exercised in the use of higher-strength steels, in reduced sections, in order to limit the deflection to within reasonable bounds. However, as experience has been gained, further advantage has been taken of the higher-strength grades. Currently, Lloyd's rules provide K_L factors for the determination of the hull girder section modulus in higher-strength steels and the values are shown in Table 2.13. For mild steel (235 N/mm^2 min. YS), the K_L factor = 1.0 and therefore significant reductions in the section modulus are now permitted

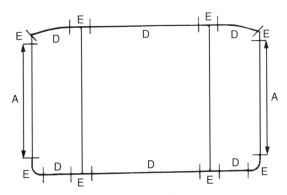

Figure 2.18 *Midships section showing distribution of Grades A, D and E*

Table 2.12 *Steel selection in merchant ships*
(a) Material classes

Structural member	Material class		
	Within 0.4L amidships	*Between 0.4L and 0.6L amidships*	*Outside 0.6L amidships*
Where *L* > 250 m: Sheerstrake or rounded gunwale Stringer plate at strength deck	V	III	II
Where *L* ≤ 250 m: Sheerstrake or rounded gunwale Stringer plate at strength deck			
Bilge strake Deck strake in way of longitudinal bulkhead	IV	III	II
Strength deck plating Bottom plating including keel Continuous longitudinal members above strength deck	III	I	I
Upper strake of longitudinal bulkhead Upper strake of topside tank			
Deck plating, other than above, exposed to weather Side plating Lower strake of longitudinal bulkhead	II	I	I
External plating of rudder horn	–	–	III
Sternframe Internal components of rudder horn Rudder Shaft bracket	–	–	II

(b) Steel grades

Thickness, t (mm)	Class									
	I		II		III		IV		V	
	Mild steel	*H.T. steel*	*Mild steel*	*H.T. steel*	*Mild steel*	*H.T. steel*	*Mild steel*	*H.T. steel*	*Mild steel*	*H.T. steel*
t ≤ 15	A	AH	A	AH	A	AH	A	AH	D	DH
15 < *t* ≤ 20	A	AH	A	AH	A	AH	B	AH	E	DH
20 < *t* ≤ 25	A	AH	A	AH	B	AH	D	DH	E	EH
25 < *t* ≤ 30	A	AH	A	AH	D	DH	E	DH	E	EH
30 < *t* ≤ 35	A	AH	B	AH	D	DH	E	EH	E	EH
35 < *t* ≤ 40	A	AH	B	AH	D	DH	E	EH	E	EH
40 < *t* ≤ 50	B	AH	D	DH	E	EH	E	EH	E	EH
t > 50	B	AH	D	DH	E	EH	E	EH	E	EH

After Lloyd's Register of Shipping, *Rules and Regulations for the Classification of Ships*.[19]

Table 2.13 K_L *factors for the determination of hull girder section modulus in higher-strength steels*

Specified minimum yield stress $-$N/mm^2	K_L
315	0.78
340	0.74
355	0.72
390	0.68

Notes
1. Intermediate values by interpolation
2. For mild steel, $K_L = 1.0$

After Lloyd's Register of Shipping, *Rules and Regulations for the Classification of Ships.*[19]

in higher-strength steels. For local scantling (thickness) requirements, a higher-strength steel factor, K, is determined as follows:

$$K = \frac{235}{\sigma_o} \text{ or } 0.66, \text{ whichever is the greater}$$

where σ_o = minimum yield stress in N/mm^2.

For mild steel, $K = 1.0$.

This would permit the use of steels with a minimum yield stress of up to 355 N/mm^2. However, the rules state that special consideration will also be given to steels with minimum yield stress values greater than 355 N/mm^2.

Offshore structures

Since the early 1970s, the UK has exploited its natural gas and oil resources and offshore platforms have become symbols of achievement in terms of design, materials and construction. One might therefore hold the view that offshore structures of this type were developed specifically for operation in the North Sea whereas such platforms were first constructed in the Gulf of Mexico in the 1940s. However, the conditions in the North Sea are considerably more severe with operating depths of 170 m compared to 7 m in the Gulf and with very much colder and rougher climatic conditions. Offshore platforms have now been constructed in large numbers and in 1981 it was reported[20] that more than 10 000 structures were in operation world-wide.

Although most offshore structures have been constructed in steel, several very large structures were built in reinforced concrete in the 1970s. This followed the tradition, established in the late 1880s, for the use of concrete in port and harbour installations, such as piers and jetties. Such facilities rank amongst the largest man-made structures and, for example, the Statfjord C platform had a float-out weight of more than 600 000 tonnes.[21] Concrete structures rely on their sheer mass to maintain their position on the seabed and for this reason are called gravity structures. Therefore they require good foundations and a thorough investigation of the seabed conditions to guard against long-term settlement and tilting. However, in spite of satisfactory performance, construction in reinforced concrete has remained relatively rare, apparently for economic reasons.

Design considerations

In the UK, the design and construction of offshore structures must comply with Guidance Notes[22] prepared by the Department of Energy. However, the Department of Energy has authorized the following organizations to issue *Certificates of Fitness* for offshore structures, in a similar manner to that for ships:

- American Bureau of Shipping
- Bureau Veritas
- Det Norske Veritas
- Germanischer Lloyd
- Offshore Certification Bureau
- Lloyd's Register of Shipping

The Guidance Notes illustrate very clearly the complex loading situation in offshore rigs by specifying the need to take account of the following loads in the primary structure:

- Dead loads – the weight of the structure and other fixed items
- Imposed or variable loads – loads from drilling operations, consumable stores, crew members, berthing and landing loads
- Hydrostatic loads – acting in a direction normal to the contact surface
- Environmental loads – wind, wave and current which act horizontally and, in extreme conditions, exert large overturning and sliding forces on fixed installations

Structural members in the splash zone are also subjected to major impact loads termed *slamming*. Whereas hydrostatic loads can be determined to a high degree of accuracy, environmental loads are random and the extreme magnitudes have to be predicted on a probability basis.

The most common type of steel platform is the welded tubular frame construction which is illustrated in Figure 2.19. In such platforms, the most critical areas are the *nodes* which represent the intersection of a *chord* (large-diameter member) and a *brace* (small-diameter member). Nodes can be constructed in various geometries, including T, K and T–K configurations, and they constitute areas of high stress concentrations. As such, they have a very marked effect on fatigue behaviour and therefore very careful consideration is given to the design and quality of welds in these locations. The first indication that fatigue could pose a major problem in offshore structures occurred in December 1965 when *Sea Gem*, a jack-up barge, collapsed in the North Sea, killing 13 people. It was shown that this failure was due to fatigue cracking which was probably initiated from brittle fracture, originating in welded components. Detailed consideration has therefore been given to the fatigue performance of nodal joints and design S–N curves

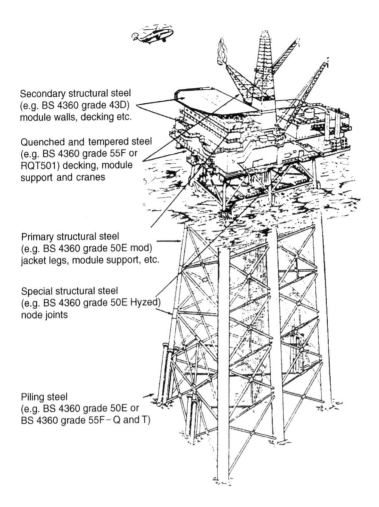

Secondary structural steel
(e.g. BS 4360 grade 43D)
module walls, decking etc.

Quenched and tempered steel
(e.g. BS 4360 grade 55F or
RQT501) decking, module
support and cranes

Primary structural steel
(e.g. BS 4360 grade 50E mod)
jacket legs, module support, etc.

Special structural steel
(e.g. BS 4360 grade 50E Hyzed)
node joints

Piling steel
(e.g. BS 4360 grade 50E or
BS 4360 grade 55F – Q and T)

Figure 2.19 *Types of steel for North Sea structures (After Billingham[21])*

have been produced which take account of the various stress concentration factors induced by different node geometries and plate thicknesses.

The Guidance Notes state that the calculated tensile stress in various members should not exceed 60% of yield stress under normal operating conditions and 80% of yield stress under extreme loading conditions.

Steel selection

Steels for tubular frame platforms can be grouped into three classes, depending on their location and duty:

1. Special structural steel.
2. Primary steel.
3. Secondary steel.

As illustrated in Figure 2.19, special structural steel is used for all the main nodes and also in the transition area to the shafts. This requires the highest grade of steel, namely BS 4360 Grade 50E Hyzed. As discussed earlier in this chapter, such steels are made to low sulphur contents and with alloy additions which modify the shape of the non-metallic inclusions so as to provide high levels of ductility in the through-thickness direction of the plate. Primary steel is used in all other structural members, such as jacket legs and topside module supports, and the steel is BS 4360 Grade 50E Mod. Secondary steel is used in lightly stressed areas such as module walls, decking, walkways and ladders and the specified material is BS 4360 Grade 43D.

Particular attention is paid to resistance to brittle fracture and the Guidance Notes lay down detailed impact and CTOD test requirements for parent plate, HAZ and weld materials. The background to these requirements has been published by Harrison and Pisarski[23] and involved a great deal of work on the correlation of Charpy V, wide plate and fracture toughness data. In the Guidance Notes, the impact test requirements are based on a minimum design temperature of $-10°C$. However, for other design temperatures, the Charpy test temperature is altered by $0.7°C$ for each $1°C$ that the design temperature differs from $-10°C$.

McLean and Oehrlein[24] have reported that the weight of steel involved in a typical tubular frame jacket is of the order of 13 000 tonnes and, together with the separately installed module support frame (2500 tonnes), supports the work area above sea level. The lower ends of the four external legs of the jacket each penetrate through a *bottle* unit, weighing 770 tonnes. These units transfer the service loads from the jacket to the piles which maintain the platform in position on the seabed. The topside modules, which house the equipment, work area and accommodation, weigh a total of 6600 tonnes. As indicated in a later chapter, austenitic stainless steels are now being used to clad the walls of the topside modules.

Cast steel nodes

In the early 1980s, significant effort was devoted to the development of cast steel nodes as alternatives to welded fabrications. The main technical incentive for this development was that castings could provide smooth, generous radii in these critical areas, compared with weld connections, and so reduce the stress concentration factor.[25] Typical compositions for these cast steel nodes are given in Table 2.14.

According to Billingham,[21] cast steel nodes have enjoyed only limited commercial exploitation but in fact they were used at fatigue-prone locations in the Hutton Field.

Table 2.14 *Composition of cast steel nodes*

	C%	Mn%	S%	Ni%	Nb%	V%	Mo%	Cr%	Cu%
Type I	0.14	1.30	0.005	0.43	0.025	0.05	0.11	0.09	0.04
Type II	0.15	0.85–1.70	0.01 max.	0.09–1.2					

Reinforcing bars

Large amounts of steel are used for the reinforcement of concrete in buildings, bridges and marine structures and reinforcing bars therefore constitute a competitor to structural steel plates and sections. At one time, reinforcing bars were regarded as low-grade, undemanding steel products and were often produced from diverted casts that were out of specification for the originally intended order. However, with the trend towards higher-strength steels and the requirement for good fabrication characteristics, reinforcing steels are now made to high quality standards. Whereas some reinforcing bars are supplied in the form of plain carbon steel with a yield strength of 250 N/mm^2, extensive use is now made of higher-strength steels with yield strengths up to 500 N/mm^2.

Standard specifications

The UK standard for steel reinforcement is BS 4449: 1988 *Carbon steel bars for the reinforcement of concrete*. It covers grades with minimum yield strength levels of 250 and 460 N/mm^2 in the form of plain rounds and *deformed* (ribbed) bars respectively. The chemical compositions of these grades are only broadly defined, as indicated in Table 2.15.

In order to provide a reasonable level of weldability, maximum carbon equivalent values of 0.42% (Grade 250) and 0.51% (Grade 460) must be observed, based on the following formula:

$$CE = C + \frac{Mn}{6} + \frac{Cr + Mo + V}{5} + \frac{Ni + Cu}{15} \ wt\%$$

To ensure adequate ductility, minimum elongation values of 22% (Grade 250) and 12% (Grade 460) are required but the materials must also be capable of being bent through 180° around formers of the following proportions:

- Grade 250 – 2 × nominal bar diameter
- Grade 460 – 3 × nominal bar diameter

The specification also includes a rebend test requirement which was introduced initially to determine the susceptibility of the materials to strain age embrittlement. This involves bending bars initially through 45° around formers of the following proportions.

- Grade 250 – 2 × nominal bar diameter
- Grade 460 – 5 × nominal bar diameter

Table 2.15

	Grade 250	Grade 460
C%	0.25 max.	0.25 max.
S%	0.060 max.	0.050 max.
P%	0.060 max.	0.050 max.
N%	0.012 max.	0.012 max.

The bars are then immersed in boiling water for at least 30 minutes. On cooling to ambient temperature, the bars must be capable of being bent back towards their original shape through an angle of at least 23°.

Traditional reinforcing steels

In the UK, high-strength reinforcing steels have been produced traditionally by:

1. The cold twisting of plain carbon steels.
2. The use of vanadium-bearing micro-alloy steels.

At one time, the former was specified in a separate standard (BS 4461), but the cold-twisted product was incorporated in BS 4449 when this standard was revised in 1988. Although these steels employ different strengthening mechanisms, namely work hardening as opposed to precipitation strengthening, their properties are sufficiently similar to be covered by a single set of requirements for tensile and bend–rebend properties.

Whitely[26] reports that the cold-twisted product can be welded, with little loss of strength, by employing high heat inputs for short periods of time. Such practices restrict the area of the heat-affected zones and the quenching effect of adjacent material leads to the formation of strong, low-temperature transformation products. Whitely also indicates that the cold-twisted product provides adequate elevated-temperature tensile properties which are pertinent to the fire resistance of reinforced concrete structures.

Although not specified in BS 4449, the impact strength of reinforcing bars is of interest in relation to the potential damage that can be caused by the crashing of vehicles into the supporting columns and deck parapets of concrete road bridges. In collaboration with the Department of Transport, extensive work on this topic was carried out by British Steel Technical.[27] Using a large pendulum impact machine, with an impact velocity of 7.6 m/s, impact transition temperature curves were developed on full-section cold-twisted bars. It was shown that unnotched bars remained fully ductile at temperatures down to at least −65°C but, with the introduction of a sharp notch, the impact transition temperature was raised significantly. However, when impact tests were carried out on large reinforced concrete beams, the effect of notching the bars was shown to be significant only in the case of cold-twisted material. This is illustrated in Figure 2.20, which shows that very little loss of impact strength was observed in mild steel (Grade 250) bars and hot-rolled (micro-alloy, Grade 460) bars.

Controlled-cooled bars

In the mid-1970s, the CRM laboratories at Liège in Belgium published details of an in-line heat treatment process for the production of high-strength reinforcing bars.[28] Designated the *Tempcore* process, the application of controlled cooling after rolling results in the formation of an outer layer of martensite which is *temp*ered subsequently by conduction of heat from the *core* of the bars. The main aim of the process is to produce weldable, high-strength reinforcement

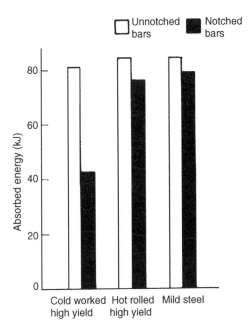

Figure 2.20 *The effect of notching reinforcing bars on the energy absorbed in concrete beams (After Armstrong et al.[27])*

cheaply by eliminating the costs associated with cold twisting or micro-alloy additions. The Tempcore process has been very successful, both technically and commercially, and has been licensed throughout the world.

The Tempcore process is illustrated schematically in Figure 2.21. On leaving the last finishing stand, the bar passes through a water cooling station which cools the outer region of the bar sufficiently quickly for the formation of martensite. At the end of the cooling process, the bar has an austenite core surrounded by a mixture of austenite and martensite, the amount of martensite increasing towards the surface of the bar. On leaving the cooling station, the bar is exposed to the atmosphere and the temperature gradient between the core and quenched surface begins to equalize. This leads to the tempering of the martensitic rim, providing an adequate balance between strength and ductility. During this second stage in the process, untransformed austenite in the outer layers of bars also transforms to bainite. The final stage of the process takes place as the bars lie on the cooling bed, namely the transformation of austenitic core. Depending upon factors such as composition, finishing temperature and cooling rate, the core can transform to ferrite and pearlite or to a mixed microstructure which includes some bainite. Thus the *Tempcore* process produces a variety of microstructures throughout the cross-section, ranging from tempered martensite in the outer layers to a region which is essentially ferrite and pearlite in the core, with an intermediate zone which may be predominantly bainitic.

Depending on the size of the bars, water cooling may be applied before the bars enter the finishing stands in order to reduce the length of the cooling station. This

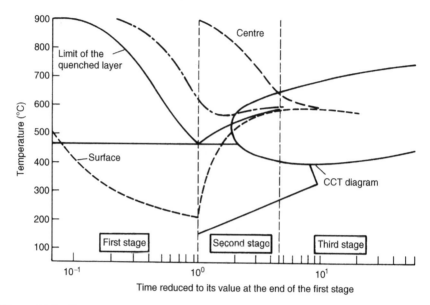

Figure 2.21 *Tempcore process – thermal evolution in different parts of a 25 mm bar, finished rolled at 900° C (After Economopoulos et al.[28])*

applies particularly to large-diameter bars which are finished at a high temperature and it is claimed[28] that the length of the cooling facility can be reduced by 70% by reducing the finishing temperature from 1050 to 900°C.

The Tempcore process is relatively cheap and provides the required mechanical properties in steels of low carbon equivalent. For these reasons, it is gradually superseding the cold-twisted and micro-alloyed products.

Steel bridges

In the UK, bridge construction is dominated currently by reinforced concrete and only about 15% of bridges are based on steel, compared to about 80% in Japan. According to Simpson,[29] one reason for this situation is that the design codes for steel bridges are complex and not easy to use. However, Simpson also points out that the failure during construction of three box girder bridges in the 1970s has also contributed to the lack of steel construction. In order to regain market share, steel suppliers and fabricators are attempting to improve the attraction of steel by providing cheaper and more effective forms of corrosion protection and by developing methods of construction that will allow faster erection schedules.

Design against brittle fracture

In the UK, the design and construction of bridges is covered by BS 5400: Part 3: 1982 *Steel, concrete and composite bridges*. This code involves the use of the complex *limit state* approach for the calculation of design stresses and therefore this discussion on the use of steel in bridges will be confined to that part of the code dealing with avoidance of brittle fracture.

As a safeguard against brittle fracture, BS 5400: 1982 specifies the maximum thickness of steel that can be used in bridge tension members, with respect to the various grades of steel listed in BS 4360 (Weldable structural steels) and the minimum operating temperature of the bridge. The various stages in the determination of this temperature are outlined below:

1. The first stage is to determine the *minimum shade air temperature* for the location of the bridge from isotherm maps, based on Meteorological Office data.
2. This initial value is then adjusted for height above sea level by subtracting 0.5°C per 100 m. Additionally, there may be the need to take account of locations where the minimum temperatures diverge from published data, e.g. in frost pockets where the minimum may be substantially lower than the published value or in coastal and some urban areas where the minimum temperature may be higher.
3. The *minimum effective bridge temperature* (MEBT) is then derived from a table in which the minimum shade air temperature is adjusted to take account of the type of bridge construction, e.g. steel or concrete decking.
4. Finally, the U value for the bridge is determined by rounding down the MEBT to the next impact test temperature in BS 4360, i.e. a value of $-17°C$ would be rounded down to $-20°C$.

Each part of a bridge which is subjected to applied stress must be classified according to the following criteria:

- Type 1 – Any part subjected to an applied stress greater than 100 N/mm² and which has either
  (a) any weld connection
  (b) weld repair not subsequently inspected
  (c) punched holes not reamed.
- Type 2 – All parts subjected to applied stresses which are not of Type 1.

Stress calculations will have provided information on the combination of steel thickness and yield strength that will satisfy the required design strength and the appropriate sub-grade of steel in BS 4360 that provides the required level of impact strength can then be derived in two ways:

1. From a table in BS 5400 which provides a correlation between the limiting thickness for various grades and the U value – the minimum effective bridge temperature – differentiating between Type 1 and Type 2 stress conditions.
2. Directly from BS 4360, having calculated the required impact strength requirements in the following manner:

for Type 1 $C_v \geq \dfrac{\sigma_y}{355} \dfrac{t}{2}$

for Type 2 $C_v \geq \dfrac{\sigma_y}{355} \dfrac{t}{4}$

where C_v = Charpy V notch energy value in joules at the minimum
effective temperature
σ_y = yield strength in N/mm^2
t = thickness in mm

It should be noted that the divisor of 355 in the above equations corresponds
to the minimum yield strength in N/mm^2 of BS 4360 Grade 50 steels.

Where severe stress concentrations occur, BS 5400 calls for more stringent toughness requirements and the impact energy value is calculated from:

$$C_v \geq \frac{\sigma_y}{355}[0.3t(1 + 0.67k)]$$

where k = stress concentration factor

As indicated earlier in this chapter, BS 5400 has not yet been updated to reflect the
fact that BS 4360 has now been withdrawn and superseded by European standards
for structural steels. However, it is understood that BS 5400 is currently being
revised to accommodate the changes in steel nomenclature. In the meantime, the
Department of Transport has issued document BD 13/90 *Design of Steel Bridges:
Use of BS 5400: Part 3: 1990*, Crown Copyright 1990, which provides guidance
on this matter.

Steel in multi-storey buildings

During the 1980s, there was a dramatic increase in the UK in the use of structural
steelwork in multi-storey buildings, primarily at the expense of *in situ* concrete.
The reasons cited[30] for this situation are:

1. The reduction in the price of structural steel relative to *in situ* concrete.
2. The reduced costs and improved methods of fire protection for structural steel.
3. The shorter site and total construction periods achieved through the use of
   steel.

All three aspects are cost or revenue related, the last item minimizing the period
during which the capital for the investment has to be financed and decreasing the
time at which the owner of a building begins to receive a return on investment.

In order to improve its share of the construction market, British Steel first
analysed the contribution of the various items to the total cost of a steel building
frame. This analysis is shown in Figure 2.22[31] and, whilst the cost of the basic
steel was very significant, items such as corrosion protection, fire protection and
fabrication also contributed very considerably to the overall cost. Each of these
items was given detailed attention and, as illustrated in Figure 2.22 the cost of
steel building frames has decreased significantly, in real value terms, over the
11–12-year period. Whereas the costs of corrosion protection and fire protection
were reduced substantially, it should be noted that the reduction in the cost of
steel played an even greater part in increasing the cost competitiveness of steel
frames.

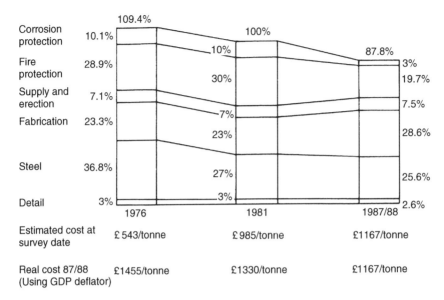

Figure 2.22   *Cost breakdown of multi-storey building frames (at 1987/88 prices)*

Building code requirements

In the UK, steel-framed buildings are specified in BS 5950: 1985 *Structural use of steelwork in buildings*. This refers to the use of structural steels specified in BS 4360 (Weldable structural steels) and the selection of a minimum level of toughness in relation to the yield strength of the steel, its thickness and service conditions.

The design strength P_y may be taken as $1.0 \times Y_s$ but cannot be greater than $0.84 \times U_s$ where Y_s and U_s are the minimum yield strengths and tensile strengths specified in BS 4360. The main types of steel used in building construction are Grades 43, 50 and 55 and, for convenience, BS 5950 tabulates the permissible design strengths for these grades, according to material thickness. This table is reproduced as Table 2.16.

Brittle fracture needs to be considered in locations which are subjected to tensile stresses in service. In this respect, the first step is the determination of the appropriate k factor, according to the level of tensile stress and location of material. This is determined from Table 2.17.

Having determined the thickness of material required from a consideration of design strength (Table 2.16), the required grade of steel can then be selected from a table in BS 5950, reproduced here as Table 2.18. This table includes the weathering grades (WR50 A, B and C) and differentiates between situations with k factors of 1 and 2, the limiting thickness/grade requirement being more severe when $k = 1$. This table also differentiates between the internal and external situations in buildings and applies providing the temperatures concerned do not fall below $-5°C$ and $-15°C$ respectively. When the steel is subjected to lower temperatures or where the steel grade or thickness is not covered in this table,

Table 2.16 *Design strength data for BS 4360 steels in building applications*

BS 4360 Grade	Thickness ($\leq$ mm)	Design strength (N/mm²)
43 A, B and C	16	275
	40	265
	100	245
50 B and C	16	355
	63	340
	100	325
55 C	16	450
	25	430
	40	415

After BS 5950: Part 1: 1985.

Table 2.17

Tensile stress due to factored load at the location	Welded location	Unreamed punched holes	Non-welded location	Drilled or reamed holes
$\leq$100 N/mm²	2	2	2	2
>100 N/mm²	1	1	2	2

then the toughness/grade requirements are determined by calculation. Thus the required impact strength at the service temperature is determined from:

$$C_v \geq \frac{Y_s t}{710k}$$

where C_v = Charpy V notch energy in joules
Y_s = the minimum yield strength of the material in N/mm²
t = the thickness of the material in mm
k = the factor determined from tensile stress/location considerations

Given that the design strength is based on yield strength, then substantial benefit is gained in building constructions from the substitution of carbon steel by higher strength structural steels, providing excessive deflection does not occur.

As illustrated in Figure 2.22, substantial savings have been made in the cost of corrosion protection and, in essence, this is related to the fact that paint systems were either over-specified or, in many cases, were simply not required.

In the previous section, it was stated that the minimum Charpy energy requirement for steel bridges was based on the following expression for Type 1 stress situations:

$$C_v \geq \frac{\sigma_y}{355} \left(\frac{t}{2} \right)$$

Apart from differences in nomenclature for yield strength and formula presentation, this expression is identical to that shown above for the minimum Charpy value for steel frame buildings, namely:

$$C_v \geq \frac{Y_s t}{710k}$$

for situations where $k = 1$.

In keeping with the situation with the bridge code BS 5400, BS 5950 has not yet been updated to reflect the withdrawal of BS 4360 and the introduction of European specifications for structural steels. However, amendments to Table 2 of BS 5950: Part 2 and Table 4 of BS 5950: Part 1 are being prepared to reflect the changes in steel nomenclature.

Steels for pipelines

Pipelines are very efficient for the mass transportation of oil and gas and can extend over vast distances. For example, the Alaskan National Gas Transportation System involves more than 4000 miles of linepipe in diameters ranging from 42 to 56 in. In Europe, the pipelines from northern Russia to Germany and Austria extend for more than 3100 miles and are predominantly in 56-in diameter linepipe. In either case, the operating pressures are up to 1450 lb/in.[2] Over the past 30 years, the trend in pipeline design has been to larger sizes and higher operating pressures in order to increase the efficiency of transportation. This has been accomplished through the provision of steels with progressive increases in yield strength coupled with good weldability and sufficient toughness to restrict crack propagation. However, other trends in the extraction of oil and gas have imposed further requirements on the performance of linepipe steels. These include the need for high-strength, heavy-wall pipe to prevent buckling during pipeline installation in deep water, e.g. in depths of 170 m in the North Sea. The operation of pipelines in arctic regions, coupled with the transportation of liquid natural gas (LNG), have also imposed further demands for improved levels of toughness at low operating temperatures. However, a particularly important trend has been the exploitation of sour oil and gas reserves which has required the development of linepipe steels with resistance to hydrogen-induced cracking (HIC). Thus in addition to higher strength and toughness, developing pipeline technologies have required improved resistance to corrosion which has been met with specific alloy additions and special control over non-metallic inclusions. Of major importance in meeting the property requirements of the oil and gas industries have been the developments in the thermomechanical processing of steel and the bulk of linepipe steels are now supplied in the controlled-rolled condition.

Specifications and property requirements

Most linepipe specifications in the world are based on those issued by the American Petroleum Institute (API) which cover high test linepipe (5LX series) and spiral weld linepipe (5LS series). The API specification for high test linepipe was introduced in 1948 and at that time included only one grade, X42, with a yield strength of 42 ksi. Since that time, higher strength steels have been developed and the specification now includes grades up to X80 (80 ksi = 551 N/mm^2).

Most specifications use yield strength as the design criterion, and as the yield strength is increased, the wall thickness can be reduced proportionately, using the same design safety factor. Thus the substitution of X70 by X80 grade can lead to a reduction in wall thickness of 12.5%, which illustrates very clearly the incentive for the use of higher-strength steels. However, the measurement of yield strength in linepipe has been an area of contention between steelmakers and linepipe operators for some time because of the change in strength that occurs between as-delivered plate and pipe material. Tensile specimens are cut from finished pipe and these are cold flattened prior to testing. Due to the method of preparation, the yield strength measured in such test specimens can be significantly lower than that obtained on undeformed plate. The differential is due to the well established *Bauschinger effect* which leads to a decrease in yield strength when tensile testing is preceded by stressing in the opposite direction, e.g. during pipe unbending and flattening. In X70 pipe, the Bauschinger effect can result in a reduction in yield strength of 69 to 83 N/mm^2 (10 to 12 ksi) and therefore plate material has to be supplied with extra strength so as to compensate for this apparent loss in yield strength. However, when the yield strength of pipe material is measured in a ring tension test, the value is much closer to that measured in plate material. The Bauschinger effect in linepipe materials is particularly marked in traditional ferrite–pearlite steels which exhibit discontinuous yielding in the tensile test. The effect is reduced in steels containing small amounts of bainite or martensite, and in steels containing a significant amount of lower temperature transformation products, the unbending and flattening operation can lead to an increase in yield strength (or 0.2% proof stress) compared with undeformed plate. These materials exhibit continuous stress–strain curves and the high rate of work hardening compensates for the loss in strength due to the Bauschinger effect.

Toughness is a major requirement in linepipe materials and detailed consideration has been given to both fracture initiation and propagation. To design against fracture initiation, the concept of *flow stress dependent critical flaw size* is employed. This predicts the critical flaw size, relative to the Charpy toughness level, for specific pipe dimensions and operating pressures. Above this critical size of defect, the toughness level required to prevent fracture initiation becomes infinite and depends solely on the flow stress. The crack opening displacement test has also been used to determine fracture initiation in linepipe materials, particularly in relation to the heat-affected zone.

When fracture occurs in an oil pipeline, fluid decompression takes place very rapidly and therefore the driving force for crack propagation dissipates very quickly in time and space. However, this is not the case in gas pipelines and therefore the possibility of developing a long-running crack is a major concern. However, the avoidance of brittle propagation is generally assured by specifying a minimum of 85% shear area at the minimum service temperature, in full-thickness specimens in the Battelle Drop Weight Tear test. Various formulae have also been derived to specify the minimum Charpy level which will ensure the arrest of a propagating ductile fracture. These indicate that higher Charpy energy values are required for higher operating pressures and for pipes with higher strengths, larger diameters and heavier wall thicknesses. However, there still appears to be some concern about the adequacy of Charpy values for the prediction of crack arrest behaviour.

The weldability of linepipe materials is important, firstly in relation to pipe fabrication and secondly with regard to the girth welds that are used in the field for pipeline construction. Obviously the latter represent the more arduous welding requirements, particularly in low-temperature environments, but, in general, the use of carbon equivalent formulae appears to be adequate in ensuring crack-free welds. Traditionally, the International Institute of Welding (IIW) formula has been used to assess the weldability of materials:

$$\text{Carbon equivalent} = C + \frac{Mn}{6} + \frac{Cr + Mo + V}{5} + \frac{Cu + Ni}{15}$$

However, there is the general feeling that the IIW formula is not adequate to define the behaviour of modern steels with low carbon contents and the following relationship by Ito and Bessyo is sometimes preferred:

$$\text{Carbon equivalent} = C + \frac{Si}{30} + \frac{Mn + Cu + Cr}{20} + \frac{Ni}{60} + \frac{Mo}{15} + \frac{V}{10} + (B \times 5)$$

As indicated earlier, thermomechanical processing has permitted the development of high-strength steels with low carbon contents and this has contributed greatly to improved weldability in linepipe steels.

Linepipe manufacturing processes

Japan is a leading producer of linepipe and a summary of the processes and size ranges available in that country in 1981 is given in Figure 2.23. This indicates that linepipe is produced as seamless and welded tubing, the former being restricted to relatively small-diameter, thick-walled tubing. Welded pipes are produced by electric resistance welding (ERW) and submerged arc welding (SAW), the latter being used for both longitudinal and spiral welded pipe. ERW pipes are produced in sizes up to 600 mm (24 in) in diameter and up to 19 mm (0.75 in) wall thickness. Longitudinal welded pipes are produced mainly by the U-O process and account for most of the pipes used for oil and gas transmission lines. However, a small amount of longitudinal welded pipe is also produced by roll bending. As illustrated in Figure 2.23, very large diameter pipe is produced by the spiral welded process in wall thickness up to 25 mm (1 in).

Steel compositions for linepipe

A feature of the API 5LX specification is that it lays down very broad requirements for chemical composition, specifying only the maximum permitted levels for carbon, manganese, sulphur and phosphorus. On the other hand, customer specifications are much more restrictive in composition so as to obtain high levels of toughness and weldability at a specific level of yield strength. Even so, steelmakers can still exercise various options in terms of composition and

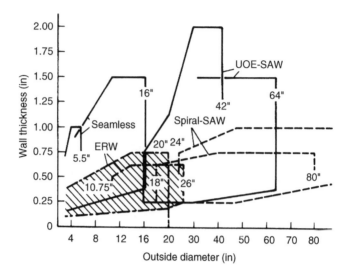

Figure 2.23 *Size ranges available with main Japanese linepipe manufacturing processes (After Nara et al.[32])*

thermomechanical processing and therefore a wide range of compositions is used to satisfy the property requirements of individual grades within the API specification. Sage[33] has produced a useful guide to the types of composition used for linepipe and this is shown in Figure 2.24. It will be noted that this chart differentiates between steels that exhibit the Bauschinger effect and those that experience a slight increase in yield strength during pipe making. The detailed chemical compositions used in various countries to satisfy API requirements are illustrated in Table 2.19. This includes some very low carbon, boron-treated steels[34] which develop a bainitic microstructure. In the production of these steels, the slab reheating temperature is in the range 1000–1150°C and the presence of fine particles of TiN inhibit austenite grain growth. The steels are finished at a temperature of about 700°C.

In general, X60 grade is satisfied by controlled-rolled ferrite–pearlite steels containing about 0.03% Nb. For X65 and X70, it is usual to supplement the grain-refining effect of niobium with dispersion strengthening from vanadium, and steels 3 and 5 in Table 2.19 are typical of this practice. In order to achieve X80 properties, small additions of nickel or molybdenum are made to the Nb–V steels and the steels are also subjected to a very severe controlled-rolling practice.

Pipeline fittings

Pipelines use a variety of fittings, such as valves, bends and tees, which have the same property requirements as linepipe, namely high strength, toughness and weldability. However, because of their method of production, the development of higher strengths in fittings has tended to lag behind that in linepipe material, which relies heavily on thermomechanical processing for the achievement of properties. Fittings are produced from forgings or as fabrications from plate and,

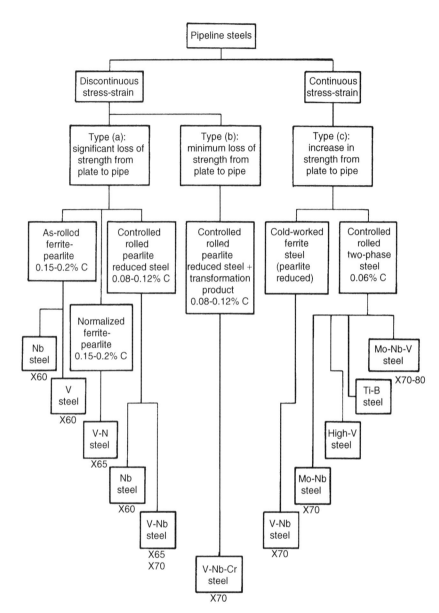

Figure 2.24 *Summary of steel pipes in production for high-strength steel pipelines in Europe, the Americas and Japan (After Sage[33])*

in either case, the hot-working operations are not conducive to the generation or preservation of the fine-grained structure, characteristic of controlled-rolled linepipe. Pipeline fittings are therefore supplied mainly in the normalized or quenched and tempered condition. However, by limiting the hot-working temperature to about 750°C, fittings can be produced from controlled-rolled plate with little loss in properties.

Table 2.18   *Selection of steels for buildings*

| BS 4360 Grade | Maximum thickness of parts subjected to applied tensile stress | | | |
| | Sections (except hollow sections) and flat bars | | Plates, wide flats and universal wide flats | |
	Internal (mm)	External (mm)	Internal (mm)	External (mm)
Values for $k = 1$				
43A	25	15	25	15
43B	30	20	30	20
43C	50	40	50	40
43D	50	50	75	75
43E	50	50	75	75
50A	16	10	16	10
50B	20	12	20	12
50C	40	27	75	55
50D	40	40	75	75
55C	19	16	19	16
55E	63	63	63	63
WR50A	12	12	12	12
WR50B	45	27	45	27
WR50C	50	50	50	50
Values for $k = 2$				
43A	50	30	50	30
43B	50	40	50	40
43C	50	50	50	50
43D	50	50	75	75
43E	50	50	75	75
50A	32	20	32	20
50B	40	25	40	25
50C	40	40	75	75
50D	40	40	75	75
55C	19	19	19	19
55E	63	63	63	63
WR50A	12	12	12	12
WR50B	50	50	50	50
WR50C	50	50	50	50

After BS 5950: Part 1: 1985.

Rogerson and Jones[39] have reviewed the topic of steels for high-pressure gas fittings and their summary of steel compositions is shown in Table 2.20. Thus, proprietary C–Mn–V steels, such as *Hyplus 29* and *Creuselso 42*, have been used for these applications which produce yield strength values up to 450 N/mm^2, depending upon section size. *Nicuage IN787* has been used successfully in the Alaskan oil pipelines in strength levels that satisfy X80. This Mn–Mo–Cu steel also has the type of composition that is resistant to sour gas/oil environments, as discussed in the next section.

Table 2.19 Chemical composition of linepipe steels

Steel no.	Grade	WT (mm)	C%	Si%	Mn%	P%	S%	Ni%	Mo%	Cu%	Nb%	V%	Ti%	B%	Ref.
1	X65	16	0.02	0.14	1.59	0.018	0.003				0.04		0.017	0.001	34
2	X65	25	0.03	0.16	1.61	0.016	0.003	0.17			0.05		0.016	0.001	34
3	X65	25	0.06		1.35	0.025	0.005	0.25		0.33	0.04	0.07			35
4	X70	20	0.03	0.14	1.91	0.018	0.003				0.05		0.018	0.001	34
5	X70	19.6	0.08		1.6						0.04	0.07			36
6	X80	12	0.07		1.65		0.002	0.38	0.22		0.05	0.075			37
7	X80	20	0.02	0.26	1.95	0.022	0.003		0.31		0.04		0.019	0.001	34
8	X80	19	0.08	0.1	1.5						0.052	0.076			38
9	X80	19	0.036	0.1	1.6			0.35	0.29		0.64				38

Table 2.20 *Compositions of high-strength pipeline fittings*

							Composition %										Type
		C	Si	Mn	S	P	Al	V	Nb	Ti	Ni	Cr	Mo	Cu	Co	N	
1	Hyplus 29	0.19	0.37	1.67	0.008	0.018	0.021	0.121	0.005	0.012	0.074	0.089	0.016	0.156	0.017	0.019	Tube
2	Controlled rolled	0.16	0.23	1.31	0.012	0.018	0.027	0.049	0.028	0.005	0.027	0.035	0.018	0.056	0.011	–	Plate
3	Controlled rolled	0.16	0.29	1.23	0.011	0.011	0.024	0.052	0.044	0.005	0.027	0.029	0.008	0.075		–	Bend
4	V–N–Ni steel	0.14	0.38	1.46	0.015	0.018		0.1			0.55	0.07	0.03	0.15		0.006	Plate
5(i)	V–N–Ni steel	0.19	0.4	1.63	0.004	0.008	0.01	0.108	0.005	0.005	0.55	0.166		0.099		–	Bend
(ii)	V–N–Ni steel	0.17	0.39	1.56	0.015		0.011	0.169		0.005	0.507	0.16	0.029	0.099		0.017	Bend
6	Acicular ferrite	0.06	0.26	1.59	0.009	0.022	0.043	0.018	0.029	0.005	0.213	0.005	0.341	0.008	0.005	–	Plate
7	Nicuage (IN787)	0.04	0.28	0.46	0.012	0.008	0.044	0.038	0.052	0.005	0.812	0.716	0.205	1.03	0.021		Plate
8	Modified Nicuage	0.05	0.26	1.23	0.008	0.015	0.027	0.031	0.071	0.005	1.1	0.018	0.206	1.18	0.012		Plate
9	BSC Mn–Mo–Cu	0.14	0.31	1.39	0.007	0.01	0.028	0.098	0.027	0.005		0.06	0.22	0.48		0.013	Plate

After Rodgerson and Jones.[39]

Steels for sour gas service

In recent years, the demand has increased for linepipe with
ronmental fracture due to the exploitation of sour gas wells, i
significant levels of H_2S and CO_2. The National Association (
neers (NACE) has determined that a fluid is designated *sour*
greater than 0.0035 atm. partial pressure of H_2S. Both H_2S a
corrosive in the presence of moisture and the corrosivity of natu $_{gas}$ is determined solely by the levels of these compounds. Although gas pipelines do not
normally operate under corrosive conditions, a temperature drop in the gas to
below its dewpoint or failure in dehydration plant can lead to the introduction of
moisture. Similarly, other operational measures such as desulphurization and the
use of inhibitors are not considered to be totally adequate and therefore there is
a need for steels with inherent resistance to sour environments.

Two types of fracture can be introduced by H_2S, namely *hydrogen-induced
cracking* (HIC) and *sulphide stress corrosion cracking* (SSCC). The latter
form of attack is generally confined to steels with yield strengths greater than
about 550 N/mm^2 and therefore does not feature prominently in linepipe steels.
However, the fact that high hardness levels can be generated in the heat-affected
zones of welds should not be overlooked since SSCC can result in catastrophic
brittle fracture. On the other hand, HIC results in a form of blistering or
delamination and can take place in the absence of stress. Atomic hydrogen is
generated at cathodic sites under wet sour conditions and diffuses into the steel,
forming molecular hydrogen at the interface between non-metallic inclusions
and the matrix. When the internal pressure due to the build-up of molecular
hydrogen exceeds a critical level, HIC is initiated. Long elongated inclusions
such as Type II MnS are particularly favourable sites for crack initiation but
planar arrays of globular oxides are also effective. Cracking can proceed along
segregated bands containing lower temperature transformation products such as
bainite and martensite. Cracking tends to be parallel to the surface but can be
straight or stepwise.

The susceptibility to HIC is assessed by immersion of unstressed coupons in a
synthetic solution of seawater, saturated in H_2S with a pH of 5.1–5.3 (BP test)
or in the more aggressive solution of 0.5% CH_3COOH + 5% $NaCl$ + H_2S sat. at
a pH of 3.5–3.8 (NACE test). In either case, the test duration is 96 h and the
test parameters include crack length, crack width or blister formation.

From the remarks made earlier, it could be anticipated that the control of
non-metallic inclusions would feature prominently in the development of HIC-
resistant steels. The sulphur content is therefore generally reduced to below
0.01% and additions of calcium or rare earth metals are made to produce a
globular sulphide morphology. In some Japanese practices, the sulphur contents
are reduced to 0.001–0.003% with calcium additions of 0.0015–0.0035%.

Segregation effects can be minimized by restricting the levels of elements such
as carbon, manganese and phosphorus and fortunately the use of controlled rolling
enables high strengths to be obtained at low carbon levels. On the other hand,
manganese is beneficial in improving toughness by refining the ferrite grain size

...herefore a minimum level of manganese must be maintained, particularly in ...nepipe for arctic service.

Certain alloy additions are also effective in improving the resistance to HIC, notably copper and nickel. At pH levels above 5, these elements form protective films which prevent the diffusion of hydrogen into the steel. However, at pH levels below 5, these protective films are not formed and therefore these alloy additions show little advantage in the NACE test, outlined earlier. On the other hand, copper additions of 0.2–0.3% are extremely beneficial in preventing HIC, as determined by the BP test.

In summary, linepipe steels with resistance to HIC embody the following features:

1. Low levels of carbon, sulphur and phosphorus.
2. Additions of calcium or rare earth metals to globularize the sulphide inclusions.
3. Low levels of oxide inclusions.
4. Freedom from segregation so as to avoid bainitic or martensitic bands.
5. Additions of small amounts of copper or chromium.

References

1. Hall, E.O. *Proc. Phys. Soc. Series B*, **64**, 747 (1951).
2. Petch, N.J. *Proc. Swampscott Conf.*, MIT Press, p. 54 (1955).
3. Pickering, F.B. *Physical Metallurgy and the Design of Steels*, Applied Science Publishers.
4. Pickering, F.B. and Gladman, T. *ISI Special Report 81* (1961).
5. Irvine, K.J., Pickering, F.B. and Gladman, T. *JISI*, **205**, 161 (1967).
6. Gladman, T., Dulieu, D. and McIvor, I.D. In *Proc. MicroAlloying 75*, (Washington, 1975), Union Carbide, p. 25 (1975).
7. Vanderbeck, R.W. *Weld J.*, **37** (1958).
8. Sellars, C.M. In *Proc. HSLA Steels: Metallurgy and Applications* (eds Gray, J.M., Ko, T., Zhang Shouhua, Wu Baorong and Xie Xishan) (Beijing, 1985) ASM.
9. Tamura, I., Ouchi, C., Tanaka, T. and Sekine, H. *Thermomechanical Processing of High Strength Low Alloy Steels*, Butterworths (1988).
10. Cuddy, L.J. In *Proc. Thermomechanical Processing of Microalloyed Austenite* (eds Deardo, A.J., Ratz, G.A. and Wray, P.J.) (Warrendale Pa, 1982) Met. Soc. AIME (1982).
11. Cohen, M. and Hansen, S.S. In *Proc. HSLA Steels: Metallurgy and Applications* (eds Gray, J.M., Ko, T., Zhang Shouhua, Wu Baorong and Xie Xishan) (Beijing, 1985) ASM (1985).
12. United States Steel Technical Report – *Corrosion Performance of Mn–Cr–Cu–V Type USS Cor-Ten High Strength Low Alloy Steel in Various Atmospheres*, February 1965, Project No. 47.001-001 (7) and 47.001-003 (3).

13. United States Steel Technical Report – *A Study of the Initial Atmospheric Corrosion of Carbon Steel, USS Cor-Ten A Steel and USS Cor-Ten B Steel*, May 1966, Project No. 47.004-002 (1).

14. Horton, J.B. *The Rusting of Low Alloy Steels in the Atmosphere*, Paper to Pittsburgh Regional Technical Meeting of AISI, November 1965.

15. Hudson, J.C. and Stanners, J.F. *JISI*, July (1955).

16. Larrabee, C.P. and Coburn, S.K. In *Proc. First International Congress on Metallic Corrosion* (London, 1961) Butterworths (1961).

17. Pircher, H. and Klapdar, W. In *Proc. MicroAlloying '75* (Washington, 1975), Union Carbide Corporation (1977).

18. Boyd, G.M. and Bushell, T.W. *Trans. Royal Institution of Naval Architects*, July (1961).

19. Lloyd's Register of Shipping *Rules and Regulations for the Classification of Ships*, 1995.

20. Graff, W.J. *Introduction to Offshore Structures*, Golf Publishing (1981).

21. Billingham, J. *Metals and Materials*, August, 472 (1985).

22. Department of Energy *Offshore Installations – Guidance on Design and Construction*, HMSO, London (1984).

23. Harrison, J.D. and Pisarski, H.G. *Background to New Guidance on Structural Steel and Steel Construction Standards in Offshore Structures*, HMSO, London (1986).

24. McLean, A. and Oehrlein, R. In *Proc. Performance of Offshore Structures*, Autumn Review Course Series 3, No. 7, The Institution of Metallurgists, p. 65 (1976).

25. Webster, S.E. *Journal of Petroleum Technology*, October, p. 1999 (1981).

26. Whitely, J.D. *Concrete*, December, 28 (1981) and January, 30 (1982).

27. Armstrong, B.M., James, D.B., Latham, D.J., Taylor, V.E., Wilson, C. and Heighington, K. In *Proc. First International Conference on Concrete for Hazard Protection* (Edinburgh, 1987) The Concrete Society, p. 241 (1987).

28. Economopoulos, M., Respen, Y., Lessel, G. and Steffes, G. *Application of Tempcore Process to the Fabrication of High Yield Strength Concrete – Reinforcing Bars*, CRM No. 45, December 1975.

29. Simpson, R.J. *Metals and Materials*, October, 598 (1989).

30. Constructional Steelwork Economic Development Committee *Efficiency in the Construction of Steel Framed Multi-Storey Buildings*, National Economic Development Office.

31. Preston, R.R. Private communication.

32. Nara, Y., Kyogoku, T., Yamura, T. and Takeuchi, I. In *Proc. Steels for Line Pipe and Pipeline Fittings* (London, 1981) The Metals Society/The Welding Institute, p. 201 (1981).

33. Sage, A.M. In *Proc. Steels for Line Pipe and Pipeline Fittings* (London, 1981) The Metals Society/The Welding Institute, p. 39 (1981).

34. Nakasugi, H. *et al.* In *Proc. Steels for Linepipe and Pipeline Fittings* (London, 1981) The Metals Society/The Welding Institute, p. 94 (1981).

35. Cavaghan, N.J. *et al.* In *Proc. Steels for Linepipe and Pipeline Fittings* (London, 1981) The Metals Society/The Welding Institute p. 200 (1981).
36. Shiga, C. *et al.* In *Proc. Steels for Linepipe and Pipeline Fittings* (London, 1981) The Metals Society/The Welding Institute, p. 134 (1981).
37. Coolen, A. *et al.* In *Proc. Steels for Linepipe and Pipeline Fittings* (London, 1981) The Metals Society/The Welding Institute, p. 209 (1981).
38. Lander, H.N. *et al.* In *Proc. Steels for Linepipe and Pipeline Fittings* (London, 1981) The Metals Society/The Welding Institute, p. 146 (1981).
39. Rogerson, P. and Jones, C.L. In *Proc. Steels for Linepipe and Pipeline Fittings* (London, 1981) The Metals Society/The Welding Institute, p. 271 (1981).

3 Engineering steels

Overview

The term *engineering steels* applies to a wide range of compositions that are generally heat treated to produce high strength levels, i.e. tensile strengths greater than 750 N/mm². These steels are subjected to high service stresses and are typified by the compositions that are used in automotive engine and transmission components, steam turbines, bearings, rails and wire ropes. As well as carbon and low-alloy grades, engineering steels also embrace the *maraging* compositions that are generally based on 18% Ni and which are capable of developing tensile strengths greater than 2000 N/mm².

Engineering steels are concerned primarily with the generation of a particular level of strength in a specific section size or *ruling section*. This introduces the concept of *hardenability* which is concerned with the ease with which a steel can harden in depth rather than the attainment of a specific level of hardness/strength. In turn, this relates to the effects of alloying elements on hardenability and the influence of cooling rate on a specific composition or section size. Much of the information that is available today on hardenability concepts and the metallurgical factors affecting hardenability was generated in the United States in the 1930s with names such as Grossman, Bain, Grange, Jominy and Lamont featuring prominently in the literature. This period also coincided with the introduction of isothermal transformation diagrams which paved the way to the detailed understanding of the decomposition of austenite and a qualitative indication of hardenability.

Up until the late 1940s, engineering steels often contained substantial levels of nickel and molybdenum, the concept being that these elements were required in order to provide a good combination of strength and toughness. Whereas these alloy additions certainly fulfilled this objective, what was to change in subsequent years was the generation of quantitative data on the actual level of toughness that was required in engineering components. This paved the way to the substitution of nickel and molybdenum by cheaper elements such as manganese, chromium and boron and the more economical use of alloy additions for particular hardenability requirements. The theme of cost reduction was also pursued very vigorously in the 1970s and 1980s with the introduction of medium-carbon, micro-alloy steels for automotive forgings. As illustrated later in the text, these steels offer the potential of major savings over traditional quenched and tempered alloy grades through lower steel costs, the elimination of heat treatment and improved machinability.

Steel cleanness and the reduction of non-metallic inclusions have also been of major concern to users of engineering steels, particularly in applications with the potential for failure by fatigue. Bearing steels are a typical example and the fatigue performance of these steels has been improved progressively over the years with the adoption of facilities such as vacuum degassing (1950s), argon shrouding of the molten stream (1960s) and vacuum steelmaking (1970s). However, major improvements in cleanness have also been obtained in bulk steelmaking processes for bearing steels with the introduction of secondary steelmaking facilities.

The machining of automotive components can account for up to 60% of the total cost and therefore major effort has been devoted to the development of engineering steels with improved machinability. In the main, these developments have been focused on the traditional resulphurized grades but with the addition of elements such as calcium and tellurium for sulphide shape control and improved transverse properties.

In summary, the author's overall perception of this sector has been one of continuing effort to achieve cost reduction, initially through the use of cheaper alloying elements but latterly via the concept of lower through costs and involving a reduction in the cost of heat treatment and machining.

Underlying metallurgical principles

Whereas strip and structural steels are based primarily on ferrite or ferrite–pearlite microstructures, engineering grades are generally heat treated to provide high strengths via the generation of lower-temperature transformation products, such as bainite or martensite. In order to form such products, the alloy content of the steels must be high enough to suppress the formation of ferrite and pearlite, under the cooling conditions employed from the austenitic state. Although water quenching would reduce the critical alloy content required to achieve a martensitic structure in a particular section size, this might generate severe internal stresses or quench cracking in a component and therefore oil quenching is generally employed.

In their simplest form, engineering steels are based on C–Mn compositions which are only effective in producing a martensitic structure in small section sizes. Such steels are described as having low *hardenability* in the context of through-hardening but can develop high levels of surface hardening through induction heating and quenching which develops a martensitic structure in the outer fibres of the component. However, engineering steels are generally produced via the basic electric arc furnace in which the addition of large amounts of scrap steel introduces a substantial amount of *residual elements*, such as copper and nickel. Such elements can be present at total levels of about 0.5% and therefore contribute significantly to the hardenability of carbon or low-alloy engineering grades.

For larger section sizes, steels with higher alloy content and hardenability are required in order to generate a substantially martensitic structure. However, whereas most of the common alloying elements will increase hardenability and promote the formation of martensite, due consideration must be given to the cost of these elements and, more particularly, to their specific cost in achieving a given increment in hardenability. Thus elements such as manganese, chromium and boron may be used in preference to nickel and molybdenum, provided that hardenability is the prime consideration. However, the excessive addition of one particular element can lead to segregation at the casting stage which may not be removed completely on subsequent processing and heat treatment. Therefore components of large section size, which require a high level of hardenability to achieve through-hardening, will be based on multiple alloying additions, some of which may also be incorporated to confer tempering resistance.

The carbon content of engineering steels is important in that it controls the strength of martensitic structures and also contributes significantly to the hardenability of the steels. Very broadly, the carbon content of most engineering steels falls into two categories, namely a level of around 0.2% in carburizing grades and about 0.4% in through-hardening grades. In the former, the carbon content in the surface regions of a component is raised to about 0.8% by gaseous diffusion in the austenitic state and on oil quenching, a high-carbon martensitic *case* is developed on a lower-carbon martensitic *core*. Such a process therefore develops a duplex microstructure in which a hard, fatigue-resistant case is supported by a lower-strength, ductile core. Carburized steels are used extensively in automotive transmission components, such as gears and back axles, which must be produced to very accurate dimensions in order to avoid misalignment, overloading and premature failure. This introduces the problem of *distortion*, namely the changes in shape that accompany heat treatment and transformation and which are exacerbated by the large temperature gradients developed under fast cooling rates from the austenitic state.

Having produced a martensitic microstructure on quenching from a temperature about 20°C above Ac_3, engineering steels are tempered in order to provide a good combination of strength and ductility/toughness. The tempering temperatures may be around 200°C in carburized components, which introduces stress relief, and up to 650°C in the higher-carbon, through-hardening grades which produces a substantial change in structure. In turn, this invokes consideration of the effect of alloying elements on tempering resistance and, in particular, the benefit that will be obtained from adding expensive elements, such as vanadium and molybdenum. These elements produce more stable carbides than iron, manganese and chromium, retarding the degeneration of the martensitic structure. Vanadium and molybdenum therefore feature prominently in ferritic creep-resisting grades for power generation applications and also in high-speed cutting steels, in order to provide high strength during elevated temperature service.

Nickel may be regarded as a common alloying element in engineering steels but, in fact, it has relatively little effect on hardenability. Additionally, it has little affinity for carbon and therefore is ineffective in retarding the tempering of martensite. However, nickel is perceived to be a toughening agent and is incorporated in some of the higher-strength, martensitic grades, notably maraging steels, and these will be discussed later in this chapter.

Engineering steels are used for the production of bearings in the carburized condition or from through-hardened steels of the 1% C–Cr type. In this respect, detailed consideration must be given to steel cleanness in order to alleviate the adverse effects of non-metallic inclusions on fatigue resistance. Bearing steels are therefore subjected to stringent deoxidation practices and may be vacuum degassed or vacuum melted in order to eliminate hard, angular inclusions which are particularly detrimental.

Although the bulk of steels for engineering applications are based on tempered martensitic structures, there are notable examples in which the required strength is developed by aircooling from the austenitic region in order to generate a predominantly pearlitic microstructure. These include rail steels, micro-alloy automotive forgings and high-carbon, wire rod. Again these materials will be discussed in more detail, later in the text, but in general they are based on high carbon contents with manganese levels up to about 1.5% which augments carbon in the formation of pearlite. Very clearly, such steels involve completely different concepts, in terms of underlying metallurgy, from those present in martensitic grades and the strengthening mechanisms are those involving grain size, volume fraction of ferrite and pearlite, and the interlamellar spacing of pearlite. Consideration must also be given to solid solution hardening effects due to manganese, silicon and free nitrogen. The structure–property relationships of these medium-carbon, pearlitic grades are therefore similar to those involved in the lower-carbon, strip and structural grades, but of course with greater emphasis on the contribution from the higher volume fraction of pearlite. Micro-alloy forging grades are also similar in concept and indeed represent an extension to high-strength, low alloy (HSLA) grades of strip and structural steels, albeit with a higher carbon content and higher volume fraction of pearlite. Medium carbon, micro-alloy forging grades generally incorporate an addition of 0.05–0.20% V which is soluble at the reheating temperature and which results in the precipitation of vanadium carbonitride in both the proeutectoid ferrite and the ferrite lamellae of the pearlite, on cooling from the forging operation. This results in tensile strengths in the range 800–1100 N/mm^2 over a wide range of section size, which are comparable to those achieved in conventional, quenched and tempered martensitic grades.

Machining is a very important stage in the production of most engineering components and, in many automotive transmission parts, can account for up to 60% of total production costs. Due consideration must therefore be given to optimizing the machinability of engineering steels, consistent with other property requirements. Whereas normalizing or carbide-spheroidizing heat treatments may be applied in order to provide an easily machined microstructure, most engineering components must be machined in the fully heat-treated, high-strength condition. In this case, small amounts of elements such as sulphur and lead may be added to the steels which improve the machining performance very dramatically. This derives from the presence of MnS inclusions or discrete particles of lead which exude into the tool–chip interface, acting as a lubricant and also forming a protective deposit on the tool.

High sulphur additions impair the transverse ductility of steels, particularly when MnS is present as long elongated inclusions. However, inclusion-modifying agents, such as calcium, can be added to the steel which improve the transverse properties. Lead is generally present as globular particles or as tails to the MnS inclusions. As such, it causes little detriment to the mechanical properties but major hygiene precautions must be taken during steelmaking to minimize lead fume and the associated toxicity effects.

Heat treatment aspects

Isothermal transformation diagrams

Isothermal transformation diagrams were first published by Bain and Davenport[1] in the United States in 1930 and paved the way to the detailed understanding of the effects of alloying elements on the heat treatment response in steels. A steel is first heated to a temperature in the austenitic range, typically 20°C above Ac_3, and then cooled rapidly in a bath to a lower temperature, allowing isothermal transformation to proceed. The progress of transformation can be followed by dilatometry, the degree of transformation depending upon the holding time at temperature. As illustrated schematically in Figure 3.1, the start and finish of transformation to ferrite, pearlite, bainite and martensite are then shown on a diagram as a function of temperature and time. Isothermal transformation diagrams, also known as TTT (time-temperature-transformation) diagrams, are simple in concept but are not representative of the majority of commercial heat treatments which involve a continuous-cooling operation. However, *martempering* and *austempering* are examples of heat treatment that employ isothermal sequences.

A schematic illustration of martempering is given in Figure 3.2. Following a conventional austenitizing treatment, a component is cooled rapidly to a temperature just above the M_s temperature and held at this temperature long enough to ensure that it attains a uniform temperature from surface to centre. The material is then air cooled through the M_s–M_f temperature range to form martensite and is subsequently tempered. Martempering is therefore not a tempering operation but a treatment that leads to low levels of residual stresses and minimizes distortion and cracking.

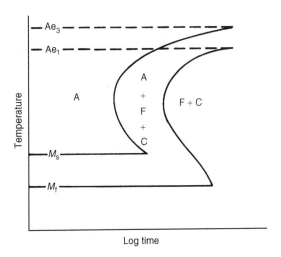

Figure 3.1 *Isothermal transformation diagram*

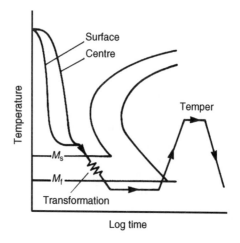

Figure 3.2   *Martempering treatment*

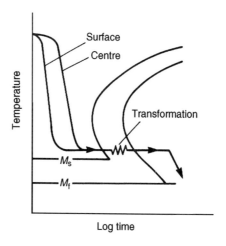

Figure 3.3   *Austempering treatment*

In *austempering*, Figure 3.3, a component is quenched into a salt bath, again at a temperature above M_s, but the material is held at this temperature long enough to allow complete isothermal transformation to bainite. The component is then quenched or air cooled to room temperature. Again no tempering is involved in the conventional sense but austempering represents a short, economical heat treatment cycle that provides a good combination of strength and toughness. The cooling rate must be fast enough to avoid the formation of pearlite and the isothermal treatment must be long enough to complete the transformation to bainite.

Although they proved to be extremely useful in gaining a better understanding of austenite decomposition and transformation kinetics, it became apparent that

isothermal transformation diagrams were only of limited value in the evaluation of continuous-cooling processes. Although attempts were made to convert
isothermal transformation diagrams to a continuous-cooling format, these were
only partially successful. Attention turned therefore to the direct determination
of continuous-cooling transformation diagrams.

Continuous-cooling transformation (CCT) diagrams

CCT diagrams are generated from a series of temperature–length curves. A specimen is heated slowly into the austenitic range and, as illustrated in Figure 3.4,
the heating curve provides the facility for the determination of the Ac_1 and Ac_3
temperatures. The specimen is then cooled at a prescribed rate and the start and
finish of transformation can be determined respectively from the initial deviation
from the cooling curve and the subsequent conformity to the heating curve. This
exercise is repeated for a series of cooling rates, ranging from the simulation of
water quenching in a small-diameter rod to that experienced in furnace cooling.

CCT diagrams can be presented in two ways:

1. Temperature–time plots in which the cooling time is plotted horizontally on
   a log scale.
2. Temperature–bar diameter plots, the latter representing different bar sizes
   cooled at rates simulative of air, oil and water cooling.

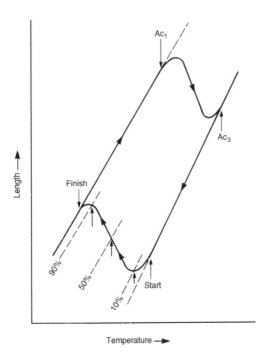

Figure 3.4 *Length changes in heating and cooling*

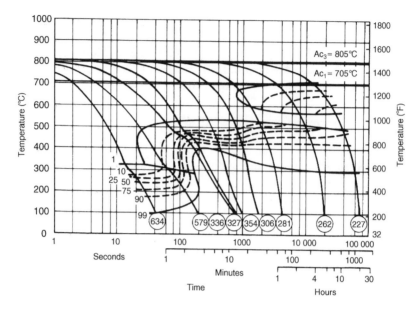

Figure 3.5   *Continuous-cooling transformation diagram for a 0.39% C 1.45% Mn 0.49% Mo steel (After Cias[2])*

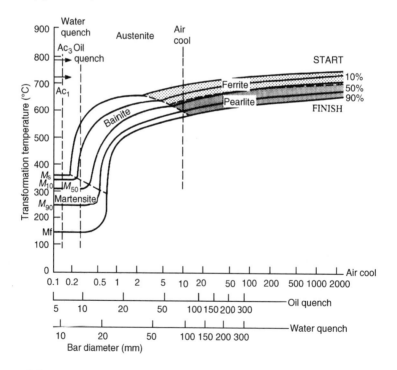

Figure 3.6   *Continuous-cooling transformation diagram for a 0.38% C 0.6% Mn steel; the dotted vertical lines indicate the transformations that occur in a 10 mm bar after air cooling, oil quenching and water quenching respectively (After Atkins[3])*

Examples of both types of presentation are shown in Figures 3.5 and 3.6. A feature of the temperature–time presentation is the insertion of the cooling curves used to generate the CCT diagrams and the hardness developed after each cooling rate is shown in a circle at the end of the cooling curves. This type of presentation was favoured by Cias[2] of Climax Molybdenum who compiled a series of CCT diagrams for medium-carbon alloy steels. Temperature–bar diameter plots were used by Atkins[3] of British Steel in a major publication of CCT diagrams, covering a wide range of carbon and alloy steels.

Hardenability testing

The term *hardenability* is used in various ways to describe the heat treatment response of steels, employing one or other of interrelated parameters such as hardness and microstructure. When evaluated by hardness testing, hardenability is often defined as the capacity of a steel to harden *in depth* under a given set of heat treatment conditions. What must be emphasized in this definition is the fact that hardenability is concerned with the depth of hardening or the hardness profile in a component, rather than the attainment of a specific level of hardness. Using microstructure as the control parameter, Siebert *et al.*[4] have defined hardenability as:

> 'the capacity of a steel to transform partially or completely from austenite to some percentage of martensite at a given depth when cooled under some given condition'

Whereas the hardenability of a steel can be determined from continuous-cooling transformation diagrams, the construction of such diagrams is both time-consuming and expensive and therefore more economical methods were required for the measurement of hardenability. A number of hardenability tests were developed in the 1930s but the best known and most widely used is the *Jominy end quench* test. Developed in imperial units, the Jominy specimen is a cylinder 102 mm (4 in) long × 25.4 mm (1 in) diameter with a flange at one end. It is usual to normalize the material to be tested prior to machining the specimen in order to eliminate variations in the hardening response that might be introduced by differences in microstructure in the as-rolled condition. The specimen is heated to the appropriate austenitizing temperature and then transferred quickly to a fixture which suspends the specimen above a tube through which a column of water is directed against the bottom face. The arrangement is shown in Figure 3.7 and it should be emphasized that the water flow is tightly specified and controlled in order to produce a consistent quenching effect. Whereas the quenched end of the specimen experiences a rapid rate of cooling, the effect diminishes along the length of the specimen to give a value approaching air cooling at the other end. When the quenching operation is complete, flats are ground at diametrically opposed positions on the specimen to a depth of 0.38 mm to remove decarburized material and provide a suitable surface for hardness testing. This can involve either Vickers (HV) or Rockwell (HRC) hardness testing at intervals of about 1.5 mm for alloy steels or 0.75 mm for carbon steels. A typical Jominy hardenability curve is shown in Figure 3.8, which reflects

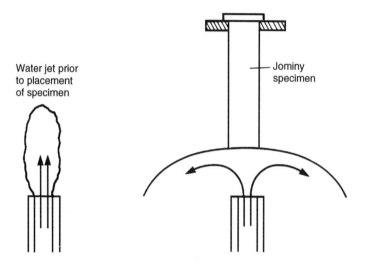

Figure 3.7   *Jominy hardenability testing*

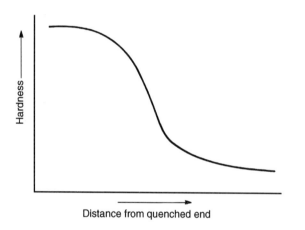

Figure 3.8   *Schematic Jominy hardenability curve*

the martensitic hardening that can be achieved at the end quenched position in a steel of reasonable hardenability and transformation to bainite and ferrite–pearlite at the slower rates of cooling.

Although it remains predominant, the Jominy hardenability test has been criticized for not being sufficiently discriminating between steels of low hardenability, i.e. those involving a very rapid decrease in hardness just beyond the quenched end. For such steels, the *SAC* test is deemed to be more appropriate, although it has been used very rarely in the UK. In this test, the specimen is again a cylinder but measures 140 mm long × 25.4 mm diameter. After normalizing and austenitizing at a suitable temperature above Ac$_3$, the specimen is quenched overall in water. After quenching, a cylinder 25 mm long is cut from the test specimen and the end faces are ground very carefully to remove any tempering

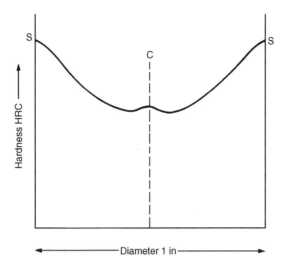

Figure 3.9   *Schematic SAC hardenability curve*

effects that might have been introduced during cutting. Rockwell (HRC) hardness measurements are then made at four positions on the original cylinder face and the average hardness provides the surface (S) value. Rockwell testing is then carried out along the cross-section of the specimen from surface to centre and provides the type of hardness profile illustrated in Figure 3.9. The total area under the curve provides the area (A) value in units of Rockwell-inch (using the original imperial unit) and the hardness at the centre gives the C value. The SAC value of a steel might be reported as 65-51-39, which would indicate a surface hardness of 65 HRC, an area value of 51 Rockwell-inch and a centre hardness of 39 HRC. One interesting feature of the SAC test is that it can reveal central segregation in a bar as indicated by a hardness peak at the centre position.

Factors affecting hardenability

Grain size

In a homogeneous austenitic structure, the nucleation of pearlite occurs almost exclusively at the grain boundaries and therefore the larger the grain boundary surface area, the greater are the nucleation sites for pearlite formation. Thus the hardenability of a given composition will increase with increasing austenitizing temperature and austenite grain size. The major effect of austenitizing temperature on the hardenability of a 0.55% C 1% Cr 0.2% Mo steel is shown in Figure 3.10, which is based on the work of Grange and presented by Grossmann.[5]

Although grain coarsening could be employed as a cheap method of achieving high hardenability, this approach is rarely adopted because the toughness and ductility are impaired. Instead, most commercial engineering steels are made to an aluminium-treated, fine-grain practice in order to produce microstructures that will provide a good combination of strength and toughness/ductility.

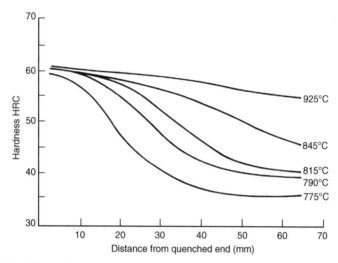

Figure 3.10   *Effect of austenitizing temperature on the Jominy hardenability of 0.55% C 0.84% Mn 0.92% Cr 0.21% Mo steel (After Grange[5])*

Alloying elements

Because of their very distinct effects on hardenability, it is convenient to consider alloying elements in three separate groups:

1. Carbon.
2. General group – Cr, Mn, Mo, Si, Ni, V, etc.
3. Boron.

Carbon must be placed in a special category because it is the element that controls the hardness of martensite and therefore defines the maximum hardness that can be achieved in a given steel composition. The relationship between carbon content and the hardness of martensite is shown in Figure 3.11. The effect is reasonably

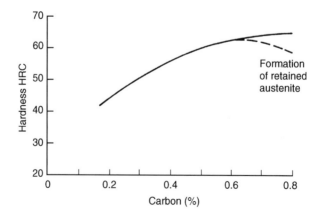

Figure 3.11   *Effect of carbon on the hardness of martensite*

linear at carbon contents up to about 0.6% but, depending upon the alloy content of the steel, the hardness then reaches a plateau and declines at higher carbon contents. This is due to the depression of the M_f temperature below room temperature, leading to incomplete transformation and the presence of retained austenite. However, because of the marked effect of carbon on hardness, it is important to bear this point in mind when attempting to compare the hardenability of steels of different carbon content.

In addition to its effect on hardness, carbon also has a significant effect on hardenability. This is illustrated in Figures 3.12 and 3.13, which show the progressive effect of carbon in base steels containing 0.8% Mn and 0.5% Ni 0.5% Cr 0.2% Mo (SAE 8600) respectively.[6] The effect of carbon on the hardness of martensite is evident in both types of steel but the effect on hardenability is very much more pronounced in the Ni–Cr–Mo steel. However, carbon is rarely used as a hardenability agent because of its adverse effect on toughness and its tendency to promote distortion and cracking. In addition, high-carbon steels are hard and difficult to cut or shear in the annealed condition.

With the exception of cobalt, small additions of all alloying elements will retard the transformation of austenite to pearlite and thereby increase hardenability. However, the elements that are most commonly used for the promotion of hardenability are manganese, chromium and molybdenum but nickel and vanadium are frequently incorporated for additional purposes. A considerable amount of work has been carried out, particularly in the United States, to quantify the effects of the major alloying elements on hardenability and, despite some complex and interactive effects, the general order of potency has been established. The published data on this topic have been reviewed very thoroughly

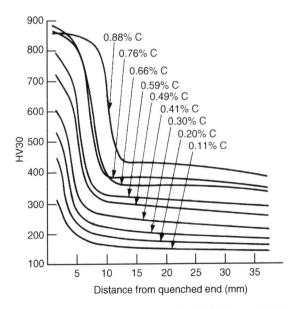

Figure 3.12 *Effect of carbon on hardenability of a 0.8% Mn steel (After Llewellyn and Cook[6])*

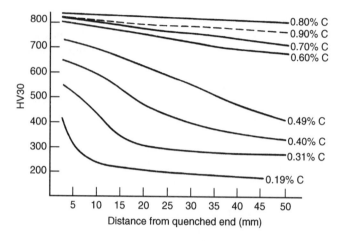

Figure 3.13 *Effect of carbon on hardenability of SAE 8600 steels (after Llewellyn and Cook[6])*

by Siebert *et al.*,[4] who include information on the effects of elements such as copper, tungsten and phosphorus as well as the five common alloying elements.

Work by deRetana and Doane[7] evaluated the effects of the major alloying elements on the hardenability of low-carbon steels of the type used for case carburizing. Their information is shown in Figure 3.14, where the change in hardenability is expressed by means of a multiplying factor, calculated as follows:

$$\text{Multiplying factor} = \frac{\text{hardenability of (base steel} + \text{alloying element)}}{\text{hardenability of base steel}}$$

The data shown in Figure 3.14 were derived from Jominy tests on a variety of commercial and experimental casts of carburizing grades in which only one element was varied in the initial part of the work. The multiplying factors were tested subsequently in multi-element steels and modified empirically to provide more widely applicable, averaged factors. However, the authors were unable to develop a single factor for molybdenum due to interactive effects and therefore separate factors are shown for this element for use in low- and high-nickel steels.

Although the effects can vary significantly with the carbon content and base composition, a guide to the potency of elements in the general group in promoting hardenability is shown below:

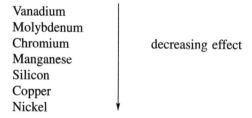

Vanadium
Molybdenum
Chromium
Manganese   decreasing effect
Silicon
Copper
Nickel

However, carbon could be placed at the top of this list and elements such as phosphorus and nitrogen, although present in small amounts, appear to produce

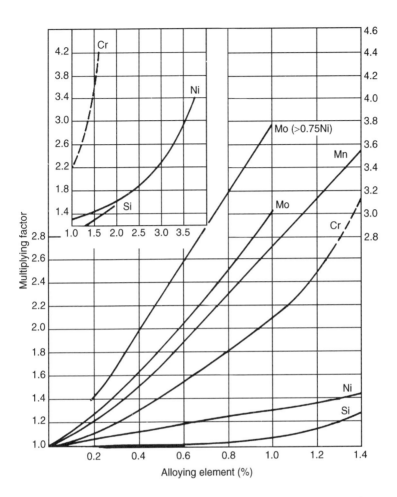

Figure 3.14 *Average multiplying factors for several elements in alloy steels containing 0.15 – 0.25% carbon (After deRetana and Doane[7])*

hardenability effects of a similar magnitude to carbon. Although vanadium has a powerful effect on hardenability, it has a low solubility in steels due to the formation of vanadium carbide. Therefore the level of addition is generally small and, as illustrated shortly, may be used more to retard the tempering process than as a hardenability agent.

Boron

In the context of low-alloy engineering steels, boron is a unique alloying element from the following standpoints:

1. The addition of 0.002–0.003% **B** to a suitably protected base composition produces a hardenability effect comparable with that obtained from 0.5% Mo, 0.7% Cr or 1.0% Ni.

2. The effect of boron on hardenability is relatively constant provided a minimum level of soluble boron is present in the steel.
3. The potency of boron is related to the carbon content of the steel, being very effective at low carbon contents but decreasing to zero at the eutectoid carbon level.

Because of its high affinity for oxygen and nitrogen, boron is added to steel in conjunction with even stronger oxide- and nitride-forming elements in order to produce metallurgically active, soluble boron. In electric arc steelmaking, this involves additions of about 0.03% Al and 0.03% Ti, either separately or in the form of proprietary compounds containing the required levels of boron, aluminium and titanium. Without these additions, boron would react with oxygen and nitrogen in the steel and form insoluble boron compounds which have no effect on hardenability.

Gladman[8] has shown that the location of boron in steels is very dependent on the heat treatments that are applied. In low-alloy steels, boron was distributed uniformly throughout the microstructure of a 25.4 mm bar after water quenching from the austenitic range. However, in air-cooled samples, the grain boundaries were enriched in boron compared with the body of the grains. Ueno and Inoue[9] also investigated the presence of boron in a 0.1% C 3.0% Mn steel and showed that boron first segregated to and then precipitated at the grain boundaries, according to a typical 'C' curve pattern. They also showed that an increase in boron content decreased the incubation periods for segregation and precipitation, and in a steel containing 0.002% B, solution treated at 1350°C, quenching in iced brine was required in order to prevent the segregation of boron to the grain boundaries. It would appear therefore that under normal heat treatment conditions, involving oil quenching from temperatures of 820–920°C, boron segregates to the austenite grain boundaries and suppresses the formation of high-temperature transformation products.

Llewellyn and Cook[6] carried out a detailed investigation of the metallurgy of boron-treated engineering steels containing a wide range of carbon contents. The effect of boron content on hardenability was studied in a base composition of 0.2% C 0.5% Ni 0.5% Cr 0.2% Mo (SAE 8620) and the following Jominy hardenability criteria were examined:

1. Jominy distance to [hardness at J 1.25 mm−25 HV].
2. Jominy distance to 350 HV, i.e. near the inflexion in the hardenability curve.
3. Hardness at J 9.8 mm, equivalent to the cooling rate at the centre of an oil-quenched, 28 mm bar.

As illustrated in Figure 3.15, each of these criteria reaches a maximum at a soluble boron content of about 0.0007%. Further additions produce a reduction in hardenability but a reasonably steady-state condition is achieved at boron contents in excess of 0.0015%. A similar pattern of results was also observed by Kapadia *et al.*[10] and, in commercial practice, it is usual to aim for soluble boron contents of 0.002–0.003%, accepting a slight loss in hardenability in favour of a consistent hardenability effect.

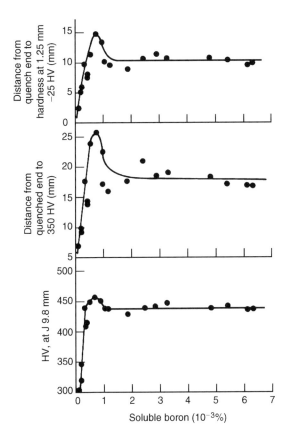

Figure 3.15   *Effect of boron on hardenability (After Llewellyn and Cook[6])*

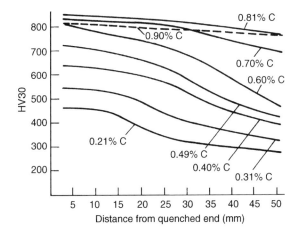

Figure 3.16   *Effect of carbon on hardenability of SAE 86B00 steels (After Llewellyn and Cook[6])*

The interaction between boron and carbon was also investigated in SAE 8600 base steels and the Jominy hardenability curves for boron-free and boron-bearing versions of these steels are shown in Figures 3.13 and 3.16. These figures also indicate the formation of retained austenite at carbon contents in excess of 0.8%. However, the interaction between boron and carbon is illustrated in Figure 3.17 where the hardenability criteria examined are the Jominy distances which coincide with decreases in hardness of 50, 100 and 125 HV below that obtained at the quenched end of the specimen (H_{max}). This figure shows that boron produces a marked increase in hardenability at a carbon content of 0.2% but the effect is steadily reduced to zero as the carbon content is increased to a level of 0.53–0.54%. Above this carbon level, boron has a detrimental effect on hardenability.

The above type of experiment was also carried out in base steels containing 0.8% Mn and the results are summarized in Figure 3.18. In this case, the efficiency of boron as a hardenability agent is expressed by means of a multiplying factor BF, calculated on the basis of:

$$BF = \frac{\text{hardenability of (base steel} + \text{boron)}}{\text{hardenability of base steel}}$$

Thus the effect of boron on hardenability is again reduced with increase in carbon content but, in this particular base composition (0.8% Mn), reaches a value of zero (BF = 1.0) at a carbon content of about 0.85%.

From the above data, it is evident that the boron effect varies not only with carbon content but also with the alloy content of the base steel. Llewellyn and Cook have proposed that the critical carbon content may well correspond to the eutectoid level for the compositions concerned and that bainite might be promoted at the expense of martensite in boron-treated, hyper-eutectoid steels.

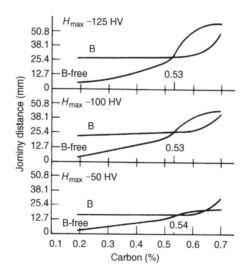

Figure 3.17   *Effect of boron on hardenability at various carbon levels (After Llewellyn and Cook[6])*

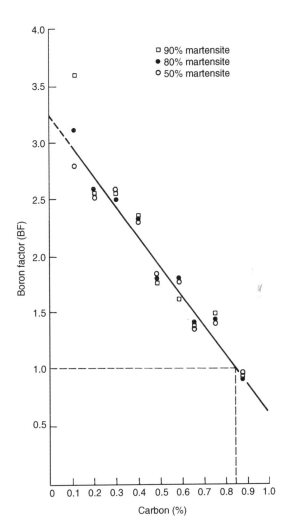

Figure 3.18   *Effect of carbon on boron multiplying factor (After Llewellyn and Cook[6])*

A consequence of the above effect is that the case hardenability of a carburized boron-treated steel is lower than that of a boron-free steel of comparable core hardenability. However, the limiting section sizes that will provide adequate case depths are probably well in excess of those that present a problem with regard to core hardenability.

Tempering resistance

Although the objective of quenching a steel is generally to produce a martensitic structure of high strength, steels are rarely put into service in the as-quenched condition because this represents a state of high stress, low toughness and poor ductility. After quenching, components are therefore tempered at an elevated temperature in order to obtain a better combination of properties. In carburized gears, the tempering treatment might be carried out at a temperature as low as 180°C, the purpose being to achieve relief of internal stresses without causing any significant softening in either the case or core of the components. On the other hand, medium-carbon steels containing about 0.4% C might be tempered at temperatures up to 650°C which results in a significant decrease in strength compared with the as-quenched condition but which is necessary in order to produce adequate toughness and ductility.

The mechanism of tempering involves the partial degeneration of martensite via the diffusion of carbon atoms out of solid solution to form fine carbides. If the tempering treatments are carried out at sufficiently high temperatures for long periods, then the breakdown of martensite can be complete, the microstructure consisting of spheroidized carbide in a matrix of ferrite. However, treatments that are carried out to achieve this type of microstructure would be termed *annealing* rather than tempering.

Alloying elements affect the tempering process by retarding or suppressing the formation of Fe_3C, either by stabilizing the ε-carbide ($Fe_{2.4}C$) which is formed initially in the breakdown of martensite or by forming carbides that are more stable and grow more slowly than Fe_3C. Alloying elements therefore provide the opportunity of tempering at higher temperatures in order to obtain a higher ductility for a given strength level. Alternatively, improved tempering resistance might be employed to allow a component to operate at a higher temperature without softening.

Smallman[11] has produced the information given in Table 3.1 on the effect of alloying elements on tempering. In this table, the negative value ascribed to carbon indicates an acceleration in the tempering process which is due presumably to the increased supersaturation/driving force effect.

Work by Grange and Baughman[12] established the following rank order of potency in promoting tempering resistance:

Vanadium	
Molybdenum	
Chromium	decreasing effect
Manganese	
Silicon	
Copper	
Nickel	

Thus there is broad agreement between the two sets of data, although Smallman indicates a relatively higher effect for silicon.

Although alloying elements such as vanadium and molybdenum are effective in promoting both hardenability and tempering resistance, they are expensive and the design of composition in engineering steels is dictated as much by costs as property requirements. Thus hardenability is achieved most cheaply through additions of manganese, chromium and boron but with little contribution to tempering resistance. However, such a condition may be perfectly acceptable in a case-carburized component where the tempering treatment is carried out at a low temperature. On the other hand, when the property requirements or

Table 3.1

Element	Retardation in tempering per 1% addition
C	−40
Co	8
Cr	0
Mn	8
Mo	17
Ni	8
Si	20
V	30
W	10

operating conditions dictate the use of high tempering temperatures, resort is made to molybdenum and vanadium additions, but at relatively modest levels. At levels greater than about 1%, molybdenum induces a *secondary hardening* reaction.

In addition to its effect on tempering resistance, molybdenum is also added to steels to suppress temper embrittlement. This involves the co-segregation of alloying elements such as manganese and silicon and impurity elements such as antimony, arsenic, tin and phosphorus to the prior austenite grain boundaries in steels with bainitic or martensitic microstructures. Temper embrittlement occurs during exposure in the temperature range 325–575°C, either from operating within this range or by slow cooling through the range from a higher tempering temperature. This topic will be discussed further when dealing with *Steels for steam power turbines*, later in this chapter.

Whereas nickel is a well known alloying element in the context of engineering steels, it is expensive and not very effective in enhancing either hardenability or tempering resistance. However, it is perceived to be a ferrite strengthener and is used in engineering steels in order to promote toughness.

Surface-hardening treatments

Case carburizing

Case carburizing involves the diffusion of carbon into the surface layers of a low-carbon steel by heating it in contact with a carbonaceous material. The objective is to produce a hard, wear-resistant case with a high resistance to both bending and contact fatigue, whilst still maintaining the toughness and ductility of the low-carbon *core*. Carburizing is carried out at temperatures in the range 825–925°C in solid, liquid or gaseous media but, in each treatment, the transport of carbon from the carburizing medium takes place via the gaseous state, usually CO.

Carburizing in a solid, granular medium, such as charcoal, is termed *pack carburizing* and has been practised since ancient times. However, whereas the early treatments relied solely on the reaction between charcoal and atmospheric oxygen for the generation of CO_2 and CO, *energizers* such as sodium or barium carbonate are now added to the carburizing compound. During the heating-up period, the energizer breaks down to form CO_2 which then reacts with carbon in the charcoal to form CO:

$$BaCO_3 \rightarrow BaO + CO_2 \tag{1}$$

$$CO_2 + C \rightarrow 2CO \tag{2}$$

In turn, the CO reacts at the steel surface to form atomic carbon which defuses rapidly into the austenitic structure:

$$2CO \rightarrow C + CO_2 \tag{3}$$

The CO_2 produced in this reaction then reacts with charcoal, reproducing reaction (2), and the cycle is repeated.

Pack carburizing is generally carried out at a temperature of 925°C and a case depth of about 1.5 mm can be obtained after carburizing for eight hours at this temperature. After this treatment, the component is removed from the carburizing compound and heat treated by various forms of quenching which will be described later.

Carburizing in liquid media generally takes place in molten salts (*salt bath carburizing*) in which the active constituent is sodium cyanide (NaCN), potassium cyanide (KCN) or calcium cyanide $Ca(CN)_2$. Oxygen is available at the salt bath–atmosphere interface and carburizing again takes place via the generation of CO. However, nitrogen is also liberated from the cyanides and diffuses into the steel. The amount of carbon and nitrogen absorbed by the steel is related to the temperature and cyanide content of the bath. With a NaCN content of 50%, the surface concentrations of carbon and nitrogen are of the order of 0.9% and 0.2% respectively, after treatments involving $2\frac{1}{2}$ hours at 900°C. However, salt bath carburizing is mainly used for small parts and treatments for $\frac{1}{2}$–1 hour produce a case depth of about 0.25 mm.

Gas carburizing is carried out in hydrocarbon gases such as propane (C_3H_8) or butane (C_4H_{10}) in sealed furnaces at temperatures of about 925°C. Major strides have been made in the control technology of the process and, since the 1950s, gas carburizing has become the most important method of case hardening. During the heating-up stage, the component is surrounded by an inert or reducing atmosphere. This atmosphere is referred to as the *carrier gas*, which is commonly an endothermic gas but may also be nitrogen based. On reaching the carburizing temperature, the furnace atmosphere is enriched to the required carbon level (*carbon potential*) by the addition of hydrocarbons which generate CO and carbon is absorbed at the surface of the steel. The carbon potential is controlled by varying the ratio of hydrocarbon to carrier gas but a surface carbon content of 0.8–0.9% is generally employed. A four-hour treatment at 925°C will produce a case depth of about 1.25 mm.

Following the carburizing operation, the components are subjected to hardening heat treatments, involving different forms of quenching:

- *Direct quenching* – quenching directly from the carburizing temperature
- *Single quenching* – allowing the component to cool to a temperature of about 840°C from the carburizing temperature before quenching

- *Reheat quenching* – cooling to room temperature and then reheating to a temperature of about 840°C before quenching
- *Double reheat quenching* – reheating first to about 900°C and quenching to produce a fine-grained core, followed by a second quenching treatment from a lower temperature, such as 800°C, in order to produce a fine-grained case region.

In case carburizing, single quenching is generally preferred to direct quenching because it reduces the thermal gradients in the steel, thereby minimizing dimensional movement or *distortion*. This topic will be discussed in the next section. Reheat quenching is employed in pack carburizing and also in gas carburizing if machining operations must be carried out before the final hardening treatment. Double reheat quenching is now virtually obsolete since satisfactory grain refinement can now be obtained in the core and case of modern fine-grained steels in the shorter quenching practices.

The type of quenching medium employed, e.g. oil or polymer, depends on the mechanical properties required but, as illustrated shortly, the quenching rate may have to be controlled very carefully in order to minimize dimensional movement. Following the quenching operation, carburized components are generally tempered in the range 150–200°C for periods of 2–10 hours in order to produce some stress relief in the high-carbon martensitic case.

Carburizing is used extensively in the automotive industry for the treatment of shafts and gears. It is also an important process in the production of large bearings in which the required level of hardness and fatigue resistance cannot be achieved in through-hardening grades such as 1.0% C 1.5% Cr (SAE 52100).

Nitriding

As its name suggests, *nitriding* involves the introduction of nitrogen into the surface of a steel but, unlike carburizing, it is carried out in the ferritic state at temperatures of the order of 500–575°C. However, like carburizing, it can be performed in solid, liquid or gaseous media but the most common is that involving ammonia gas (*gas nitriding*) which dissociates to form nitrogen and hydrogen:

$$2NH_3 \rightarrow 2N + 3H_2$$

Nascent, atomic nitrogen diffuses into the steel, forming nitrides in the surface region.

In *salt bath nitriding*, mixtures of NaCN and KCN are employed and the holding times are rarely longer than two hours, compared to periods of 10–100 hours in gas nitriding. A variation of salt bath nitriding is the *Sulfinuz* process which involves the addition of sodium sulphide (Na_2S) to the bath. This results in the absorption of sulphur into the steel as well as the introduction of carbon and nitrogen which is characteristic of conventional salt bath nitriding. The presence of sulphur in the surface of the component improves the anti-frictional behaviour and also the corrosion resistance of the steel.

Nitriding is carried out on steels containing strong nitride-forming elements such as Al, Cr and V and in BS 970 (*Wrought steels for mechanical and allied*

Table 3.2

Grade	C%	Si%	Mn%	Cr%	Mo%	V%	Al%
709M40	0.4	0.25	0.85	1.05	0.3		
722M24	0.24	0.25	0.55	3.25	0.55		
897M39	0.39	0.25	0.55	3.25	0.95	0.2	
905M39	0.39	0.25	0.55	1.6	0.2		1.1

purposes), the grades of steel shown in the Table 3.2 are identified specifically as being suitable for nitriding.

Given the low temperature involved in the process, nitriding can be carried out after the conventional hardening and tempering treatments have been applied to through-hardened steels.

Carbonitriding (nitro-carburizing)

Carbonitriding can be regarded as a variant of gas carburizing in which both carbon and nitrogen are introduced into the steel surface. This is achieved by the introduction of ammonia gas into the carburizing atmosphere which cracks, liberating nascent nitrogen. In some respects, the term *carbonitriding* may be misleading in that it is completely different from the nitriding process that takes place at low temperatures in the ferritic state and the term *nitro-carburizing* might be more appropriate. However, the process is carried out at temperatures of the order of 870°C and the case depths are lower than those produced by gas carburizing, e.g. a four-hour carbonitriding treatment at this temperature produces a case depth of about 0.75 mm.

The introduction of nitrogen produces a significant increase in the hardenability of the case region such that high surface hardness levels can be produced in steels of relatively low alloy content.

Induction hardening

In the processes described above, the hardness of the surface is increased by modifying the chemical composition of this region. In *induction hardening*, the composition of the material is unchanged but the surface is hardened by selective heat treatment. This is achieved by induction heating but a less controlled effect can be produced by the direct impingement of an oxy-acetylene torch in the process of *flame hardening*.

This process is generally applied to steels containing 0.30–0.50% C which give hardness values in the range 50–60 HRC. The steels may be C–Mn or low-alloy grades and induction hardening is carried out in the normalized or quenched and tempered condition, depending upon the section size or the properties required in the core.

Relative merits of surface treatments

In presenting a detailed review of gas carburizing, Parrish and Harper[13] examined the benefits of this process in relation to other surface-hardening treatments such as nitriding and induction hardening. Their main findings can be summarized as follows:

- *Carburizing* – capable of producing a wide range of case depths and core strengths and providing good resistance to bending and contact fatigue. The main drawback of the process is the distortion that occurs due to the thermal gradients induced by quenching from the austenitic range.
- *Nitriding* – produces relatively shallow hardening, e.g. a highly alloyed grade such as 897M39 ($3\frac{1}{4}$% Cr–Mo–V) has an effective case depth (at 500 HV) of only 0.35 mm after nitriding for 80 hours. However, nitriding is a distortion-free process that produces a surface which is resistant to scuffing and adhesive wear.
- *Induction hardening* – capable of producing a wide range of case depths using a range of compositions. Capable of producing similar contact fatigue and wear resistance to case carburizing and the process produces little distortion.

Distortion in case-carburized components

When transformable steels are heat treated, the volume changes that occur during heating and cooling (quenching) are not completely complementary and a component will exhibit a small change in shape compared with its original, pre-heat-treated condition. The term *distortion* is widely used to describe such changes in shape and represents a significant problem in the production of precision engineering components such as automotive gears. In such components, slight inaccuracies in shape lead to irregular tooth contact patterns which can result in problems ranging from a high level of noise in a gearbox or back axle to an overload situation which produces premature fatigue failure. The effect is therefore very important commercially but, given that dimensional change is inevitable under fast-cooling conditions, the approach to the problem is one of control and consistency of response rather than elimination.

In the mid-1960s, Murray[14] published work on the effects of composition on distortion in carburizing steels, using the *Navy C* specimen. This consisted of a split ring in which the dimensional change is measured by the degree of gap opening after heat treatment. Whereas some interesting results were obtained in both the UK and United States with this specimen, it had two major limitations:

1. The unrestrained nature of the split ring could result in gap openings of up to 1.2 mm which were very much larger than the dimensional changes that occur in automotive gears.
2. The effective section size of the specimen was small which made it unsuitable for the investigation of steels with medium to high hardenability.

Llewellyn and Cook[15] therefore designed a new specimen for the investigation of distortion which was washer-shaped with an outer diameter (OD) of 132 mm, bore diameter (BD) of 44 mm and a thickness (T) of 22 mm. These dimensions represented a compromise between the section size that was considered typical of a medium-size truck gear and the limiting size of specimen that could be produced from 50 kg experimental casts of steel. The specimens were carburized at 925°C for $5\frac{1}{2}$ hours at a carbon potential of 0.8%, furnace cooled to 840°C and then cooled at various rates to room temperature.

It was shown that hardenability has a marked effect on dimensional change, as illustrated in experimental casts of steel containing 1–3% Cr. The Jominy hardenability curves for these steels are shown in Figure 3.19 and the dimensional changes that occurred on oil quenching are illustrated in Figure 3.20. As the hardenability of the steels is increased, progressive contractions occur in the OD and BD and these are compensated by increases in thickness. This pattern of results was also repeated in other series of steels of different alloy content. As the alloy content and hardenability are increased, the transformation temperatures are depressed from those involving upper bainite to lower bainite and, ultimately, martensite. With regard to distortion, the consequence of this progression is the increasing volume change on transformation. Therefore, the progressively greater movement that results from increasing hardenability is associated with the progressive increase in volume on transformation. However, such an explanation does not address the reason for the irregular changes in shape that occur in these specimens but these are obviously associated with the complex stress situation that arises from the differential cooling rates that take place in a disc or component of limited symmetry.

One departure from the general trend between hardenability and distortion was observed in the case of boron-treated steels which behaved in an entirely different manner from boron-free steels of comparable hardenability. A satisfactory explanation for the unique behaviour of boron steels has not been proposed but they differ from boron-free steels in the following respects:

1. The case hardenability of a boron steel is lower than that of a boron-free steel of the same core hardenability.
2. Boron-treated steels will tend to have a higher M_s–M_f transformation range than boron-free steels of comparable hardenability.

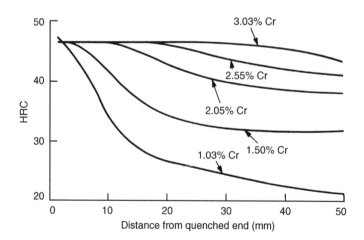

Figure 3.19 *Effect of chromium on Jominy hardenability (After Llewellyn and Cook[15])*

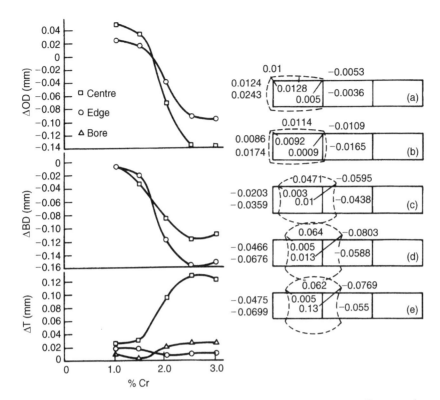

Figure 3.20 *Effect of chromium on dimensional change. Left top: outer diameter change. Left centre: bore diameter change. Left bottom: thickness change. Right: (a) 1.03% Cr; (b) 1.5% Cr; (c) 2.05% Cr; (d) 2.55% Cr; (e) 3.03% Cr (After Llewellyn and Cook[15])*

Llewellyn and Cook showed that carburizing reduces the amount of movement that occurs in low-alloy/low-hardenability steels and related the effect to the depression of the M_s–M_f transformation range in the case-carburized region of a component. It was observed that carburizing had a smaller effect on dimensional change in highly alloyed steels and this may be related to the fact that these steels have low transformation ranges.

Cooling rate from the austenitic range was shown to have a dramatic effect on dimensional movement and water quenching can virtually eliminate the

Table 3.3 *Steels used to investigate the effect of quenching rate on distortion*

Grade	C	Si	Mn	P	S	Cr	Mo	Ni	Al	B	Ti
637H17 (En 352)	0.18	0.28	0.81	0.011	0.037	0.92	0.1	1.1	–	–	–
822H17 (En 355)	0.18	0.34	0.6	0.025	0.016	1.47	0.2	2	0.056	–	–
835H15 (En 39B)	0.14	0.35	0.42	0.012	0.018	1.13	0.21	3.9	0.03	–	–
CM60	0.14	0.26	0.98	0.008	0.049	0.31	0.12	0.15	0.014	0.003	ND[a]
CM80	0.19	0.35	1.37	0.009	0.045	0.28	0.12	0.17	0.056	0.003	0.051

[a]ND = not determined.

influence of other important factors such as carburizing treatment and alloy content/hardenability. This was illustrated in samples of commercial steels which were air cooled, slow oil quenched, fast oil quenched and water quenched, following a carburizing treatment at 925°C and furnace cooling to 840°C. The compositions of these steels are given in Table 3.3 and the hardenability curves are shown in Figure 3.21. The mean dimensional changes that took place after heat treatment are illustrated in Figure 3.22. In this figure, the various quenching treatments are shown in terms of the H value, an expression of the severity of the quenching rate. After air cooling, relatively little movement occurs in any of the steels, but as the quenching rate is increased to an H value of 0.8, most of the steels exhibit a significant decrease in OD and BD and a complementary increase in thickness. However, when the specimens are water quenched, a marked increase takes place in OD and BD and, despite the marked variation in hardenability, each of the steels tends to exhibit similar levels of movement.

From the foregoing remarks, it can be appreciated that a number of measures can be taken to reduce or control the degree of distortion that occurs in carburized components:

1. In some cases, quenching presses can be used to restrain the amount of movement that occurs.
2. Given the marked effect of hardenability, automotive manufacturers in particular will generally order steels to restricted hardenability bands, i.e. to only a half or even a third of the band width of the standard specification. This will minimize the variation in distortion such that the changes can be predicted and accommodated in the pre-heat treatment geometry.
3. The major effect of cooling rate is widely recognized in the automotive industry. Special quenching oils or warm oils are sometimes used in order to reduce temperature gradients during quenching, thereby reducing the degree of irregular dimensional change.

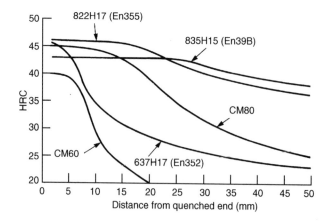

Figure 3.21   *Jominy hardenability of commercial steels (After Llewellyn and Cook[15])*

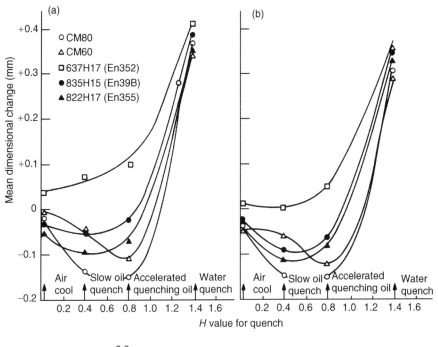

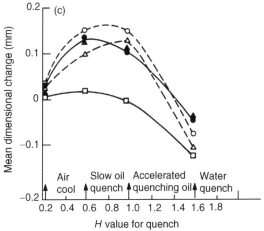

Figure 3.22 *Effect of quenching rate on mean dimensional change in (a) outer diameter, (b) bore diameter and (c) middle thickness (After Llewellyn and Cook[15])*

Standard specifications

The major UK standard for engineering steels is BS 970: Part 1: 1996 *Wrought steels for mechanical and allied engineering purposes*. It deals with steels of the following type in the form of blooms, billets, slabs, bars, rods and forgings:

- As-rolled, as-rolled and softened, and micro-alloyed C and CMn steels (Section 2).
- Through hardening boron steels (Section 3).
- Case hardening steels (Section 4).
- Stainless and heat resisting steels (Section 5).

Acknowledging that the standard contains a large number of grades, this edition of BS 970 has continued the practice of separating the steels into two categories. Category 1 steels are the recommended series of steels for use in new designs and in established designs whenever possible and are printed throughout the standard in normal (upright) type. Category 2 steels make up the remainder and are shown in *italic* (sloping) type.

The old En numbering system which became established after the Second World War was withdrawn some years ago and replaced by a logical, if complex, system which is outlined below.

In all designations, apart from those for stainless steels, the last two digits indicate the mean of the specified carbon content.

Each designation carries a letter which defines the supply requirements:

- M – for steels to specified mechanical properties
- H – for supply to hardenability requirements
- A – when supplied only to a specified analysis

In carbon and carbon – manganese steels, the first three digits indicate the mean of the manganese content and examples are shown in Table 3.4.

Free-cutting steels are now included in BS 970: Part 3: 1991 *Bright bars for general engineering purposes*. In these grades, the first three digits are in the range 200 to 240 and the second and third digits indicate the minimum or mean of the sulphur content multiplied by 100 (Table 3.5).

In the case of stainless steels, the designations reflect the well established AISI system. Thus the first three digits are in the 300 or 400 ranges, representing the

Table 3.4

Grade	C%	Mn%	Supplied to
060A62	0.6–0.65	0.5–0.7	Analysis only
080H46	0.43–0.5	0.6–1	Hardenability requirements
150M36	0.32–0.4	1.3–1.7	Mechanical properties

Table 3.5

Grade	C%	Mn%	S%	Supplied to
226M44	0.4–0.48	1.3–1.7	0.22–0.3	Mechanical properties

Table 3.6

Grade	C%	Cr%	Ni%
420S29	0.14–0.2	11.5–13.5	–
420S37	0.2–0.28	12–14	–
431S29	0.12–0.2	15–18	2–3
304S15	0.06 max.	17.5–19	8–11
310S31	0.15 max.	24–26	19–22

austenitic and martensitic/ferritic grades respectively. The interventing letter S also signifies a stainless steel but the last two digits, chosen arbitrarily in the range 11 to 99, indicate variants within the main type. Examples are shown in Table 3.6.

Quenched and tempered grades are no longer included in BS 970, following the publication of the following European standards:

BS EN 10083-1: 1991   *Quenched and tempered steels (special steels)*
BS EN 10083-2: 1991   *Quenched and tempered steels (unalloyed quality steels)*

In the former, a useful comparison is provided of grades in the new European standard and equivalent grades in previous national standards. This information is shown in Table 3.7.

For many years, BS 970 adopted a system whereby a single letter was used to specify a particular range of tensile strength in the quenched and tempered condition. This system is still widely used in the UK and the full range of letters and the associated tensile ranges are shown in Table 3.8.

Former versions of BS 970: Part 1 contained an appendix which provided a guide to the selection of Category 1 through hardening steels, based on tensile strength and limiting ruling section. This guide was prepared for applications where the most important criterion was the level of mechanical properties required in the finished part and with the aim of assisting in the selection of the most cost-effective grade for a given tensile strength and section size. Although many of the steel designations are now obsolete with the introduction of European standards, it was considered worthwhile to maintain this information in the present text and this is shown in Table 3.9. The first column in this table (*Heat treatment condition*) lists the various ranges of tensile strength according to the lettering system P–Z defined above. Along each row of the table, the ruling section increases and the various grades of steel that will satisfy the required combination of tensile strength and section size are identified. Thus a simple C–Mn steel (080M30) is recommended for an application calling for a tensile strength of 625–775 N/mm^2 (Q condition) in small section sizes. On the other hand, a 2.5% Ni–Cr–Mo steel (826M40) is recommended when a tensile strength of 1075–1225 N/mm^2 (W condition) is required in section sizes greater than 150 mm.

Table 3.7 Comparison of steel grades specified in BS EN 10083-1: 1991, ISO 683-1: 1991, ISO 683-1 and other steel grades previously standardized nationally

EN 10083-1	ISO 683-1: 1987[1]	Germany[1] Alpha-numeric Name	Germany[1] Material number	Finland	United Kingdom[1]	France[1]	Sweden SS steel	Spain Name	Spain Number
2 C 22	–	(Ck 22)	(1.1151)	–	(070M20)	[XC 18]	–	–	–
3 C 22	–	(Cm 22)	(1.1149)	–	–	[XC 18 u]	–	–	–
2 C 25	(C 25 E4)	Ck 25	1.1158	–	(070M26)	[XC 25]	–	C25K	F1120
3 C 25	(C 25 M2)	Cm 25	1.1163	–	–	[XC 25 u]	–	C25K-1	F1125(1)
2 C 30	(C 30 E4)	Ck 30	1.1178	–	(080M30)	[XC 32]	–	–	–
3 C 30	(C 30 M2)	Cm 30	1.1179	–	–	[XC 32 u]	–	–	–
2 C 35	(C 35 E4)	Ck 35	1.1181	C 35	(080M36)	[XC 38 H 1]	1572	C35K	F1130
3 C 35	(C 35 M2)	Cm 35	1.1180	–	–	[XC 38 H 1 u]	–	C35K-1	F1135(1)
2 C 40	(C 40 E4)	Ck 40	1.1186	–	(080M40)	[XC 42 H 1]	–	–	–
3 C 40	(C 40 M2)	Cm 40	1.1189	–	–	[XC 42 H 1 u]	–	–	–
2 C 45	(C 45 E4)	Ck 45	1.1191	C 45	(080M46)	[XC 48 H 1]	1672	C45K	F1140
3 C 45	(C 45 M2)	Cm 45	1.1201	–	–	[XC 48 H 1 u]	–	C45K-1	F1145(1)
2 C 50	(C 50 E4)	Ck 50	1.1206	–	(080M50)	–	1674	–	–
3 C 50	(C 50 M2)	Cm 50	1.1241	–	–	–	–	–	–
2 C 55	(C 55 E4)	Ck 55	1.1203	–	(070M55)	[XC 55 H 1]	–	C55K	F1150
3 C 55	(C 55 M2)	Cm 55	1.1209	–	–	[XC 55 H 1 u]	–	C55K-1	F1155(1)
2 C 60	(C 60 E4)	Ck 60	1.1221	–	(070M60)	–	–	–	–
3 C 60	(C 60 M2)	Cm 60	1.1223	–	–	–	–	–	–
28 Mn 6	(28 Mn 6)	28 Mn 6	1.1170	–	(150M19)	–	–	–	–
38 Cr 2	–	38 Cr 2	1.7003	–	–	(38 C 2)	–	–	–
38 CrS 2	–	38 CrS 2	1.7023	–	–	(38 C 2 u)	–	–	–
46 Cr 2	–	46 Cr 2	1.7006	–	–	–	–	–	–
46 Cr S 2	–	46 CrS 2	1.7025	–	–	–	–	–	–

34 Cr 4	34 Cr 4	1.7033	–	(530M32)	(32 C 4)	–	–	–
34 CrS 4	34 CrS 4	1.7037	–	–	(32 C 4 u)	–	–	–
37 Cr 4	37 Cr 3	1.7034	–	(530M36)	(38 C 4)	–	38 Cr4	F1201
37 CrS 4	37 CrS 4	1.7038	–	–	(38 C 4 u)	–	38 Cr4-1	F1206(1)
41 Cr 4	41 Cr 4	1.7035	–	(530M40)	42 C 4	–	42 Cr4	F1202
41 CrS 4	41 CrS 4	1.7039	–	–	42 C 4 u	2245	42 Cr4-1	F1207(1)
25 CrMo 4	25 CrMo 4	1.7218	25 CrMo 4	(708M25)	25 CD 4	2225	–	–
25 CrMoS 4	25 CrMoS 4	1.7213	–	–	25 CD 4 u	–	–	–
34 CrMo 4	34 CrMo 4	1.7220	34 CrMo 4	(708M32)	(34 CD 4)	2234	–	–
34 CrMoS 4	34 CrMoS 4	1.7226	–	–	(34 CD 3 u)	–	–	–
42 CrMo 4	42 CrMo 4	1.7225	42 CrMo 4	(708M40)	42 CD 4	2244	40 CrMo4	F1252
42 CrMoS 4	42 CrMoS 4	1.7227	–	–	42 CD 4 u	–	40 CrMo4-1	F1257(1)
50 CrMo 4	50 CrMo 4	1.7228	–	(708M50)	–	–	–	–
36 CrNiMo 4	36 CrNiMo 4	1.6511	–	(817M37)	–	–	–	–
34 CrNiMo 6	(34 CrNiMo 6)	(1.6582)	34 CrNiMo 6	(817M40)	–	2541	–	–
30 CrNiMo 8	(31 CrNiMo 8)	1.6580	30 CrNiMo 8	[823M30]	30 CND 8	–	–	–
36 NiCrMo 16	–	–	–	[835M30]	35 NCD 16	–	–	–
51 CrV 4	[51 CrV 4]	1.8159	50 Cr V 4	[735A50]	(50 CV 4)	–	51 CrV4	F1430

Notes:

[1]If a steel grade is given in round brackets, this means that the chemical composition differs only slightly from EN 10083-1. If it is given in square brackets, this means that greater differences exist in the chemical composition compared with EN 10083-1. If there are no brackets around the steel grade, this means that there are practically no differences in the chemical composition compared with EN 10083-1.

After BS EN 10083-1: 1991.

Table 3.8

Reference symbol	Tensile strength (N/mm^2)
P	550–700
Q	625–775
R	700–850
S	775–925
T	850–1000
U	925–1075
V	1000–1150
W	1075–1225
X	1150–1300
Y	1225–1375
Z	1550 min.

Steel prices

Because of the very large number of steel grades covered by BS 970, one can only be very selective in attempting to provide an introduction to the cost structure of engineering steels. Purely on an arbitrary basis, the selection was narrowed down to some of the grades in the 19–29 mm ruling selection column in the Category 1 steels tabulation (Table 3.9). However, this selection was augmented by one further grade, namely 826M40, which is recommended for the attainment of the highest strength (1075–1225 N/mm^2) in the largest ruling section (150–250 mm) in this table.

The prices of the six steels concerned, in the form of large billets, are shown in Figure 3.23. This indicates a steady increase in price which reflects the progressive addition of alloying elements such as chromium, molybdenum and nickel.

Machinable steels

Machining is an important stage in the production of most engineering components and, in many automotive transmission parts, machining can account for up to 60% of the total production costs. It is not surprising therefore that the engineering industries have called for steels with improved and consistent levels of machinability, whilst still maintaining the other properties that ensure good service performance. The common machining processes include turning, milling, grinding, drilling and broaching and several of these operations might be carried out on an automatic lathe in the production of a single component. Each of these processes differs in terms of the metal cutting action and involves different conditions of temperature, strain rate and chip formation. Therefore the machining performance of a steel cannot be defined by means of a single parameter.

Table 3.9 *Guide to the selection of category 1 through hardening steels*

Heat treatment condition	Tensile strength range (N/mm²)[a]	HB range	Ruling Section						
			≤13 mm	>13 mm ≤19 mm	>19 mm ≤29 mm	>29 mm ≤63 mm	>63 mm ≤100 mm	>100 mm ≤150 mm	>150 mm ≤250 mm
Q	625–775	179–229	080M30	080M30	080M36	080M40 150M19	120M36	135M44 150M36 080M50[b]	708M40
R	700–850	201–255	212M36[c] 080M36 212M36[c] 226M44[c]	212M36[c] 080M40 216M44[c] 226M44[c]	212M36[c] 120M36 150M19 080M46[b] 216M44[c] 226M44[c]	212M36[c] 135M44 150M36 080M50[b] 216M44[c] 226M44[c]	080M46[b] 216M44[c] 135M44 530M40 070M55[b] 216M44[c] 606M36[c]	216M44[c] 708M40 605M36	708M40
S	775–925	223–277	120M36 080M46[b] 216M44[c] 226M44[c]	120M36 080M50[b] 216M44[c] 226M44[c]	135M44 150M36 080M50[b] 216M44[c] 226M44[c]	530M40 070M55[b] 606M36[c]	708M40 605M36	709M40	709M40

continued overleaf

Table 3.9 (*continued*)

Heat treatment condition	Tensile strength range (N/mm²)[a]	HB range	Ruling Section						
			≤13 mm	>13 mm ≤19 mm	>19 mm ≤29 mm	>29 mm ≤63 mm	>63 mm ≤100 mm	>100 mm ≤150 mm	>150 mm ≤250 mm
T	850–1000	248–302	135M44[d] 080M50[b,d] 216M44[c,d] 226M44[c,d]	135M44[d] 070M55[b,d]	530M40	708M40 605M36	709M40	817M40	817M40
U	925–1075	269–331	135M44[d]	606M36[c] 708M40 605M36	606M36[c] 708M40 605M36	709M40	817M40	720M32	720M32
V	1000–1150	293–352	708M40 605M36	708M40 605M36	709M40	817M40	720M32	720M32	826M40
W	1075–1225	311–375	708M40[d,e] 817M40[e]	709M40[d,e] 817M40[e]	817M40[e]	720M32[e]	720M32[e]	826M40[e]	826M40[e]

[a] 1 N/mm² = 1 MPa.
[b] No specified impact properties.
[c] Free cutting steel.
[d] Full mechanical properties may not always be obtained by bulk heat treatment but the properties can be achieved by the appropriate heat treatment of die forgings and components.
[e] Often ordered in the softened condition for machining and subsequent heat treatment.

After BS 970; Part 1: 1983

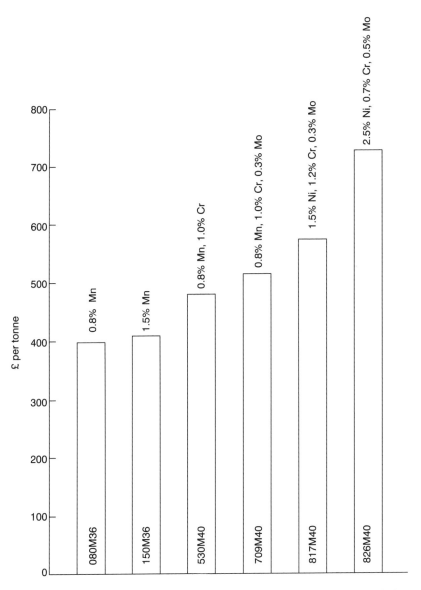

Figure 3.23 *Basis prices of engineering steel billets as of 23 September 1996*

However, some of the features that are often invoked as measures of machinability are:

- Tool life
- Production rate (e.g. components/hour)
- Power consumption
- Surface finish of component
- Chip form
- Ease of swarf removal

Following a brief description of machinability testing, the role of free cutting additives will be discussed. Attention will then be turned to the composition and machining characteristics of:

1. Low-carbon free cutting steels.
2. Medium-carbon steels.
3. Low-alloy steels.
4. Stainless steels.

Machinability testing

Many organizations have developed their own individual test methods for evaluating machinability and these can range from tests involving tool failure in only a few minutes of metal cutting to those simulating commercial practices and lasting several hours. However, reproducibility is of absolute importance and, regardless of the type of test employed, the cutting conditions such as *speed* (e.g. peripheral bar speed in m/min), *feed* (rate of travel of cutting tool in mm/rev) and *depth of cut* (metal removed per cut in mm) must be carefully controlled.

A widely adopted laboratory test for machinability is the Taylor Tool Life test in which the life of the tool is determined at various cutting speeds. As illustrated in Figure 3.24, plots of tool life (T) against cutting speed (V) provide a straight line giving the equation:

$$VT^n = C$$

where C and n are constants.

One particular parameter of tool life which features prominently in machining evaluations is V20, the cutting speed that will provide a tool life of 20 minutes. However, such short-term tests may not necessarily provide an accurate guide to the performance in longer term industrial machining operations.

Although simple turning tests such as those described above are useful in steel development or for quality control purposes, steel users often call for longer term tests involving multi-machining operations. Thus up to two tonnes of bright drawn bar stock might be consumed over a period of about seven hours in tests involving the production of the type of test piece shown in Figure 3.25. This involves turning, drilling, plunge cutting and parting operations and the cutting rates are adjusted so as to maintain a specific dimensional tolerance in the component over a simulated shift period. The machinability rating of a given steel sample can then be expressed in terms of components per hour or production time per component.

In each of the laboratory tests for machinability, rigorous attention must be given to maintaining standard characteristics in the cutting tools in terms of grade, hardness and tool geometry.

Role of free cutting additives

Various elements are added to steel in order to improve the machining performance. The main free cutting additives and their perceived action in improving machinability are as follows.

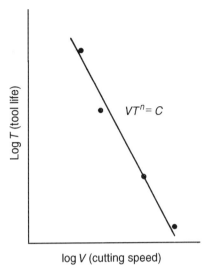

Figure 3.24   *Taylor tool life curve*

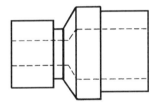

Figure 3.25   *Machinability test piece (By permission of British Steel plc)*

Sulphur

Sulphur is the cheapest and most widely used free cutting additive in steels. Whereas most specifications for engineering steels restrict the sulphur content to 0.05% max., levels up to about 0.35% S are incorporated in free cutting steels. Sufficient manganese is also added to such steels to ensure that all the sulphur is present as MnS rather than FeS, which causes *hot shortness* (cracking) during hot working. The MnS inclusions deform plastically during chip formation into planes of low strength which facilitate deformation in the primary shear zone. The MnS inclusions also exude into the tool–chip interface, acting as a lubricant and also forming a protective deposit on the tool. The net effect is a reduction in cutting forces and temperatures and a substantial reduction in the tool wear rate. This is illustrated in Figure 3.26, which relates to low-carbon free cutting steels and shows the marked decrease in flank wear rate with increase in the volume fraction of MnS inclusions. Although relatively little improvement in wear rate is achieved at sulphide volume fractions greater than 1.5%, the chip form and surface finish continue to improve with further additions of sulphur.

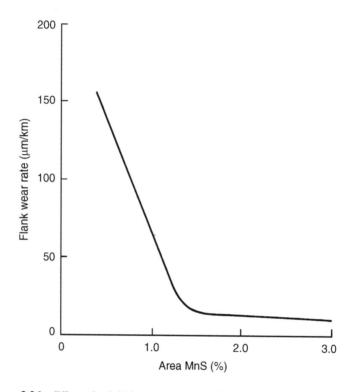

Figure 3.26   *Effect of sulphide content on machinability (After Wannel* et al.[16])

For many years, it has been postulated that the morphology of the MnS inclusions plays a major role in the machining performance of free cutting steels, large globular inclusions being far more effective than thin elongated inclusions. However, sulphide morphology is influenced markedly by the state of deoxidation of the steel, a heavily killed steel promoting the formation of MnS inclusions that are easily deformed into elongated inclusions during hot working. In addition, killed steels tend to contain hard, abrasive oxide inclusions which have an adverse effect on machinability. Therefore it is extremely difficult to differentiate between the effects that might be due to sulphide morphology and those that are clearly due to the presence of hard, abrasive oxide inclusions.

One of the disadvantages of high sulphur contents is that they impair the transverse ductility of steels, particularly when MnS is present as long elongated inclusions. As indicated later, inclusion-modifying agents can be added to high-sulphur steels in order to promote a more favourable sulphide morphology.

Lead

After sulphur, lead is the next most common additive and additions of 0.15–0.35% Pb are incorporated in free cutting steels. Such levels are soluble in molten steel but are precipitated as discrete particles of lead during solidification.

Due to its high density, major precautions have to be taken in the production of leaded steels in order to avoid segregation effects. Again lead reduces the frictional effects at the tool–chip interface, where it becomes molten at the elevated temperatures generated during cutting. Lead is also thought to have an embrittling effect in the primary shear zone, thereby shortening the chips and improving surface finish. The lead particles are often present as tails to the MnS inclusions.

Lead has little effect on the mechanical properties of steel at ambient temperature since it is generally present as a globular constituent.

The toxic effects of lead are well known and care has to be taken to contain lead fumes during the production of leaded steels. Similar measures are also taken during the drop forging of leaded steels but the author is not aware that hygiene problems arise during the machining of these steels.

Tellurium

Tellurium is an efficient but relatively expensive addition in free cutting steels and is generally restricted to a maximum level of 0.1%. In larger amounts, it leads to cracking during hot working. Tellurium is generally present as manganese telluride, which is a low melting point compound and should therefore act in a similar way to lead. The high surface activity of tellurium is also considered important in its action as a free machining additive. Small additions of tellurium, e.g. 0.01%, are also added to engineering steels in order to produce more globular MnS inclusions and promote better transverse properties.

Like lead, tellurium presents hygiene problems during steelmaking (*garlic breath*) and requires the operation of fume extraction systems.

Selenium

Selenium additions of 0.05–0.1% are made to low-alloy steels, but in free machining stainless steels many specifications call for a minimum of 0.15% Se. Selenium is present as a mixed sulphide–selenide and, like tellurium, small additions of this element are also effective in promoting more globular sulphide inclusions.

Bismuth

Bismuth is closely related chemically to lead and its free cutting properties have been known for many years. However, commercial interest in bismuth has only been significant since about 1980, as greater anxieties have been voiced about the use of lead. Levels of up to 0.1% Bi are typical in free cutting grades and bismuth is present as tails to the MnS inclusions. Its action in promoting improved machining characteristics appears to be similar to that of lead.

Calcium

As indicated earlier, oxide inclusions are hard and abrasive and detract from the machinability of steels. This is particularly the case with alumina inclusions which are formed in engineering steels due to the practice of adding aluminium as a deoxidant or grain-refining element. However, the adverse effects of alumina can be reduced with the addition of calcium, which results in the formation of calcium aluminate. These inclusions soften during high-speed machining and form protective layers on the surface of carbide tools. Calcium is also effective in reducing the projected length of MnS inclusions, thereby improving the transverse properties of resulphurized steels.

Low-carbon free cutting steels

For some engineering components, the mechanical property requirements are minimal and by far the most important requirement is a high and consistent level of machinability. Hose couplings and automotive spark plug bodies are examples of such components which are mass produced at high machining rates. Sulphur contents in the range 0.25–0.35% are typical but, as illustrated in Figure 3.27, substantial improvements in machinability are obtained by the addition of lead to resulphurized steels. Where extremely high rates of machining are required, a low-carbon free cutting steel might be treated with sulphur, lead and bismuth, e.g. 0.25%, S, 0.25% Pb, 0.08% Bi.

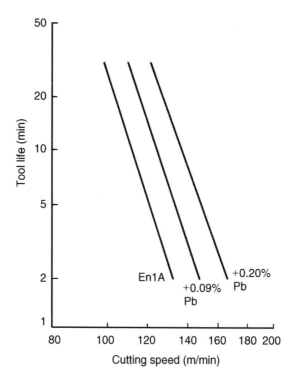

Figure 3.27   *Effect of lead on machinability (After Wannel et al.[16])*

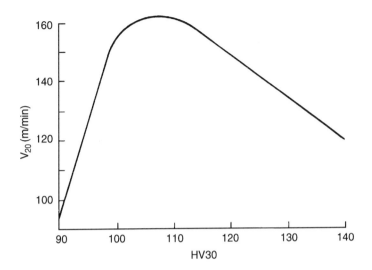

Figure 3.28 *Effect of hardness on machinability (After Wannel* et al.[16]*)*

The hardness of low-carbon free cutting steels is important and the effect appears to be associated with the achievement of an optimum embrittling effect for chip formation. This is illustrated in Figure 3.28, which indicates an optimum hardness of 105–110 HV in the normalized condition. Therefore the composition of these steels, including residual elements such as chromium, nickel and copper, needs to be controlled in order to achieve the optimum hardness. Additions of phosphorus and nitrogen may also be employed in achieving the desired embrittlement effects. Above the critical hardness level, embrittlement and ease of chip formation give way to increased abrasion and reduced tool life.

As indicated earlier, oxide inclusions are particularly damaging to machinability and the deoxidation practices adopted for low-carbon free cutting steels have to be controlled very carefully in order to minimize the level of these inclusions. With the move from ingot production to continuous casting, this problem has become more acute as the steels need to be more heavily deoxidized in order to minimize gas evolution during solidification.

Medium-carbon free cutting steels

Medium-carbon steels, containing 0.35–0.5% C and up to 1.5% Mn, are often used in the normalized condition for engineering components requiring a tensile strength of up to about 1000 N/mm^2. The free cutting versions of these steels are generally based on sulphur contents of 0.2–0.3% but, because of their higher strength, they are significantly harder to machine than low-carbon free cutting steels. Therefore benefit can be gained in selecting a grade of steel with the lowest level of carbon consistent with achieving the required strength in the end product.

Whereas silicate inclusions impair the machinability of low-carbon free cutting steels, they are not particularly damaging in medium-carbon steels when machined with carbide tools. This is due to the fact that higher temperatures are generated in the cutting of these steels, leading to the softening of the silicate inclusions. However, alumina particles are again particularly detrimental, reducing the machining performance of both plain and resulphurized medium-carbon steels.

Machinable low-alloy steels

In high-strength automotive transmission components, a high level of toughness and ductility may be required in the transverse direction. For this reason, free machining versions of low-alloy engineering steels were based traditionally on the more costly lead additions rather than sulphur. However, as indicated earlier, the addition of inclusion-modifying agents can reduce the anisotropy in mechanical properties and provide improved service performance in resulphurized steels. This is illustrated in Figure 3.29, which shows the improvement in transverse impact properties obtained in SAE 4140 steel[17] with calcium treatment. Bearing in mind that high-strength engineering steels are generally fine grained with aluminium, calcium treatment will also lead to an improvement in machinability with carbide cutting tools due to the formation of calcium aluminate rather than alumina.

Machinable stainless steels

Although out of context in terms of chapter heading, it was thought that benefit might be gained in continuity of technology by dealing with machinable stainless steels at this stage.

Austenitic stainless steels have high rates of work hardening and this results in poor machinability. However, this problem has been exacerbated to some extent

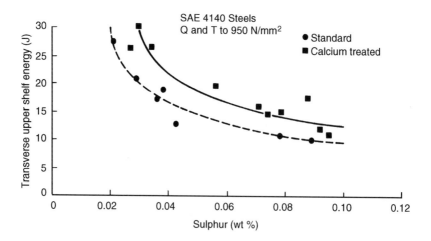

Figure 3.29   *Effect of calcium treatment on transverse upper shelf energy of SAE 4140 steels (After Pickett et al.[17])*

by modern steelmaking practices, such as arc-AOD, which produce low levels of sulphur, e.g. often less than 0.01%. To overcome this particular problem, the steels can be resulphurized to a level just below the maximum generally permitted in standard specifications, e.g. 0.03% max.

Like carbon and low-alloy steels, large additions of sulphur are also made to stainless steels in order to provide free cutting grades. Thus AISI 303 is an 18% Cr 9% Ni austenitic stainless steel containing 0.15% S min. However, the large volume fraction of MnS inclusions in this grade impairs the corrosion properties such that the performance is significantly worse than that of the standard 18% Cr 9% Ni (Type 304) stainless steel. Where this presents a significant problem, free cutting properties coupled with improved corrosion resistance can be obtained in Type 303Se, the 18% Cr 9% Ni grade containing 0.15% Se min.

Steels for gas containers

High-pressure gas cylinders have been in use since the 1870s when they were first introduced for the transportation of liquefied carbon dioxide for the aerated drinks industry. Since that time, their use has extended enormously to deal with the conveyance of a variety of *permanent gases*, such as air, argon, helium, hydrogen, nitrogen and oxygen, and other *liquefiable gases*, such as butane, propane, nitrous oxide and sulphur dioxide. Given the explosive, flammable or toxic nature of some of these gases, it is not surprising that the manufacture and utilization of gas containers are the subject of major scrutiny by the Home Office and also by international bodies. However, whilst still maintaining very high standards of safety, the use of higher strength steels has been permitted which has greatly increased the carrying efficiency of gas cylinders in terms of their gas capacity per weight.

In the UK, the relevant British Standard is BS 5045 *Transportable gas containers*, of which Parts 1 and 2 are the main sections for steel containers:

- Part 1: Specification for seamless gas containers above 0.5 litre water capacity
- Part 2: Steel containers up to 130 litre water capacity with welded seams

This topic is the subject of reviews by Irani[18] and Naylor.[19]

Steel compositions

BS 5045: Part 1: 1982 specifies four grades of steel for seamless cylinders, covering C, C–Mn, Cr–Mo and Ni–Cr–Mo compositions, and details are shown in Table 3.10. The most popular compositions are the C–Mn and Cr–Mo grades, which are hardened and tempered to provide a minimum yield strength value of 755 N/mm^2. However, as indicated later, the yield strength is restricted to lower levels for certain types of hydrogen containers.

In relation to welded containers, BS 5045: Part 2: 1989 permits the use of C or C–Mn steels but, as illustrated in Table 3.11, these grades provide minimum yield strength values in the range 215–350 N/mm^2, i.e. significantly lower than that attained in seamless containers.

Table 3.10 *Seamless steel gas containers to BS 5045: Part I: 1982*

Material	Code	C%	Si%	Mn%	Cr%	Mo%	Ni%	YS[a] (N/mm^2) min.	TS[a] (N/mm^2)
Carbon steels	M	0.15–0.25	0.05–0.35	0.4–0.9	–	–	–	250	430–510
	C	0.35–0.45	0.05–0.35	0.6–1	–	–	–	310	570–680
C–Mn steels	Mn	0.4 max.	0.1–0.35	1.3–1.7	–	–	–	445	650–760
	Mn H							755	890–1030
Cr–Mo alloy steel	CM	0.37 max.	0.1–0.35	0.4–0.9	0.8–1.2	0.15–0.25	0.5 max.	755	890–1030
Ni–Cr–Mo alloy steel	NCM	0.27–0.35	0.1–0.35	0.5–0.7	0.5–0.8	0.4–0.7	2.3–2.8	755	890–1030

[a]For containers for use in hydrogen trailer service, the minimum yield strength shall not exceed 680 N/mm^2 and the tensile strength shall be within the range 800–930 N/mm^2.

After BS 5045: Part I: 1982 Amendment No. 1, August 1986.

Table 3.11 *Welded steel gas containers to BS 5045: Part 2: 1989*

Chemical and physical properties	Type A	Type B	Type C	Type D	Type E	Type F	Type G
Carbon % max.	0.2	0.18	0.2	0.2	0.15	0.18	0.25
Silicon % max.	0.3	0.25	–	0.3	0.3	0.4	0.35
Manganese							
% min.	–	0.4	0.7	–	–	0.5	0.6
% max.	0.6	1.2	1.5	0.6	1.2	1.4	1.4
Phosphorus % max.	0.05	0.04	0.05	0.05	0.025	0.04	0.03
Sulphur % max.	0.05	0.04	0.05	0.05	0.03	0.04	0.045
Grain-refining elements % max.	–	–	– [a]	–	0.3[a]	–	0.7
Yield stress (N/mm^2) min.	215	275	310	200	350	285	250
Tensile strength (N/mm^2) min.	340	400	430	320	430	430	430
Tensile strength (N/mm^2) max.	430	490	585	420	650	510	550
Elongation % min.[b]							
$L_0 = 50$ mm	28	24	21	29	21	–	21
$L_0 = 5.65\sqrt{S_0}$	33	29	20	35	25	20	25

[a]Grain-refining elements are limited to: niobium 0.08%, titanium 0.2%, vanadium 0.2%, niobium plus vanadium 0.2%.
[b]Where any other non-proportional gauge lengths are used, conversions are in accordance with BS 3894: Part 1.
Note 1. L_0 is the original gauge length.
 S_0 is the original cross-sectional area.
Note 2. Type A and type C are equivalent to grades of BS 1449: 1962 (withdrawn); type E is equivalent to grade 43/35 of BS 1449: Part 1; type D is similar to Euronorm 120; type G is equivalent to type 151 grade 430 of BS 1501: Part 1.

After BS 5045: Part 2: 1989.

Design and manufacture

The manufacture of a seamless steel cylinder is shown schematically in Figure 3.30, which illustrates the high degree of metal forming involved in this complex hot-working operation. The process is all the more remarkable when it is realized that the production of seamless cylinders from thick plate or billets was developed in the 1880s and 1890s and has changed very little to the present day.

 The cylinder wall is the thinnest region of the vessel and is therefore subjected to the highest stresses. However, under fatigue conditions, the higher stresses are generated at the junction of the relatively thin wall and the thicker concave base. BS 5045: Part 1 details the following formulae for the calculation of minimum wall thickness in seamless containers:

$$t = \frac{0.3 \, (\text{test pressure}) \times (\text{internal diameter})}{7 \times f_e - (\text{test pressure})} \tag{1}$$

$$\text{or } t = \frac{0.3 \, (\text{test pressure}) \times (\text{external diameter})}{7 \times f_e - 0.4 \, (\text{test pressure})} \tag{2}$$

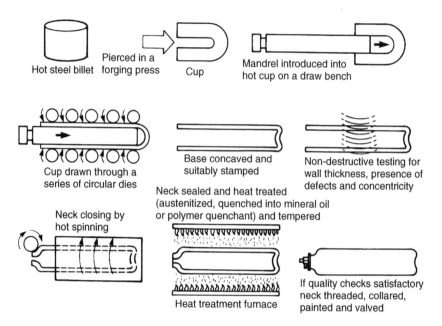

Figure 3.30   *The production of seamless steel gas cylinders (After Irani[18])*

where f_e is the maximum permissible equivalent stress at test pressure.

In general, f_e is equal to $0.75 \times$ minimum yield strength. However, a value of $0.875 \times$ minimum yield strength is allowed in special portable containers, e.g. those used in aircraft, underwater breathing apparatus and life raft inflation, thereby facilitating the use of lighter containers. For a Cr–Mo steel cylinder (230 mm OD), used for the transportation of oxygen at 200 bar (equivalent to a test pressure of 300 bar), equation (2) provides a minimum wall thickness t of 5.39 mm.[18]

BS 5045: Part 2 provides the same equations for the calculation of minimum thickness as those shown earlier for seamless containers (equations (1) and (2)). However, as indicated previously, the proof strength values of steels specified in Part 2 are significantly lower than those included in Part 1. Thus for a given test pressure, welded containers would require a greater thickness than seamless containers but, in effect, the former operate at lower pressures.

Rather strangely, BS 5045 Parts 1 and 2 make no reference to impact test requirements although Oldfield[20] reports that *leak-before-burst* philosophies are growing in acceptance and that fracture toughness tests are being developed for use with gas containers. This is particularly pertinent when handling aggressive gases, as illustrated in the following section.

Hydrogen gas containers

During the 1970s and early 1980s, almost one hundred failures occurred in Europe (20 in the UK) in gas containers that were used for the delivery of hydrogen by road trailer. Failure was due to the propagation of fatigue cracks from small manufacturing defects which were present at the inner surface of the cylinder, close to the highly stressed junction where the wall joins the concave base. The fatigue cracks propagated through to the outside wall of the cylinder, resulting in leakage of gas. The failed containers had been made by various manufacturers and conformed to the appropriate specifications at that time. Under the aegis of the Health and Safety Executive, a committee was established to investigate the failure mechanism and make appropriate recommendations. The results of this investigation are the subject of an interesting paper by Harris *et al.*[21]

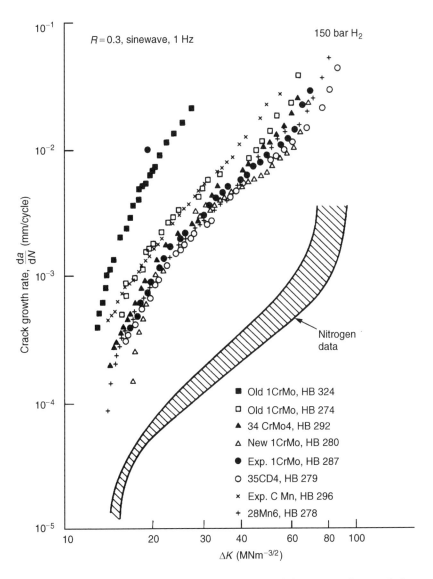

Figure 3.31 *The influence of hydrogen on the rate of fatigue crack growth in gas container steels (After Harris* et al.[21]*)*

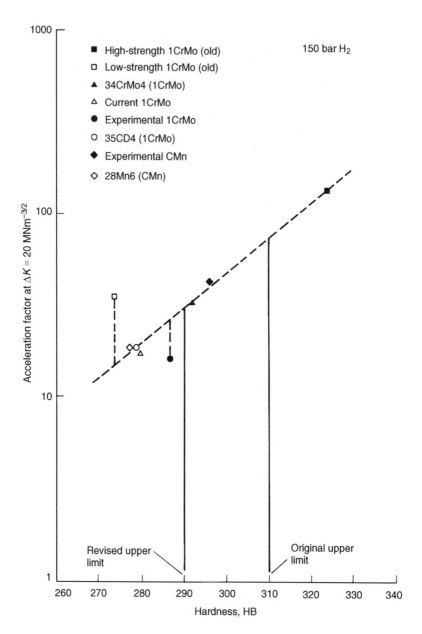

Figure 3.32 *The influence of hardness on the enhancement of fatigue crack growth by hydrogen (After Harris et al.[21])*

The problem containers in the UK were manufactured in Cr–Mo steel, which had been used for many years for the manufacture of storage containers without any serious problems. However, the transportable hydrogen containers differed from the storage containers in one important respect, namely the refilling frequency, which was the main source of fatigue cycling, the other being due to transportational stresses. Whereas the storage containers could experience up to 12 refills per annum, the transportable containers might be refilled twice daily. Harris *et al.* carried out fatigue crack growth tests on Cr–Mo and C–Mn steels in hydrogen gas at 150 bar and their results are shown in Figure 3.31. Whereas the bulk of the test data fell within a relatively narrow scatter band, the major exception was the Cr–Mo steel with the relatively high hardness of 324 HB. On the other hand, the fatigue crack growth of all the steels in hydrogen was very much greater than that induced in a nitrogen atmosphere. The acceleration factor by which the crack growth rate in hydrogen exceeds that in nitrogen at a stress intensity value of 20 $MNm^{-3/2}$ is shown as a function of hardness in Figure 3.32. This figure demonstrates very clearly the influence of the hardness (or strength) of the steel on the fatigue behaviour, the acceleration factor increasing from below 20 to above 100 as the hardness is increased from 274 to 324 HB.

Subsequent to the above work, amendments were incorporated into the standard for transportable gas containers and appear as Appendix E to BS 5045: Part 1 (Containers for use in hydrogen trailer service). This restricts the hardness range to 230–290 HB. In addition, the yield stress is not allowed to exceed 680 N/mm^2 and the tensile strength must be within the range 800–930 N/mm^2.

Higher strength steels

As indicated earlier, the carrying capacity of gas containers has been increased progressively over the years and this has been achieved primarily through the use of steels with progressively higher yield strength. This trend is continuing and Naylor[19] describes development work on gas container steels with minimum yield strength values of 950 N/mm^2. These higher strength values are achieved with micro-alloying additions of vanadium and also with higher levels of molybdenum or silicon. Naylor reports that cylinder-manufacturing trials were being carried out on these experimental compositions.

Bearing steels

Bearings constitute vital components in most items of machinery, permitting accurate movement under low frictional conditions. In addition, they are also required to transmit high loads and provide long service lives under arduous fatigue conditions. Stemming mainly from the requirement of the aeroengine industry, major effort has been devoted to improving the level and consistency of bearing fatigue performance and, with the adoption of cleaner steelmaking techniques, it is claimed that bearing life has increased by a factor of 100 since the early 1940s.[22]

The grade of steel adopted internationally for through-hardened bearings is SAE 52100, the 1.0% C 1.5% Cr composition. This material is generally solution treated at a temperature of about 850°C, followed by oil quenching and tempering in the range 180–250°C. This results in a microstructure of lightly tempered martensite, primary (undissolved) carbides and up to about 5% retained austenite. For larger bearings, carburizing grades such as SAE 4720 (2% Ni–Mo) are adopted. In the United States, M50 (0.8% C, 4.0% Cr, 4.25% Mo) is used extensively in main shaft gas turbine bearings, helicopter transmission bearings and other aerospace applications. In 1983, Bamberger[23] introduced a variant of M50 steel which was designated M-50 NiL. This is a 0.12% C 3.5% Ni 4.0% Cr 4.25% Mo 1.2% V composition which is case hardened to produce bearings with high fracture toughness and long life.

Bearing fatigue testing

Whereas bearings are required to have long endurance lives, perhaps extending over several years, laboratory evaluation tests must be accelerated such that a meaningful result can be obtained in a matter of days. As such, these tests are undertaken at higher loads and speeds than those experienced under typical service conditions. Various bearing fatigue tests have been developed which are capable of assessing the performance of bearing steels in the form of balls, washers, cones and cylinders but in the UK the Unisteel washer test[24] is still in operation following its introduction in the early 1950s. A sectional view of the machine is shown in Figure 3.33. The test washer measures 76 mm o.d. × 51 mm i.d. × 5.5 mm thick and forms the top race of a standard thrust bearing. The cage of the standard bearing is operated with only nine balls instead of the normal 18 in order to provide a maximum calculated Hertzian stress of 3725 N/mm^2 at a relatively low load. Both sides of the specimen are tested.

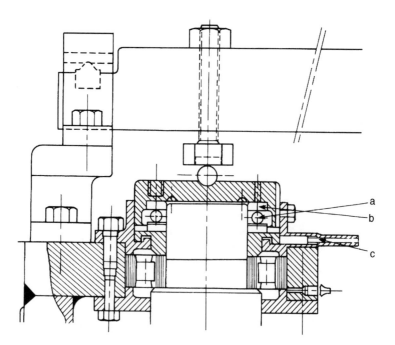

Figure 3.33 *Sectional view of Unisteel Bearing Fatigue Rig (a) balls; (b) test bearing; (c) standard thrust race (After Johnson and Sewell[24])*

Factors affecting fatigue performance

Johnson and Sewell[24] were among the earliest investigators to establish a good quantitative relationship between the bearing fatigue performance of SAE 52100 and inclusion content. The inclusions act as stress raisers, forming incipient cracks which then propagate under stress reversals until a fatigue pit (*or spall*) is formed on the surface of the component. As illustrated in Figure 3.34, Johnson and Sewell showed that oxide inclusions such as alumina and silicates have an adverse effect on fatigue performance, whereas sulphides appear to be beneficial. These authors showed that a better correlation between fatigue performance and inclusion content was obtained when titanium nitride was included in the inclusion count such that the inclusion parameter was based on:

$$\text{Number of alumina} + \text{silicate} + \frac{\text{TiN}}{2}$$

Oxides are considered to be detrimental because they are brittle and, as illustrated by Brooksbank and Andrews,[25] they also become surrounded by tensile stresses on cooling from elevated temperatures due to differences in the thermal expansion characteristics between the oxide particles and the matrix. The beneficial effect of sulphides is generally ascribed to the fact that they tend to encapsulate the more angular oxide inclusions, thereby reducing the detrimental tensile stresses. However, the beneficial effects of sulphides are often disputed and some hold the view that they are non-detrimental rather than positively beneficial. Other workers have indicated that TiN inclusions have a relatively small effect on bearing fatigue performance and such an effect is consistent with the tesselated stress calculations of Brooksbank and Andrews.

Given the importance of inclusion content in relation to bearing fatigue performance, major attention has been given to the development of reliable methods of

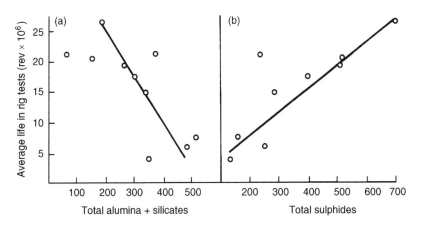

Figure 3.34 *Relationship between average life and inclusion content: counts based on total inclusions observed (×750) in 516 fields representing a total area of ≃9 mm² (After Johnson and Sewell[24])*

inclusion assessment. For many years, both the steelmakers and bearing manufacturers have employed the Jerkontoret (JK) method which is specified in ASTM Practice for Determining the Inclusion Content of Steel–E45. This rates a steel in terms of the worst field of the 100 fields that are examined against standard charts. However, as stated by Hampshire and King,[26] this technique is capable of ignoring the fact that a steel with 100 equally bad fields would be ranked the same as a steel with only one bad field. Therefore a new method was required that took into account both inclusion severity and frequency and such a procedure is the SAM counting technique. This is detailed in Supplementary Requirements S2 of ASTM A 295-84. The SAM procedure concentrates on the frequency of *Type B* aluminates and *Type D* globular oxides because these tend to be the most damaging inclusions to bearing fatigue performance. *Type A* manganese sulphides are disregarded as being insignificant to fatigue life and *Type C* silicates are ignored because they occur infrequently in bearing steels.

The SAM count is regarded as a major improvement over the JK method but major strides have been made in introducing cleaner steels and a large number of fields must now be examined in order to obtain a statistically valid assessment of inclusion content. Attention is being given to the use of automatic image analysis techniques but these require very careful polishing procedures and, again, problems are being experienced with these techniques in the accurate assessment of steels with low oxygen and sulphur contents.

Relatively little work has been carried out on the role of primary carbides in bearing steels. In general, it is accepted that these particles are very beneficial in increasing the wear resistance of the material and they are also instrumental in inhibiting grain growth during solution treatment. However, it is considered that large primary carbides, particularly when present in segregated bands, act as stress raisers and are detrimental to fatigue life.

The high carbon content of through-hardening steels such as SAE 52100 depresses the M_f temperature of the material below room temperature, resulting in the formation of retained austenite. A similar situation also exists in the case of carburized bearings, the high alloy content augmenting the effect of carbon in depressing the $M_s - M_f$ transformation range. However, the retained austenite will transform to martensite under the high Hertzian stresses induced in service and will tend to introduce dimensional changes in the bearing component. For this reason, a low level of retained austenite is desirable for critical bearing applications and is generally limited to a maximum of about 5%. The formation of strain-induced martensite will create a compressive residual stress and should therefore be beneficial to the fatigue performance. The *white bands* observed in the microstructure of SAE 52100 bearings after service are thought to be associated with the residual stresses created by the formation of strain-induced martensite.

Modern steelmaking methods

During the 1950s, vacuum degassing facilities were introduced which resulted in a marked improvement in the cleanness of bearing steels. By the early 1960s, argon shrouding of the molten stream was adopted which prevented reoxidation, leading to a further reduction in the non-metallic inclusion content. In the mid-1970s, many steelmakers introduced vacuum induction melting and vacuum arc remelting (VIM, VAR) to produce exceptionally clean steels. However, dramatic improvements in cleanness have also been obtained in bulk steelmaking processes for bearing grades, particularly with the introduction of secondary steelmaking facilities. Davies *et al.*[27] have described the facilities that have been installed at Stocksbridge Engineering Steels for the production of bearing steels and these are summarized in Table 3.12. Steels from a 100 t electric arc furnace are processed via a vacuum arc degassing (VAD) unit while melts from a 150 t furnace are transferred to a ladle furnace (LF), equipped with argon stirring, arc heating and alloying facilities. Care is also taken to minimize the inclusion content with the use of high alumina and graphitized magnesia refractories.

The SAM count rating for alumina (Type B) and globular oxides (Type D) on casts of SAE 52100, produced by Stocksbridge Engineering Steels between 1984 and 1986, have been monitored by Hampshire and King.[26] These data are shown in Figure 3.35. The major decrease in Type D inclusions in April 1985 coincides with the introduction of the ladle steelmaking facility and it is reported that individual Type D SAM counts since that time have rarely exceeded a value of 3. On the other hand, the introduction of ladle steelmaking did not have a dramatic effect on the incidence of Type B alumina inclusions, the SAM count being between 3 and 8 over the period shown. Mean oxygen contents of less than about 10 ppm can now be obtained via the ladle refining route, i.e. approaching that produced by vacuum arc remelting.

Table 3.12 *Facilities for the production of clean steels at Stocksbridge Engineering Steels*

Process route element	B Furnace route	C Furnace route
Slag-free tapping	Submerged taphole	Submerged taphole
Ladle stirring	Argon through sidewall plug	Argon through sidewall and bottom plugs
Slag control	Specific synthetic slags	Specific synthetic slags
Reheating	Arc heating in VAD	Arc heating in LF
Degassing	Integral part of VAD	Separate tank degasser
Alloying	via 13 alloy hoppers	via 13 alloy hoppers
Ladle refractories	75–85% Al_2O_3 walls and bottom 92% MgO slag line	82–85% Al_2O_3 walls and bottom 96% MgO slag line
Ingot teeming shroud	Inert gas	Inert gas
Ingot holloware	60% Al_2O_3	60% Al_2O_3
CC ladle to tundish[a]	Graphitized Al_2O_3 shroud	Not applicable
CC tundish	700-mm deep, MgO-lined	Not applicable
CC tundish to mould	Graphitized Al_2O_3 submerged entry shroud and flux, or inert gas	Not applicable

[a]CC = continuous caster.
After Davies *et al.*[27]

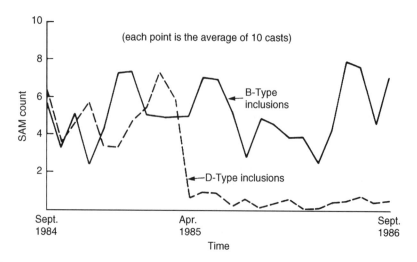

Figure 3.35   *SAM counts of steel delivered from major steel supplier (After Hampshire and King[26])*

High-speed steels

High-speed steels are so called because of their ability to maintain a high level of hardness during high-speed machining operations. They are characterized by high carbon contents, sometimes up to 1.5%, and major additions of strong carbide-forming elements such as chromium, molybdenum, tungsten and vanadium. Up to 12% Co is also included in some of the more complex grades. The constitution, manufacture and properties of high-speed steels have been reviewed very thoroughly in a book by Hoyle.[28]

Role of alloying elements

The microstructure of high-speed steels consists of primary and secondary carbides in a matrix of tempered martensite. The primary carbides are coarse particles which are not dissolved during solution treatment for the hardening operation and the secondary carbides are fine particles that are precipitated in martensite during tempering treatments. The role of the major alloying elements is outlined below.

A high carbon content is required in order to produce a hard martensitic matrix and also to form primary carbides. Both constituents are effective in providing abrasion resistance during metal cutting. However, the amount of carbon that can be accommodated in high-speed steels is limited on two counts. Firstly, carbon lowers the solidus temperature of steels substantially and therefore reduces the maximum temperature that can be employed for solution treatment prior to hardening. Secondly, carbon has a powerful effect in depressing the M_s–M_f temperature range and a very high level of carbon in solution will lead to excessive amounts of retained austenite.

The chromium content of most grades of high-speed steel is about 4%. Chromium forms carbides such as $M_{23}C_6$ and M_7C_3 but these carbides dissolve fairly readily and are taken into solution at the normal solution treatment temperatures employed for these steels, i.e. 1200–1300°C. Chromium is therefore added as a hardenability agent and promotes the formation of martensite. In addition, chromium is also beneficial in improving the scaling resistance of these steels at the high temperatures generated during machining operations.

Molybdenum and tungsten are the principal alloying elements in high-speed steels and can be present in levels up to about 10% and 20% respectively. From a metallurgical standpoint, the two elements are very similar and in most types of steel they are interchangeable on the basis of atomic weight %. Both elements are strong carbide formers and the principal carbide type is M_6C. This carbide is sometimes called eta-carbide and has a low solid solubility in steel. Therefore molybdenum and tungsten contribute little to the hardenability of high-speed steels but the small amounts that are dissolved are very effective in promoting tempering resistance and maintaining a high level of hardness at the cutting temperature (*red hardness*). Both elements also induce secondary hardening reactions in martensite, as illustrated schematically in Figure 3.36. This figure also indicates that appreciable softening does not take place until the temperature exceeds about 550°C, which therefore represents the effective maximum operating temperature for these steels.

In the early development of high-speed steels, tungsten was regarded very much as the preferred element and molybdenum the substitute element to be tolerated only in special circumstances. However, this prejudice has now largely disappeared and there appears to be no technical reason for recommending a tungsten grade in favour of a comparable molybdenum grade.

All high-speed steels contain between 1% and 5% V and as the vanadium content is increased, the carbon is generally increased by at least 0.1% for each additional 1% V. Vanadium is a very strong carbide-forming element and produces extremely hard particles of V_4C_3 (MC type). This carbide has a very low solid solubility in steel and again contributes very little to hardenability.

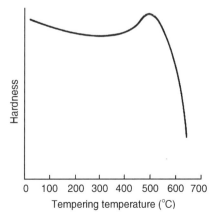

Figure 3.36 *Tempering curve for high-speed steel (schematic) (After Hoyle[28])*

However, vanadium carbide improves the abrasion resistance of high-speed steels and is also beneficial as a grain-refining agent.

Although not a standard addition, up to 12% Co is incorporated in some high-speed steels and, according to Hoyle,[28] the role of this element is less clearly defined than that of the other major additions. Cobalt does not form carbides and has a high solubility in austenite. For a given hardening temperature, it reduces the level of retained austenite and also improves the secondary hardening response. Cobalt also increases the thermal conductivity, particularly at high temperatures. The net result is that cobalt increases the red hardness of high-speed steels and improves their performance in fast-cutting operations.

Heat treatment

Given the high carbon content and complex alloy design, it might be anticipated that the heat treatment of high-speed steels would involve rather complicated procedures. However, these procedures are based on sound metallurgical principles and are designed to achieve specific microstructural features and properties. A schematic illustration of the heat treatment of high-speed steels, including the annealing, hardening and tempering cycles, is shown in Figure 3.37.

Because of their high alloy content, high-speed steels are air hardening and will form martensite on cooling from the austenite temperature range. Therefore, after forging or hot-forming operations, these steels must be annealed in order to produce:

1. A softened condition for easier machining operations.
2. The relief of internal stresses.
3. A suitable microstructure for the subsequent hardening treatment.

Annealing is carried out by heating slowly to a temperature just above Ac_1 which involves temperatures in the range 850–900°C, depending upon the particular grade of steel. The material is held at the annealing temperature for two to four hours and then furnace cooled to a temperature below 600°C. The rate

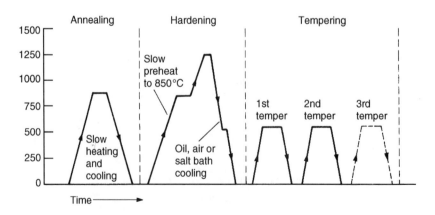

Figure 3.37   *The heat treatment of high-speed steels*

of cooling can then be increased. This treatment results in the formation of a ferrite matrix with finely dispersed carbide particles. However, the large primary carbides remain virtually unaffected by this treatment. After annealing, the hardness of the standard grades of high-speed steel is less than 300 HB.

As indicated earlier, elements such as molybdenum, tungsten and vanadium form stable carbides which have only limited solubilities in steel. However, in order to produce a high-carbon martensite, with good tempering resistance and the facility for secondary hardening, it is essential that a proportion of these carbides is taken into solution. Solution treatment temperatures very close to the solidus temperatures are therefore employed, i.e. temperatures in the range 1200–1300°C, depending upon the grade. According to Hoyle,[28] the solution of $M_{23}C_6$ (Cr-based) carbide begins at temperatures just above 900°C and is complete at about 1100°C. The solution of M_6C (Mo- and W-based) carbides begins at about 1150°C and continues until the solidus is reached. On the other hand, MC (V-based) carbide is extremely stable and little solution is achieved, even at temperatures close to the solidus.

In order to minimize thermal shock, the steel is preheated slowly to a temperature of about 850°C. This is generally carried out in one furnace and the steel is then transferred to a high-temperature furnace with a neutral atmosphere. However, the soaking time at the hardening temperature must be short, e.g. two to five minutes, in order to minimize decarburization and grain growth.

Following solution treatment, the cooling operation can be carried out in air or by quenching into oil or a salt bath. Salt bath temperatures of 500–600°C are employed and the treatment is carried out in order to reduce temperature gradients and thereby reduce distortion and the risk of cracking. However, the salt bath treatment should only be long enough to allow the material to attain a uniform temperature and the component should then be air cooled. Transformation to martensite will begin at a temperature below 200°C and will be about 80% complete at room temperature. Whereas it is important to ensure that the steel reaches ambient temperature in order to achieve a high degree of transformation, tempering treatments must take place immediately after the hardening cycle in order to prevent stabilizing effects in the residual retained austenite.

Tempering treatments are carried out at 530–570°C and serve two purposes:

1. To produce a tempered martensitic structure which will be hard but stable at elevated temperatures.
2. To destabilize the retained austenite such that martensite will form on cooling to room temperature.

As indicated in Figure 3.36, a secondary hardening reaction takes place with a peak at a temperature of about 550°C, i.e. the typical tempering temperature for high-speed steel. Therefore, as well as providing a structure which will be stable in the short term up to this temperature, the tempering treatment also produces the maximum level of hardness. During the first tempering treatment, carbide precipitation takes place in both the martensite and retained austenite that were present in the structure after the hardening treatment. In the latter case, carbide precipitation depletes the phase in carbon and alloying elements, such that the M_s-M_f temperature range is raised and the retained austenite transforms to martensite on cooling to room temperature. Therefore at the end of the first tempering treatment, the microstructure will consist of:

1. Primary (undissolved) carbides;
2. Tempered martensite containing secondary carbides;
3. Untempered martensite;

and a second tempering treatment is required in order to temper the newly formed martensite. The second tempering treatment is again carried out at 530–570°C and for most grades this achieves the required microstructure, i.e. primary carbides in a matrix of tempered martensite. However, for optimum performance and maximum dimensional stability, it may be necessary to carry out a third tempering treatment, particularly in high-cobalt grades.

Standard specifications and uses

In the UK, high-speed steels are specified in BS 4659: 1989 *Tool and die steels*, which also includes details of hot-work tool steels, cold-work tool steels and plastic moulding grades. The specified ranges for the main alloying elements in high-speed steels, together with the hardness requirements in the annealed and hardened conditions, are summarized in Table 3.13. The grade designations adopted for the various types of tool steel in this specification follow the system laid down by the American Iron and Steel Institute (AISI) except that in all cases they are preceded by the letter 'B'. Thus grade T1 in the American standard is designated BT1 in the UK standard. In both cases, the letter 'T' refers to grades in which tungsten is the principal alloying element and the letter 'M' is used for molybdenum or tungsten–molybdenum grades. In the German system, the designations have the prefix 'S' (Schnellstahl) and the numbers that follow represent the levels of tungsten, molybdenum, vanadium and cobalt in that order. The chromium content is not specified but is most likely to be of the order of 4%. Thus M2/BM2 (6% W, 5% Mo, 4% Cr, 2% V) in the American and British standards have the same alloy design as S6-5-2 in the German standard.

Although 13 grades of high-speed steels are listed in BS 4659: 1989, the three popular grades of high-speed steel are listed in Table 3.14.

In the United States, M10 (8% Mo, 4% Cr, 2% V) is also a common grade and with the three steels listed above makes up nearly 90% of the general purpose high-speed tools. However, an M10-type steel is not listed in the British standard.

Hoyle[28] has prepared a basic guide to tool selection and this is shown in Table 3.15. This selection reflects the benefit of cobalt additions in maintaining a high level of red hardness at high cutting speeds and the use of higher carbon and vanadium contents in order to increase the resistance to abrasive wear when cutting very hard materials.

Although high-speed steels constitute the major type of cutting tool, sintered carbides now feature very prominently in this market and ceramic materials, such as alumina and cubic boron nitride, are also being used for fast machining operations or for very hard workpiece materials.

Table 3.13 High-speed steels – BS 4659: 1989. Details of composition and hardness

Designation	C%	Cr%	Mo%	W%	V%	Co%	Hardness after annealing HB (max.)	Hardness after heat treatment HV (min.)
BM1	0.75–0.85	3.75–4.5	8–9	1–2	1–1.25	1 max.	241	823
BM2	0.82–0.92	3.75–4.5	4.75–5.5	6–6.75	1.75–2.05	1 max.	248	836
BM4	1.25–1.4	3.75–4.5	4.25–5	5.75–6.5	3.75–4.25	1 max.	255	849
BM15	1.45–1.6	4.5–5	2.75–3.25	6.25–7	4.75–5.25	4.5–5.5	277	869
BM35	0.85–0.95	3.75–4.5	4.75–5.25	6–6.75	1.75–2.15	4.6–5.2	269	869
BM42	1–1.1	3.5–4.25	9–10	1–2	1–1.3	7.5–8.5	269	897
BT1	0.7–0.8	3.75–4.5	0.7 max.	17.5–18.5	1–1.25	1 max.	255	823
BT4	0.7–0.8	3.75–4.5	1 max.	17.5–18.5	1–1.25	4.5–5.5	277	849
BT5	0.75–0.85	3.75–4.5	1 max.	18.5–19.5	1.75–2.05	9–10	290	869
BT6	0.75–0.85	3.75–4.5	1 max.	20–21	1.25–1.75	11.25–12.25	302	869
BT15	1.4–1.6	4.25–5	1 max.	12–13	4.75–5.25	4.5–5.5	290	890
BT21	0.6–0.7	3.5–4.25	0.7 max.	13.5–14.5	0.4–0.6	1 max.	255	798
BT42	1.25–1.4	3.75–4.5	2.75–3.5	8.5–9.5	2.75–3.25	9–10	277	912

After BS 4659: 1989.

Table 3.14

Grade	C%	W%	Mo%	Cr%	V%
T1	0.75	18		4	1
M1	0.8	2	8	4	1
M2	0.85	6	5	4	2

Table 3.15   *Selection of high-speed steels*

Application	Grade	Composition (%)					
Normal duty	M1	0.8C,	2W,	8Mo,	4Cr,	1V	
	M2	0.85C,	6W,	5Mo,	4Cr,	2V	
	T1	0.75C,	18W,		4Cr,	1V	
Higher speed cutting	M35	0.9C,	6W,	5Mo,	4Cr,	2V,	5Co
	T4	0.75C,	18W		4Cr,	1V,	5Co
	T6	0.8C,	20W,		4Cr,	1.5V,	12Co
Higher hardness	M15	1.5C,	7W,	3Mo,	5Cr,	5V,	5Co
	M42	1.3C,	9W,	3Mo,	4Cr,	3V,	10Co

Maraging steels

The term *maraging* relates to ageing (precipitation strengthening) reactions in very low carbon martensitic steels for the development of ultra-high strengths, i.e. 0.2% proof strength values of 1400–2400 N/mm^2. Maraging steels are characterized by high nickel contents and a very important feature is that they exhibit substantially higher levels of toughness than conventional high-carbon martensitic grades of equivalent strength.

Work on these steels began in the United States in the 1950s at the Bayonne Research Laboratory of the International Nickel Company and was directed primarily to the development of a high-strength material for submarine hulls. However, maraging steels proved unsuitable for this application and their main usage has been in the areas of aerospace, tooling and machinery, and structural engineering. An excellent summary of information on these steels is contained in a publication by the American Society for Metals.[29]

Metallurgy

Maraging steels generally contain about 18% Ni and the carbon contents are limited to 0.03% max. Typically, they are solution treated at a temperature of 820°C and air cooled to room temperature. This results in the formation of a martensitic structure, even in large section sizes due to the high hardenability effect conferred by the high nickel content and other alloying elements. In the solution-treated condition, the low-carbon martensitic structures provide the following range of properties:

- 0.2% proof strength, 790–830 N/mm^2
- tensile strength, 1000–1150 N/mm^2
- elongation, 17–19%
- reduction of area, 27–35%

These steels also contain substantial amounts of cobalt and molybdenum and smaller additions of titanium which promote age-hardening reactions when the solution-treated steels are aged at temperatures of about 480°C. In the 18% Ni–Co–Mo–Ti grades, the main precipitation-strengthening phases are Ni$_3$Mo (orthorhombic) and a complex sigma phase (tetragonal), based on FeTi.[30] As illustrated in Figure 3.38, the presence of cobalt intensifies the hardening effect of molybdenum but no ageing effects are produced in high-cobalt-molybdenum-free compositions.[31]

Commercial grades

Although a large number of maraging grades has been developed, including compositions containing 20% and 25% Ni, the main commercial grades are based on the five composition ranges shown in Table 3.16. In this table, the numbers ascribed to the various grades relate to the nominal proof strength values in units of N/mm^2. However, these grades are also designated according to their

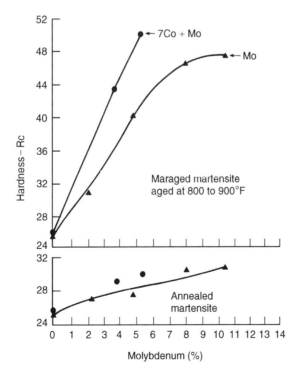

Figure 3.38 *Effect of molybdenum and cobalt on hardening response in maraging steels (After Decker* et al.[31]*)*

Table 3.16   *Composition ranges for 18% Ni−Co−Mo maraging steels*[32]

Grade	Wrought				Cast
	18Ni 1400	*18Ni 1700*	*18Ni 1900*	*18Ni 2400*	*17Ni 1600*
Nominal 0.2% proof stress:					
N/mm^2 (MPa)	1400	1700	1900	2400	1600
tonf/in^2	90	110	125	155	105
10^{3}lbf/in^2	200	250	280	350	230
kgf/mm^2	140	175	195	245	165
hbar	140	170	190	240	160
Ni	17–19	17–19	18–19	17–18	16–17.5
Co	8–9	7–8.5	8–9.5	12–13	9.5–11
Mo	3–3.5	4.6–5.1	4.6–5.2	3.5–4	4.4–4.8
Ti	0.15–0.25	0.3–0.5	0.5–0.8	1.6–2	0.15–0.45
Al	0.05–0.15	0.05–0.15	0.05–0.15	0.1–0.2	0.02–0.1
C max.	0.03	0.03	0.03	0.01	0.03
Si max.	0.12	0.12	0.12	0.1	0.1
Mn max.	0.12	0.12	0.12	0.1	0.1
Si + Mn max.	0.2	0.2	0.2	0.2	0.2
S max.	0.01	0.01	0.01	0.005	0.01
P max.	0.01	0.01	0.01	0.005	0.01
Ca added	0.05	0.05	0.05	None	None
B added	0.003	0.003	0.003	None	None
Zr added	0.02	0.02	0.02	None	None
Fe	Balance	Balance	Balance	Balance	Balance

proof strength values in ksi units, e.g. 18Ni 1700 (N/mm^2) = 18Ni 250 (ksi). As indicated in Table 3.16, the strength of the wrought grades is increased by increasing the levels of cobalt, molybdenum and titanium. Whereas the 18Ni 1400, 1700 and 1900 grades are aged for three hours at 480°C, the 18Ni 2400 grade requires three hours at 510°C or longer times at 480°C. The larger amount of titanium in 18Ni 2400 leads to a greater volume fraction of FeTi sigma phase precipitates compared with the lower strength grades.

Whereas the carbon content of these grades is restricted to low levels, the metallurgically active content is virtually zero due to the fixation of carbon by titanium. Elements such as silicon and manganese are also tightly controlled in order to promote high toughness levels. Aluminium is added primarily as a deoxidant, and although larger amounts will supplement the hardening reactions, this leads to a loss in toughness. Boron and zirconium are added in order to retard grain boundary precipitation, thereby improving toughness and stress corrosion resistance. [31]

Maraging steels are generally produced by vacuum melting and sometimes double vacuum melting (induction plus vacuum arc remelting) is employed. These procedures are designed to achieve the following:

1. Close control over the main alloying elements and the volatilization of impurities such as lead and bismuth.
2. Minimization of segregation.

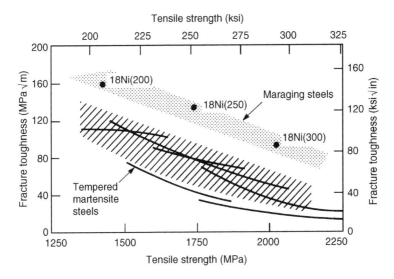

Figure 3.39 *Fracture toughness of maraging steels and other high-strength martensitic grades (After Magnee et al.[33])*

3. Low levels of oxygen, nitrogen and hydrogen.
4. Low levels of inclusions.

As indicated in Figure 3.39, the toughness of the maraging grades is significantly better than that of tempered high-carbon martensitic steels. [33]

The casting grade 17Ni 1600 is normally solution treated by homogenizing for four hours at 1150°C, followed by air cooling. Maraging then takes place at 480°C for three hours, which is similar to the conditions used for the lower strength wrought grades. However, a more complex heat treatment is sometimes employed in order to improve toughness. This involves the following:

1150°C 4 hours + 595°C 4 hours + 820°C 1 hour + 480°C 3 hours

Corrosion behaviour

The corrosion rate of maraging steels in both marine and industrial atmospheres is almost half of that shown by high-strength, low-alloy steels. In sea water, both types of steels show similar corrosion rates initially, but after six months the maraging steels corrode more slowly. The International Nickel Co.[32] have published K_{ISCC} data for exposure in aqueous environments, with and without NaCl, which show that the 18% Ni maraging steels compare favourably with other high-strength steels.

Applications

From the foregoing remarks, it is apparent that maraging steels are capable of producing very high strengths with good fracture toughness but at a cost far in excess of that of conventional, low-alloy engineering steels. They are therefore

Table 3.17 *Applications for maraging steels*[32]

Aerospace	Tooling and machinery	Structural engineering and ordnance
Aircraft forgings (e.g. undercarriage parts, wing fittings)	Punches and die bolsters for cold forging	Lightweight portable military bridges
Solid-propellant missile cases	Extrusion press rams and mandrels	Ordnance components
Jet-engine starter impellers	Aluminium die-casting and extrusion dies	Fasteners
Aircraft arrestor hooks	Cold-reducing mandrels in tube production	
Torque transmission shafts	Zinc-base alloy die-casting dies	
Aircraft ejector release units	Machine components: gears index plates lead screws	

used selectively in applications where weight saving is of paramount importance or where they can be shown to be more cost-effective than low-alloy grades. Some typical uses for maraging steels are listed in Table 3.17. As one might anticipate, these steels are used to advantage for weight reduction in aerospace and military applications but their excellent combination of properties, ease of heat treatment and dimensional stability also offer attractions in the more commercial sector of tooling and machinery.

Steels for steam power turbines

Electricity is generated on a large scale by the following sequence of operations:

1. The production of steam at high temperature and pressure in fossil-fired boilers or nuclear reactors.
2. The passage of the steam through a turbine where it impinges on the blades of a rotor, thereby creating rotational energy.
3. The transmission of the energy developed in the turbine rotor to a generator rotor which produces electricity in the windings of the stator.

The rotor shafts in the turbine and generator are produced from very large, high-integrity forgings which operate at speeds of about 3000 rev/min and over a range of temperatures, depending on their position in the power train. In a typical 660 MW coal-fired station, the turbine has three stages, namely high pressure (HP), intermediate pressure (IP) and low pressure (LP), and the last stage can involve two or three cylinders. On the other hand, the turbines for large water-cooled reactors operate at 1200 MW and involve only the HP and LP stages with very large rotors. This is illustrated in the data in Table 3.18 by Collier

Table 3.18

Component	Weight (tons)	
	660 MW Fossil unit	*1200 MW Nuclear unit*
HP rotor	15	90
IP rotor	30	–
LP rotor	50	210
Generator rotor	90	190

and Gemmill[34] for the weights of finished forgings. These authors also point out that the ingot weights for these components are usually a factor of two or three times that of the finished forgings and often involve the combined production of several steelmaking and secondary refining units in order to achieve the required ingot weights.

Attention will be focused on the rotors for turbines and generators but reference will also be made to other components such as casings, bolts and blades. A very detailed text on the materials used in both steam and nuclear power plant has been prepared by Wyatt.[35]

Turbine casings

The casings for the HP and IP stages of turbines are made from large castings, and in the UK the 0.5% Cr–Mo–V grade is favoured:

> 0.15% C, 0.5% Cr, 0.5% Mo, 0.25% V

The casings are generally cast in two halves which are bolted together longitudinally along a heavy flange. In fossil-fired plant, steam inlet temperatures of 540 or 565°C are employed and therefore the casings are subjected to high internal pressure within the creep range. In addition, the *two shifting* operation of being on load during the day and shut down overnight imposes significant thermal fatigue stresses on the casing and other components in the HP and IP stages.

Repair welding in 0.5% Cr–Mo–V casings can give rise to stress relief cracking which is associated with low creep ductility in the coarse-grained heat-affected zone of weldments. This problem is generally overcome by adopting titanium as opposed to aluminium deoxidation and this practice has proved to be beneficial to both weldability and rupture ductility.

HP and IP rotors

The following 1% Cr–Mo–V composition is used world-wide for the rotor shafts of both the HP and IP stages of turbines:

> 0.25% C, 1.0% Cr, 1.0% Mo, 0.3% V

As indicated earlier, the HP rotor is the smallest in the turbine and operates at temperatures up to the main steam temperature of 540–565°C. During operation, the rotors are subjected to centrifugal forces and also to torques which mainly affect the rotor journals due to their smaller cross-sectional area. However, given the temperature of operation, the creep and rupture behaviour of the HP rotor is of major concern.

The IP rotor is similar in design but larger than the HP rotor and again there is the need for high creep and rupture strength in the hotter regions. However, at the exhaust end of the IP stage, the blades are longer and impose greater centrifugal forces. This factor, coupled with the cooler operating conditions, therefore introduces fracture toughness as an important design parameter. In addition, transient operating phases, such as start-up and shut-down, can result in further thermal stresses and reinforce the need for adequate fracture toughness at lower temperatures. However, there tends to be an inverse relationship between rupture strength and toughness and whereas the best creep and rupture strengths in 1% Cr–Mo–V steel are obtained in an upper bainitic microstructure, such material has poor toughness. On the other hand, the lower bainitic or martensitic structures that promote good fracture toughness have poor creep strength. The final hardening treatment for 1% Cr–Mo–V rotors therefore depends very much on whether high-temperature strength or good toughness is considered to be the more important design criterion. In the UK, rotors are oil quenched for improved fracture toughness, whereas in the United States, air cooling is adopted in favour of creep strength.

Both Reynolds *et al.*[36] and Viswanathan and Jaffe[37] have reviewed the factors that might lead to an improved combination of creep strength and fracture toughness. These include the use of vacuum carbon deoxidation, electroslag remelting and modifications to the traditional 1% Cr–Mo–V composition to incorporate higher levels of chromium (1.5%) and an addition of nickel (0.7%). It has also been shown that improvements in the toughness of HP and IP rotors can be obtained by reducing the level of impurity elements and suppressing temper embrittlement. However, it is generally accepted that temper embrittlement is more pronounced in the 3.5% Ni–Cr–Mo–V LP rotor grade and this matter will receive greater attention in the next section.

LP rotors

The LP rotor is the largest in the turbine assembly and operates at temperatures between 270°C and ambient. The blades in this segment are also very long, e.g. up to 1.12 m, and impose high stresses on the rotor shaft. Therefore the main design criteria in LP rotors are a high proof strength and good fracture toughness as opposed to the major requirement for high creep strength in the HP and IP components. The 0.2% proof strength values for LP rotors are of the order of 750 N/mm^2 and the increasing demand for good fracture toughness is illustrated by the data in Table 3.19 for FATT values over the period from 1970 to 1985.

The material used for LP rotor forgings is 3.5% Ni–Cr–Mo–V steel to the following composition:

0.25% C, 3.5% Ni, 1.5% Cr, 0.5% Mo, 0.1% V

In addition, the steel is made to low levels of silicon (0.1% max.) and manganese (typically 0.2%), together with restricted levels of elements such as arsenic,

Table 3.19 *FATT Values in LP rotors*

Period	FATT°C [a]	
	Average	Range
1970–1975	+25	0 to + 50
1976–1980	+16	−18 to + 40
1981–1985	+10	−4 to + 20

[a]Fracture appearance transition temperature.

antimony, tin and phosphorus in order to minimize the effects of temper embrittlement. The evolution of the current composition will be described briefly to reflect the various changes that have taken place to accommodate the demands for increased turbine size and improved fracture toughness.

Boyle *et al.*[39] have given a very detailed account of the development work that was undertaken in the United States in the 1950s following failures in a number of turbine and generator rotors. These failures were attributed to inadequate fracture toughness and prompted an industry-wide development to improve the performance of the 2.7% Ni 0.5% Mo 0.1% V steel that was used for large turbine rotors at that time. It became evident that temper embrittlement was a significant factor in contributing to low toughness and the manganese content of the steel was reduced from 0.7% to 0.4%. In order to compensate for the loss in hardenability and to improve toughness, the nickel content was raised from 2.7% to 3.5%. Boyle *et al.* state that nickel contents as high as 5% were investigated but that the benefits derived from additions greater than 3.5% were insufficient to compensate for the difficulties introduced by the depression of the Ac_1. Presumably these difficulties related to the fact that the Ac_1 can be depressed to a temperature close to or below the nominal tempering temperature with the potential risk of reaustenitization.

Although changes to the manganese and nickel contents realized significant improvements in toughness, there was also the need to develop a composition for larger diameter rotors which required higher tensile and yield strengths, together with good toughness and ductility. This led to the addition of 1.5% Cr to the modified Ni–Mo–V to provide the current 3.5% Ni–Cr–Mo–V grade. Since the 1950s, a considerable amount of work has been carried out in the UK to improve the fracture toughness of LP and generator rotors, particularly in relation to the suppression of temper embrittlement. In addition, the original FATT data have been augmented with K_{1C} fracture toughness values and a good correlation between the two parameters has been derived via the expression:

$$K_{IC}MNm^{-3/2} = \frac{6600}{60 - B}$$

where $B = $ (test temperature − FATT value)°C.

Following homogenization treatments at temperatures around 1150°C, rotors are heated slowly to a solution treatment temperature of 840°C and then lowered vertically into an array of high-pressure water sprays. Modern LP rotors have a diameter in excess of 1700 mm and the above treatment results in the formation of a rim structure which is essentially martensitic and a core structure which is essentially bainitic. In order to avoid quench cracking, the large forgings are not quenched directly to room temperature but are held for a period at a temperature of 200°C in order to reduce internal stresses. The material is then tempered in the range 560–640°C, the lower bound avoiding the formation of isothermal temper embrittlement and the upper bound avoiding the risk of partial reaustenitization. In turn, the rate of cooling adopted from the tempering temperature represents a compromise between the fast cooling that inhibits temper embrittlement and the slow cooling that minimizes internal stress. The compromise of fan cooling is usually employed.

At one time, the requirement for very low levels of silicon and manganese would have posed major problems in deoxidation but these have been overcome by the introduction of vacuum carbon deoxidation (VCD). VCD treatment is also thought to be beneficial in improving the solidification characteristics and reducing the segregation effects in large forging ingots.

Turbine generator end rings

The copper windings in an electrical generator are not strong enough to resist the rotation stress and *end rings* are fitted to retain the windings. These components are very highly stressed and have 0.2% proof strength values greater than 1200 N/mm^2. A further requirement is that the material should be non-magnetic in order to minimize electrical losses from the generator. Up until the late 1970s, a cold-worked steel containing 18% Mn 5% Cr was used world-wide for this application but the main disadvantage of this material was its susceptibility to stress corrosion cracking (SCC). This could arise in moist conditions and the threat of catastrophic disintegration during service necessitated frequent inspection.

Because of these problems, attention was turned to the development of a more SCC-resistant material, leading to the introduction of a steel of the following composition:

   0.08%C, 0.3%Si, 18.0%Mn, 18.0%Cr, 0.60%N

This material is substantially solid solution strengthened by nitrogen and, additionally, this element also serves to preserve the austenitic (and therefore non-magnetic) structure in the presence of a high chromium content. The 18% Mn 18% Cr steel also offers the facility for attaining higher 0.2% proof strength values than 18% Mn 5% Cr steel, i.e. values up to 1400 N/mm^2. 18% Mn 18% Cr steel now represents the first-choice material for end rings and after 10 years trouble-free service has virtually eliminated the need for in-service inspection.

Turbine bolts

Large bolts or studs are used to maintain steam tightness in the flanges of high-temperature turbines. These are tightened to a tension that imposes an elastic strain of about 0.15% but at the high operating temperatures this tension reduces due to the process of creep (*stress relaxation*). Therefore bolting materials must have good creep strength in order to minimize relaxation and maintain steam tightness. Repeated relaxation and retightening will lead eventually to creep fracture and therefore the utilities limit both the operating period and the number of retightening operations so as to avoid fracture during service.

In the UK, the majority of the bolting requirements are satisfied by a range of Cr–Mo and Cr–Mo–V steels and the evolution of these steels has been described by Everson *et al.*[40] Details of the composition, heat treatment and high-temperature properties of these steels are given in Table 3.20. The Cr–Mo (Group 1) steel was introduced in the 1930s when the steam temperatures were of the order of 450°C. By the late 1940s, steam temperatures had risen to about 480°C and higher strength bolts were required. This led to the introduction of the Cr–Mo–V (Group 2) steel in which the higher strength is achieved by the formation of a fine precipitate of V_4C_3 on tempering at 700°C. However, steam temperatures continued to rise in the UK in pursuit of higher operating efficiency and in 1955 reached a level of 565°C. This change necessitated a further increase in strength and this led to the introduction of the 1% Cr 1% Mo 0.75% V (Group 5) steel. However, this material proved to have poor rupture ductility due to intergranular cracking after only short exposure at the operating temperature. A major research programme was therefore undertaken on this problem and this led to the development of the Cr–Mo–V–Ti–B (Group 6) steel. This composition represented the addition of about 0.1% Ti and 0.005% B to Group 5 steel which produced a significant improvement in rupture ductility. According to Everson, Orr and Dulieu, the principal effect is due to boron, about 50% of which is incorporated in the V_4C_3 precipitates and the remainder is dissolved in the matrix. This produces a stabilizing effect on the V_4C_3 near the grain boundary regions, making the carbides more resistant to dissolution and so reducing the rate at which denuded zones are formed. Titanium is added primarily as a nitrogen-fixing agent and so preventing the formation of boron nitride, which is metallurgically inactive. However, the formation of TiN leads to refinement of the austenite grains which also contributes to improved rupture ductility. More recently, the Cr–Mo–V–Ti–B steel has been used as boiler support rods, operating at temperatures up to 580°C.

Nimonic 80A (Ni–20.0% Cr, 2.4% Ti, 1.4% Al) is significantly stronger than the low-alloy steels and allows the use of smaller bolts and more compact flanges than are possible with ferritic bolting materials. It is also likely that Nimonic 80A would replace Cr–Mo–V (Group 6) steel as the principal turbine bolting material if steam temperatures were to be raised above the present operating level of 565°C.

Detailed studies have also shown that marked improvements in the rupture characteristics of Cr–Mo–V bolting steels can be produced by restricting the level of residual elements. Thus the rupture life of Cr–Mo–V steel can be related to the 'R' factor (P + 2.43 As + 3.75 Sn + 0.13 Cu) and Cr–Mo–V bolting steels are now made to a typical 'R' value of 0.07% compared to 0.2% in early years. This has led to a substantial increase in rupture ductility. The 'R' value can also be reduced to a level of 0.015 in VIM–VAR material to provide further improvements in creep strength and ductility.

Table 3.20 UK turbine bolting steels

SES[a] Trade name	CEGB Code	BS 1506: 1986 Type no.	Nominal composition (%)					Typical heat treatment	RT tensile strength (N/mm²)	Strength			
										Rupt.		Relax[b]	
			C	Cr	Mo	V	Others			c	d	c	d
DUREHETE 900	GP 1 (Cr–Mo)	631–850	0.4	1	0.5	–	–	870°C OQ + 660°C	850–1000	234	97	81	–
DUREHETE 950	GP 2 (Cr–Mo–V)	671–850	0.4	1	0.5	0.25	–	950°C OQ + 700°C	850–1000	324	151	83	–
–	GP 3 (3Cr–Mo–V)	–	0.3	3	0.5	0.75	–	–	–	–	–	–	–
–	GP 4 (Mo–V)	–	0.2	–	1	0.25	–						
DUREHETE 1050	GP 5 (1Cr–Mo–V)	–	0.2	1	1	0.75	–	Superseded by Durehete 1055					
DUREHETE 1050	GP 6 (1Cr–Mo–V)	681–820	0.2	1	1	0.75	Ti, B	980°C OQ + 700°C	820–1000	418	280	141	70

[a]SES, Stocksbridge Engineering Steels.
[b]Stress relaxation for 0.15% strain.
[c]10^4 h values (N/mm²) at 500°C.
[d]10^4 h values (N/mm²) at 550°C.

After Everson et al.[40]

Turbine blades

Apart from locations at the extreme ends of steam turbines, all the blades are made in 12% Cr steels. Although the metallurgy of these steels will be discussed in the next chapter, for the sake of continuity the use of 12% Cr steels in turbine blades is better described at this stage.

In the HP turbine, the blades are short and operate under the maximum steam temperature, i.e. 565°C. The creep strength of 12% Cr steels is not adequate to operate at such temperatures and therefore the first few rows of blades at the HP inlet are generally manufactured from Nimonic 80A. In the LP turbine, the blades are long, and in large turbines the exhaust blades can exceed a length of one metre. Such blades generate high centrifugal forces and again 12% Cr steels are not strong enough to cope with the conditions imposed. In such situations, the precipitation-strengthened FV520B steel may be employed, which has the composition shown in Table 3.21.

However, in recent years, interest has grown in the use of titanium alloys, typically Ti–6% Al, 4% V (Ti6Al4V), as a substitute for martensitic stainless steels in the outlet stages of turbines. The reduced weight and high resistance to corrosion fatigue of the titanium alloys allows the length of the blades to be increased substantially, compared to what can be achieved in 12% Cr steels. Thus the exhaust area of the LP turbine can be increased by about 50%, which increases the power by the same amount.

For all stages between the inlet and outlet, the grades of 12% Cr steels shown in Table 3.22 are employed. The tempering resistance and creep strength of these materials is increased progressively with the addition of molybdenum, vanadium and niobium to the 12% Cr base. As illustrated in the next chapter, nickel is added to such steels in order to preserve an austenitic structure at high temperatures, in the presence of ferrite-forming elements such as molybdenum, vanadium and niobium.

Although the 12% Cr steels have adequate resistance to attack in moist air, the formation of water droplets in the final stages of the turbine can result in severe erosion problems. This is overcome by brazing strips of erosion-resistant materials such as stellite (cobalt-based) to the leading edges of LP blades.

Table 3.21 *Composition of FV 520B*

C%	Ni%	Cr%	Mo%	Cu%	Nb%
0.05	5.5	14	1.6	1.5	0.3

Table 3.22

Type	C%	Si%	Mn%	Ni%	Cr%	Mo%	V%	Nb%
12Cr–Mo	0.1	0.3	0.3	–	12.5	0.75	–	–
12Cr–Mo–V	0.1	0.3	0.6	0.8	12	0.6	0.2	–
12Cr–Mo–V–Nb	0.13	0.5	1	0.8	11.2	0.6	0.2	0.4

Medium–high-carbon pearlitic steels

Although most of the steels that are used in engineering applications are heat treated to form a tempered bainitic/martensitic structure, there are notable examples in which the required strength is generated in air-cooled, medium–high-carbon steels with a predominantly pearlitic microstructure. These include micro-alloy forging grades, rail steels and high-carbon wire rod and these applications will be discussed in the sections that follow. However, as a precursor to these discussions, it is worthwhile to review very briefly the physical metallurgy of pearlitic steels.

In Chapters 1 and 2, the Hall–Petch relationship was introduced, namely:

$$\sigma_y = \sigma_i + k_y d^{-\frac{1}{2}}$$

where σ_y = yield strength
  σ_i = friction stress which opposes dislocation movement
    (Peierls stress)
  k_y = dislocation locking term
  d = ferrite grain size (mm)

It was stated that this basic relationship was extended later to take account of the solid solution strengthening effects of alloying elements and expressed as follows:

$$\sigma_y = \sigma_i + k' \ (\% \ \text{alloy}) \ + k_y d^{-\frac{1}{2}}$$

Following this concept, Pickering and Gladman[41] then produced the following quantitative relationships for yields and tensile strength:

$$\text{YS (N/mm}^2) = 53.9 + 32.3\% \ \text{Mn} + 83.2\% \ \text{Si} + 354 \ N_f + 17.4 d^{-\frac{1}{2}}$$

$$\text{TS (N/mm}^2) = 294 + 27.7\% \ \text{Mn} + 83.2\% \ \text{Si} + 3.85\% \ \text{pearlite} \ + 7.7 d^{-\frac{1}{2}}$$

These equations were derived by statistical analysis of steels containing up to 0.25% C and one of the interesting points to emerge was the absence of á term for carbon or pearlite in the equation for yield strength.

In the 1970s, Gladman *et al.*[42] went on to develop equations for steels containing up to 0.9% C and their equation for yield strength was presented in the following form:

$$\sigma_y = f_\alpha^n \sigma_\alpha + (1 - f_\alpha^n)\sigma_p$$

where σ_y = yield strength of the ferrite–pearlite aggregate
  σ_α = yield strength of ferrite
  σ_p = yield strength of pearlite
  f_α = volume fraction of ferrite

and by implication $(1 - f_\alpha)$ = volume fraction of pearlite.

Thus the yield strength of the aggregate is presented as the sum of the separate contributions from ferrite and pearlite and weighted according to their volume fractions. In the absence of pearlite, $\sigma_y = \sigma_\alpha$ and in the absence of ferrite, $\sigma_y = \sigma_p$. The index n in the above equation allows the yield strength to vary with pearlite content in a non-linear manner and, as indicated below, it was given the value of $\frac{1}{3}$. The full quantitative equations for yield and tensile strengths are as follows:

$$YS \ (N/mm^2) = f_\alpha^{\frac{1}{3}}[35 + 58.5\%Mn + 17.4d^{-\frac{1}{2}}] + (1 - f_\alpha^{\frac{1}{3}})[17.8 + 3.85S_0^{-\frac{1}{2}}]$$
$$+ 63.1\%Si + 426\sqrt{\%N}$$

$$TS \ (N/mm^2) = f_\alpha^{\frac{1}{3}}[247 + 1146\sqrt{\%N} + 18.2d^{-\frac{1}{2}}] + (1 - f_\alpha^{\frac{1}{3}})$$
$$[721 + 3.55S_0^{-\frac{1}{2}}] + 97.3\%Si$$

where $S_0 =$ interlamellar spacing of pearlite (mm).

Thus in high-carbon steels the volume fraction of pearlite $(1 - f_\alpha)$ and the interlamellar spacing (S_0) have a significant effect on both the yield and tensile strengths. The major difference between the two equations is the value of the constants, namely 35 and 178 N/mm^2 for yield strength and 247 and 721 N/mm^2 for tensile strength. Manganese in solid solution appears to have no significant effect on tensile strength, but this element appears in the equation for yield strength. The components of yield strength in medium-to high-carbon steels, containing 0.9% Mn, 0.3% Si and 0.007% N, are shown in Figure 3.40. This

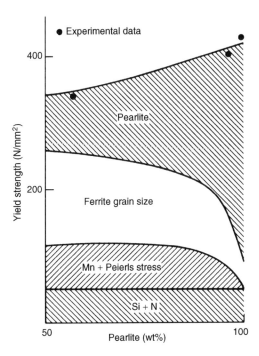

Figure 3.40 *Components of yield strength in high-carbon steels (After Gladman et al.[42])*

indicates that ferrite makes a significant contribution to the yield strength even when there is 90% pearlite in the microstructure.

Gladman *et al.*[42] also generated an equation for the impact transition temperature of high-carbon steels and this is shown below:

$$27J \text{ ITT } (°C) = f_\alpha[-46 - 11.5d^{-\frac{1}{2}}]$$

$$+ (1 - f_\alpha)[-335 + 5.6S_0^{-\frac{1}{2}} - 13.3p^{-\frac{1}{2}} + 3.48 \times 10^6 t]$$

$$+ 48.7\% \text{ Si} + 762\sqrt{\%N}$$

where d = ferrite grain (mm)
p = pearlite colony size (mm)
t = cementite plate thickness (mm)

This equation again emphasizes the importance of a fine grain size in producing a low impact transition temperature and it should be noted that the coefficient for the pearlite colony size p (13.3) is of a similar order to that of the ferrite grain size d (11.5).

Rail steels

Up until the 1970s, rails for passenger and freight trains were regarded as relatively simple undemanding products and the specifications had changed very little for a number of decades. However, investment in railway systems, the advent of high-speed passenger trains and the requirement for longer life track imposed a demand for rails of high quality, greater strength and tighter geometric tolerances. Therefore there have been major innovations in the past 20 years in terms of method of manufacture, degree of inspection and range of products.

Typically, rail steel is produced in large BOS vessels and is vacuum degassed prior to being continuously cast into large blooms. Vacuum degassing, coupled with ladle trimming facilities, permits very tight control over chemical composition. After casting, the blooms are placed in insulated boxes, whilst still at a temperature of about 600°C, and are cooled at a rate of 1°C per hour for a period of three to five days. This treatment, coupled with prior vacuum degassing, reduces the hydrogen level in the finished rail to about 0.5 ppm, thereby reducing substantially the susceptibility to hydrogen cracking. The blooms are then reheated and rolled directly to the finished rail profile. The rail produced from each bloom is hot sawn to specific lengths prior to passage through a rotary stamping machine *en route* to the cooling areas. Depending upon the properties required, the rails are either cooled normally in air or subjected to enhanced cooling for the development of high strength. On cooling to room temperature, the rails are passed through a roller-straightener machine which subjects the section to a number of severe bending reversals and emerge with a very high degree of straightness. Finally, the rails pass through a series of ultrasonic, eddy current and laser inspection stations which monitor non-metallic inclusions, external defects and the flatness of the running surface. The final operation is the cold sawing of the rail ends on high-speed machines with carbide-tipped blades. Rails are generally supplied in lengths up to 36 m.

Rail steel specifications

Historically, high-volume rail steels have been based on fully pearlitic microstructures which are characterized by high resistance to wear and plastic flow, both of which are major property requirements for good rail performance. Although the potential of martensitic structures has been evaluated, they proved to be unsuitable, possessing inadequate toughness and ductility. Therefore rail steels continue to be based on pearlitic microstructures which are generated through various combinations of carbon, manganese and other elements.

Rail steel specifications can be classified into three types, based on tensile strength:

1. Normal grades, $\sim$700 N/mm^2 min. TS
2. Wear-resisting grades, 880 N/mm^2 min. TS
3. High-strength grades, 1080–1200 N/mm^2 TS

Normal grades

Typical examples of *normal grades* are BS 11: 1985 Normal and UIC 860-O Grade 70 (Table 3.23). These are the high-tonnage grades which are used in normal service conditions in conventional railways, including high-speed passenger traffic (200 km/h), and medium-speed (100 km/h) relatively heavy axle load (25 tonne) freight. The majority of London Transport underground track is also laid in BS 11 Normal grade.

Wear-resistant grades

The hardness and wear resistance of pearlitic steels are increased by refining the pearlite lamellae. This is achieved by increasing the carbon and manganese contents as illustrated in the specifications for *wear-resisting grades* given in Tables 3.24 and 3.25.

Thus similar levels of tensile strength can be obtained from various combinations of carbon and manganese, both of which depress the temperature of transformation from austenite to pearlite and thereby refine the pearlite lamellae. These wear-resisting grades are used for heavy axle loads, high-density traffic routes or tightly curved track. However, the use of wear-resisting rails on conventional railways can also show economic advantages.

Table 3.23 *BS 11 Normal grade UIC 860-0 Grade 70*

Grade	C%	Si%	Mn%	TS min. (N/mm^2)	Elong. min. $5.65\sqrt{S_0}$
BS 11 Normal	0.45–0.6	0.05–0.35	0.95–1.25	710	9
UIC 860-0 Grade 70	0.4–0.6	0.05–0.35	0.8–1.25	680	14

Table 3.24   *BS11 wear-resisting grade A/UIC 860-0 Grade 90A*

C%	Si%	Mn%	TS min. (N/mm^2)	Elong. min. 5.65$\sqrt{S_0}$
0.65–0.8	0.1–0.5	0.8–1.3	880	8

Table 3.25   *BS11 wear-resisting grade B/UIC 860-0 Grade 90B*

C%	Si%	Mn%	TS min. (N/mm^2)	Elong. min. 5.65$\sqrt{S_0}$
0.55–0.75	0.1–0.5	1.3–1.7	880	8

High-strength grades

For the extremely arduous service conditions encountered in tightly curved track and under very high axle loads, even higher strengths are required which demand further refinement of the pearlitic structure. Up until 1985, most European railmakers produced such material by adding up to 1% Cr to the basic C–Mn composition and this increased the hardness of the rail head from 280 BHN to approximately 330 BHN with an improvement factor of about two in wear resistance. However, the gradual replacement of bolted track by welded track brought about the requirement for an adequate level of weldability, a property not readily satisfied with the high hardenability introduced by the addition of 1% Cr. Attention therefore turned to the use of accelerated cooling from the austenite range, rather than high-alloy additions, for the depression of the pearlite transformation and the development of high-strength rails. A number of railmakers have now introduced in-line cooling for the head hardening of rails and detailed accounts of the computerized facility installed by British Steel at Workington are given in publications by Preston[43] and Hodgson and Preston.[44] Brief details are as follows.

The rail leaves the finishing stands at a temperature of about 1000°C and is stood head up. The rail passes under temperature monitors and then into a 55-m-long cooling train where it is sprayed with water over all surfaces. The outgoing rail leaves the cooling station at dull red heat and the main transformation to pearlite takes place in still air. Obviously the rate of cooling has to be balanced very finely so as to provide sufficiently high rates of cooling in the centre of the rail head so as to depress the transformation of austenite to pearlite without cooling the surface of the rail at a rate which would lead to the formation of bainite or martensite. During cooling, the rail is driven through the cooling station by rollers which also maintain straightness in the rail. On leaving the cooling station, the rail is finally turned on its side and passes down the cooling banks for finishing in the normal manner. Using this facility, the hardness developed in the head of a 0.8% C 0.9% Mn rail is controlled to within a fairly narrow range of about 350–370 BHN.[44]

Flash butt welding is used to join lengths of rail into continuous track and Preston[43] has commented on the effects of this process on the hardness of enhanced cooled rail. In flash butt welding, the rail ends are heated to high temperatures by an electric arc and then squeezed together by a hydraulic force of 40–60 tonne. The rail ends fuse, excess material is extruded from the joint and material in the heat-affected zone cools through the transformation range at the rate of about 1.3°C/s. In contrast, the head of mill-hardened rail cools at a rate of about 2.5°C/s, and after flash butt welding, the hardness of enhanced cooled 0.8% C 0.9% Mn steel falls to about 300 BHN compared with 350–370 BHN in the mill-hardened condition. However, this can be avoided by increasing the alloy content of the steel such that a hardness of about 370 BHN is developed under a natural cooling rate of 1.3°C/s after welding.

Wear resistance of rails

British Steel has carried out extensive laboratory wear tests on rail steels at its Swinden Technology Centre in Rotherham. This involves the rotation of discs of railway wheel and rail materials, under a controlled contact stress and with a controlled amount of slip between the two discs. Wear is determined by weight loss on the rail test disc and expressed in terms of mg/m of slip. In this test, a very clear relationship has been established between wear rate and hardness and this relationship is shown for a range of rail steel grades in Figure 3.41.[45] Thus

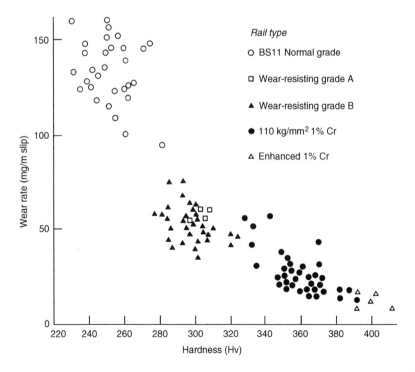

Figure 3.41   *Effect of hardness on wear rate in laboratory tests (After British Steel[45])*

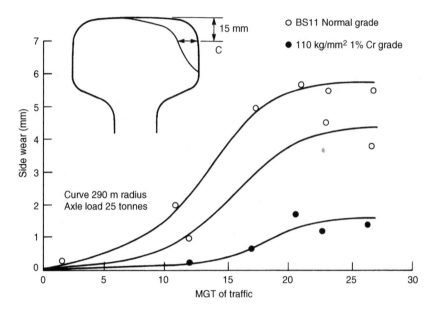

Figure 3.42 *Side wear in normal and 1% Cr rails under identical service conditions (After British Steel[45])*

BS 11 with the lowest hardness and tensile strength exhibits the highest wear rate but this falls progressively to low values as the alloy content and hardness of the materials is increased. However, whilst producing the correct order of merit, this test exaggerates the relative performance of the various steels under operating conditions and therefore laboratory rig testing has to be supplemented by measurements of rail wear under normal track service. An illustration of the development of *side wear* in BS 11 Normal and 1% Cr rails is given in Figure 3.42. This indicates that 1% Cr material has a wear resistance of three to four times that of BS 11 Normal under the particular track conditions identified in this figure. Other track tests indicate that wear-resistant grades have twice the resistance of BS 11 Normal.

Side wear of the type illustrated in Figure 3.42 is a function of axle load and track curvature, the centrifugal force causing the flange of the wheel to scrub against the gauge corner head of the high outer rail in the curve.

Austenitic 14% Mn rails

When the service conditions are such that exceptionally high rates of wear are experienced in high-strength rails, then consideration is given to the use of rails of the following composition which develop an austenitic microstructure:

C%	Si%	Mn%
0.75–0.9	0.2–0.4	13–14

The metallurgy of this material is complex but such steel has a very high resistance to wear because of its high rate of work hardening when subjected to applied stress or abrasion. This special grade of steel is made in electric arc furnaces but is rolled to rail in the same type of mill as that employed for the pearlitic grades. The mechanical properties[45] of 14% Mn rail material are given below:

TS (N/mm^2)	YS (N/mm^2)	Elong. (%)
818–973	355–386	40–60

The low YS/TS ratio illustrated above is indicative of the high rate of work hardening and material with an initial hardness of 180–210 BHN develops a hardness of over 400 BHN, and to an appreciable depth, after a short period of service. The steel can be welded, using suitably modified techniques, by either the thermit or flash butt welding processes.

14% Mn rails are used traditionally in railway points and crossings and in other situations where the extended service life justifies their higher cost compared with pearlitic grades.

At one time, it was postulated that the high rate of work hardening in this steel (also known as Hadfields Manganese Steel) was due to the formation of strain-induced martensite. However, it is now known that the hardening effect is associated with the formation of stacking faults in the austenitic structure and strain-induced martensite will only form in decarburized material or in steels of a lower alloy content.

Micro-alloy forging steels

Up until the late 1940s, the engineering steels that were used for automotive engine and transmission parts were based largely on compositions containing substantial amounts of nickel and molybdenum. The philosophy that prevailed was that these components were subjected to arduous service conditions that required high levels of strength and toughness and that this combination of properties was best achieved in Ni–Mo grades. However, during the 1950s, there was the realization that many of the steels were over-alloyed with regard to the hardenability requirements of the components and that the specified levels of strength could be achieved by steels of leaner composition. In the 1960s, the emerging technology of fracture mechanics provided greater knowedge on the level of toughness required in engineering components and indicated that satisfactory performance could be provided by steel compositions which gave lower impact energy values than the traditional Ni–Mo grades. These factors, coupled with major advances in heat treatment technology, led the way to the gradual substitution of the Ni–Mo grades by cheaper steels involving additions of manganese, chromium and boron.

By the 1970s, the opportunities for alloy reduction and substitution had largely been exhausted but competition in the automotive industry maintained the impetus for further cost reduction. Attention therefore turned to potential savings in manufacturing costs and particularly in the area of heat treatment. Traditionally, components such as crankshafts and connecting rods are cooled to room temperature after the forging operation, only to be reheated to a temperature of about 850°C prior to oil quenching. Tempering at 550–650°C then produces tensile strengths in the range 800–1100 N/mm^2. However, in the mid-1970s, German manufacturers demonstrated that these strength levels could be produced in a micro-alloy, medium-carbon steel (49 MnVS3) after air cooling from the forging operation, thus eliminating the need for heat treatment. Since that time, major effort has been devoted to the development of micro-alloy forging steels in Europe and Japan and these steels have gradually been introduced as substitutes for quenched and tempered steels in some automotive components.

Metallurgical considerations

As indicated in earlier chapters, niobium, titanium and vanadium are used as micro-alloying elements in low-carbon steels, although high soaking temperatures must be employed in order to achieve substantial solution of Nb(CN), TiC and TiN. However, vanadium has a high solubility in austenite, regardless of the carbon content, and is therefore the most suitable micro-alloying element for medium-carbon steels. On cooling from the solution treatment temperature, vanadium carbonitride precipitates in both the proeutectoid ferrite and the ferrite lamellae of the pearlite. The physical metallurgy of these steels has been reviewed by Gladman.[46]

As illustrated in Figure 3.43, the tensile properties of these grades increase progressively with vanadium content and, depending upon the levels of strength required, vanadium contents in the range 0.05–0.2% are employed. The level of precipitation strengthening is also influenced by the nitrogen content and

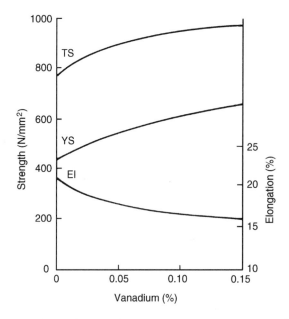

Figure 3.43 *Effect of vanadium on the tensile properties of air-cooled 0.45% C 0.9% Mn steel*

Lagnenborg[47] has shown that the tensile strength of these steels can be expressed as a function of $(V + 5 \times N)\%$. Nitrogen levels of up to 0.02% are therefore incorporated in the steels in order to intensify the strengthening effect.

One of the disadvantages of these micro-alloy steels is that they display significantly lower levels of toughness than the traditional quenched and tempered martensitic grades and this has inhibited their large-scale commercial exploitation. The low impact strength is related to the coarse pearlitic structure but this effect is exacerbated by precipitation strengthening. Whereas this problem has been overcome in structural steel plates with the use of low-temperature finishing (controlled rolling), there is little scope for the adoption of this practice in drop-forging operations due to the metal flow/die filling problems that occur at low forging temperatures.

The impact strength of these grades can be improved by lowering the carbon content and compensating for the loss in strength by increasing the manganese, vanadium and nitrogen contents. Experience in Sweden and Germany has shown that an improvement in toughness can also be obtained by increasing the silicon content of the steels. However, attention has also been given to the potential of grain-refining additions of titanium in pursuit of higher impact strength.

As indicated earlier, titanium has a low solubility in medium-carbon steels but TiN is even less soluble than TiC. Particles of TiN are therefore present at the normal soaking temperature for forging, namely 1150°C, and will refine the austenite grains provided the particles are present as a fine dispersion. According to Gladman,[46] this is achieved by restricting the titanium content to below the stoichiometric level required for reaction with nitrogen in TiN and the growth of particles is also minimized by rapid solidification from the liquid state. In practice, the titanium additions are restricted to levels of about 0.01% and the need for rapid solidification is generally satisfied by continuous casting as opposed to ingot casting.

Japanese steelmakers have expressed concern that the formation of TiN for grain refinement can reduce the level of soluble nitrogen that is available for precipitation strengthening by V(CN). However, this problem can be overcome by adjusting the nitrogen content such that the free nitrogen (total nitrogen minus nitrogen as TiN) exceeds 0.006%.

More recently, it has been reported that improved levels of toughness can be obtained in medium-carbon, micro-alloy steels by the generation of bainitic structures. However, according to Naylor[48] the benefits of this development are not yet clear-cut and the need for alloy additions to achieve a bainitic structure may detract from the viability of the approach.

Commercial exploitation

Korchynsky and Paules[49] have reviewed the various grades of micro-alloy forging steels that are produced in Europe and Japan and their lists of compositions and associated tensile properties are shown in Tables 3.26 and 3.27. Heading the list is the German grade 49MnVS3, the first medium-carbon, micro-alloy steel to be used commercially for air-cooled automotive forgings. Like the Swedish Volvo

Table 3.26 *Chemical composition of micro-alloy forging steels*

Country	Grade	C	Si	Mn	S	V	Other
Germany	49MnVS3	0.44–0.5	0.6 max.	0.7–1	0.04–0.07	0.08–0.13	
Gr. Britain	BS970-280M01	0.3–0.5	0.15–0.35	0.6–1.5	0.045–0.06	0.08–0.2	
Gr. Britain (UES)	VANARD	0.3–0.5	0.15–0.35	1–1.5	0.1 max.	0.05–0.2	
Gr. Britain (UES)	VANARD 850	0.36	0.17	1.25	0.04	0.09	0.1 Cr
Gr. Britain (UES)	VANARD 1000	0.43	0.35	1.25	0.06	0.09	0.15 Cr
Gr. Britain (Austin-Rover)	CMV 925	0.37–0.42	0.15–0.35	1.1–1.3	0.06–0.08	0.08–0.11	0.04 Mo max. Cr + Cu + Ni = 0.5 max.
Finland (OVAKO)	IVA 1000	0.47	0.5	1.1	0.05	0.13	0.5 Cr
Sweden (Volvo)	V-2906	0.43–0.47	0.15–0.4	0.6–0.8	0.04–0.06	0.07–0.1	0.2 Cr max.
Germany	44MnSiVS6	0.42–0.47	0.5–0.8	1.3–1.6	0.02–0.035	0.1–0.15	Ti optional
Germany	38MnSiVS6	0.35–0.4	0.5–0.8	1.2–1.5	0.04–0.07	0.08–0.13	Ti optional
Germany	27MnSiVS6	0.25–0.3	0.5–0.8	1.3–1.6	0.03–0.05	0.08–0.13	Ti optional
Japan (Mitsubishi-NKK)		0.32	0.25	1.45		0.06	0.01 Ti 0.01–0.016 N

After Korchynsky and Paules.[49]

Table 3.27 *Tensile properties and hardness of micro-alloy forging steels*

Country	Grade	UTS (MPa)	YS (MPa)	El (% min.)	RA (% min.)	BHN
Germany	49MnVS3	750–900	450 min.	8	20	
Gr. Britain	BS970-280M01	780–1080	540–650	18/8	20	
Gr. Britain (UES)	VANARD 850	770–930	540 min.	18	20	237–277
Gr. Britain (UES)	VANARD 1000	930–1080	650 min.	12	15	269–331
Gr. Britain (Austin-Rover)	CMV 925	850–1000	560	12	15	248–302
Finland (OVAKO)	IVA 1000	1025	750	10	20	290
Sweden (Volvo)	V-2906: <90 mm	750–900	500 min.	12		230–275
	<50 mm	800–950	520 min.	15		245–290
Germany	44MnSiVS6	950–1100	600 min.	10	20	
Germany	38MnSiVS6	820–1000	550 min.	12	25	
Germany	27MnSiVS6	800–950	500 min.	14	30	
Japan (Mitsubishi-NKK)		720–800	470–550			

After Korchynsky and Paules.[49]

Table 3.28

Grade	Steel replaced
VANARD 850	0.35% C, 1.5% Mn (216M36)
VANARD 925	0.35% C, 1.5% Mn, 0.25% Mo (605M36) 0.4% C, 0.8% Mn, 1% Cr (530M40)
VANARD 1000	0.35% C, 1.5% Mn, 0.25% Mo (605M36) 0.4% C, 0.8% Mn, 1% Cr, 0.3% Mo (709M40)
VANARD 1100	0.4% C, 0.8% Mn, 1% Cr, 0.3% Mo (709M40)

grade V2906, 49 MnVS3 has a relatively low manganese content which restricts the amount of pearlite in the microstructure, and the tensile strengths of these grades are towards the bottom of the range. In the UK, substantial effort has been devoted to the development and commercial evaluation of the VANARD range which is based on the following composition range:

0.3–0.5% C, 0.15–0.35% Si, 1–1.5% Mn, 0.05–0.2% V

Within this composition range, increasing levels of carbon, manganese and vanadium are used to provide tensile strengths in the range 850–1100 N/mm^2.

Some of the quenched and tempered, alloy steels that can be replaced by the VANARD grades are shown in Table 3.28.

Whereas the standard VANARD grades are generally made to a sulphur specification of 0.05% max., variants are produced with sulphur contents of the order of 0.08% for improved machinability. Other steels listed in Table 3.26 also specify enhanced sulphur levels. However, irrespective of the sulphur content, it is generally claimed that micro-alloy forging steels offer better machinability than traditional martensitic grades due to easier crack propagation in the predominantly pearlitic structure.

In addition to compositional effects, the mechanical properties of these steels are controlled by the soaking temperature, hot-working schedule and cooling rate to ambient temperature. The cooling rate from the finishing temperature is important since this controls the transformation temperature and the precipitation of V(CN). In conventional forging steels, the components may be placed in a bin after the forging operation and the slow cooling rates encountered in this situation would cause the precipitates to overage, resulting in a substantial loss in strength. Simple conveyer systems have therefore been introduced which enable the micro-alloy steel forgings to cool freely in air or else the cooling rate is enhanced with fan cooling.

The automotive components that are being produced in medium-carbon, micro-alloy steels include crankshafts, connecting rods, steering knuckles, axle beams and tension rods. Various manufacturers, including the Rover Group, have claimed that very substantial cost savings have been achieved by the adoption of these grades due to:

1. The lower cost of micro-alloy steels compared with the alloy grades that they replace.
2. The elimination of heat treatment costs.
3. The improved machining characteristics compared with traditional grades.

However, car makers in the UK have been reluctant to use these steels for safety-critical components due to their low impact properties and the rate of acceptance has been particularly slow in North America where manufacturers face greater threats of litigation on improper application/product liability. Therefore the further exploitation of these steels is very dependent on the development of improved toughness via controlled processing or grain-refinement techniques.

Controlled processed bars

There is a limited demand for normalized or quenched and tempered bar products that can be machined directly to the finished component form. Given that these heat treatments are expensive, there is obviously an incentive to develop the required mechanical properties in the as-rolled condition, i.e. the incentive is similar to that described in the previous section for the elimination of heat treatment in automotive forgings.

Normalized steels

Normalizing is applied to bar products in order to refine the grain size and facilitate subsequent processing operations such as machining or cold forging. The steel is reheated to just above the reaustenization temperature (Ac_3) to achieve a fine austenite grain size and the material is then allowed to cool freely in air. In order to simulate such a structure in the 'as-rolled' condition, three options can be considered:

1. Lower billet reheating temperatures.
2. Lower bar finishing temperatures.
3. Faster cooling rates after rolling.

In rolling mills that do not have inter-stand cooling or delayed rolling facilities, options (1) and (2) are interdependent and both are aimed at the refinement of the austenite grain size. Whereas the austenite grain size will affect the pearlite content and ferrite grain size, these microstructural parameters are also influenced significantly by the rate of cooling from the finishing temperature.

Japanese steelmakers have introduced the term *normalize-free* for controlled rolled products that involve billet reheating temperatures of around 1050°C and finishing at or below 900°C in order to simulate the properties of normalized bars. Whereas such finishing temperatures can be achieved by introducing delays into the rolling schedule, economic rates of production can only be sustained with the adoption of inter-stand cooling.

Quenched and tempered steels

As indicated previously, the tensile properties of quenched and tempered alloy steel forgings can be reproduced in air-cooled, micro-alloy steels but the latter tend to produce inferior impact properties. However, rolling offers a greater opportunity than forging for low-temperature finishing and therefore a greater potential for improved toughness via grain refinement.

The relative effects of vanadium and niobium on strength–toughness relationships are shown in Figure 3.44. These additions were made to base compositions containing 0.33% C but with varying manganese. Under normal

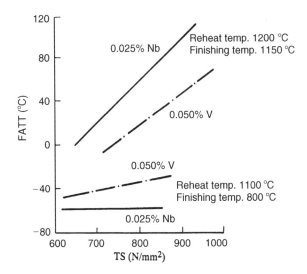

Figure 3.44   *Strength–toughness relationships in 0.33% C, micro-alloy steels*

rolling conditions (reheat 1200°C, finish 1150°C), the vanadium addition produces the better impact properties (lower FATT) and the beneficial effect appears to increase slightly with increasing strength. Controlled rolling (reheat 1100°C, finish 800°C) produces a major improvement in toughness in both types of steel but the niobium steel displays the better combination of properties, maintaining an FATT of about − 60°C over the range of tensile strength from 625 to 850 N/mm². The better performance of the niobium steel under controlled rolling conditions is probably related to the fact that niobium has a greater effect than vanadium in suppressing austenite recrystallization during hot rolling which leads to the production of a finer grain size. However, in situations where controlled rolling cannot be employed, then vanadium steels offer the more attractive properties.

High-carbon wire rod

Although tensile strengths >2000 N/mm² are normally associated with lightly tempered martensites, or with maraging grades, these strength levels can also be achieved very readily in wire products by cold-drawing rods with a fine pearlitic microstructure. The demand for such products in the UK amounts to about 230 000 tonnes/annum and some of the more important applications are listed below:

- Wire ropes
- Prestressed concrete wire
- Tyre cord reinforcement
- Bridge suspension cables
- High-pressure hose reinforcement
- Helical springs (bedding and seating)
- Core wire for electrical conductor cables
- Piano strings

Rod rolling and conditioning

All wire is produced from hot-rolled rod and in the context of *high-carbon rod* this involves steels with carbon contents in the range 0.5–0.9%. Depending upon the carbon content, this results in the formation of a mixed ferrite–pearlite or a completely pearlitic microstructure. However, natural cooling in air from a high finishing temperature results in the generation of coarse pearlite which is unsuitable for severe cold-drawing operations. Traditionally, high-carbon rod is reheated to a temperature just above Ac_3, in order to reaustenitize the material, and then quenched into a lead bath at 450–500°C. The steel is therefore allowed to transform isothermally at a relatively low temperature to form a fine lamellar pearlite. Such a treatment is termed *patenting* and develops a high-strength structure which is also capable of extensive cold drawing. Patenting is still employed to a limited extent but most high-carbon steels are now drawn directly from as-rolled rod which has been subjected to an in-line cooling operation at the end of rod rolling. This produces a microstructure which is similar to that developed by the costly patenting process, but, on average, controlled cooled rod has a tensile strength which is about 100 N/mm^2 lower than that obtained on patenting.[50]

As indicated in the previous section, the strength of pearlitic steels is influenced very markedly by the carbon content, which controls the amount of pearlite in the structure, and also by the pearlite interlamellar spacing (S_o). However, for a given carbon content, the volume fraction of pearlite can be increased and the interlamellar spacing refined by depressing the austenite to pearlite transformation temperature. This can be achieved by:

1. Increasing the prior austenite grain size.
2. Increasing the rate of cooling from rod rolling.
3. Adding alloying elements such as chromium and manganese.

Jaiswal and McIvor[50] developed the following equation to illustrate the effects of cooling rate and composition on the tensile strength of controlled cooled, plain carbon steel rod:

$$\text{Tensile strength (N/mm}^2) = [267(\log \text{CR}) - 293] + 1029(\% \text{ C})$$
$$+ 152(\% \text{ Si}) + 210(\% \text{ Mn})$$
$$+ 442(\% \text{ P})^{\frac{1}{2}} + 5244(\% \text{ N}_f)$$

This equation relates to steels with a prior austenite grain size of ASTM 7 and CR is the cooling rate in °C/s at 700°C, i.e. before the start of transformation. Although nitrogen has the largest strengthening coefficient, carbon provides a much bigger contribution to the strength of these materials because it is present in very much higher concentrations.

Wire drawing

Prior to wire drawing, the hot-rolled rods must be cleaned in order to remove scale. This may be carried out mechanically by grit blasting or by subjecting the rods to a series of bending and twisting operations. However, for the more critical applications, acid pickling is employed followed by a neutralizing wash. The cleaned rod is then coated with lime or zinc phosphate for *dry drawing* or with a layer of copper or brass for *wet drawing*.

As illustrated in Figure 3.45, pearlitic steels work harden very rapidly during wire drawing to develop tensile strengths well in excess of 2000 N/mm^2.

Micro-alloy, high-carbon rod

Jaiswal and McIvor[50] have described the use of micro-alloy additions in high-carbon rod in order to compensate for the lower tensile strengths that are achieved in controlled cooled rod compared with patented material. These authors show that additional strength can be achieved by the following mechanisms:

1. Refinement of the pearlite interlamellar spacing with the addition of chromium.
2. Solid solution strengthening using higher silicon contents.
3. Precipitation strengthening by the addition of vanadium.

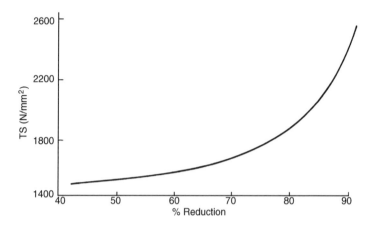

Figure 3.45 *Effect of cold drawing on the strength of high-carbon pearlitic rod*

However, micro-alloying increases the tendency for martensite formation which has a particularly damaging effect on the wire-drawing characteristics. The martensite-forming potential of various elements was therefore determined in controlled cooling experiments and the critical cooling rate at which martensite first appeared in the pearlitic matrix was determined. With a prior austenite grain size of ASTM 7, the following critical cooling rate (CCR) relationship was derived:

$$CCR \ (Ks^{-1}) = 97 - (\%Si) - 70(\%Mn) - 50(\%Cr) - 224(\%P)$$

This equation indicates that manganese is potentially more damaging than chromium but it was also shown that chromium was more effective in producing a finer interlamellar spacing. Whereas silicon had relatively little effect in promoting the formation of martensite, it was also shown to be less effective than chromium in refining the interlamellar spacing. Jaiswal and McIvor therefore propose that silicon is used in combination with chromium, or possibly vanadium, rather than as a single addition.

Given the importance of avoiding the formation of martensite, Jaiswal and McIvor recommend that chromium is better used in large-diameter rods where the cooling rate is relatively slow and therefore less likely to produce martensite. For small-diameter rods, vanadium is recommended.

In a base steel containing 0.85% C and 0.7% Mn, the above authors state that the maximum levels of strength shown in Table 3.29 can be obtained by microalloying.

Table 3.29 *Microalloyed High Carbon Rod*

Additive	Rod diameter	TS (N/mm^2)
0.8% Si, 0.25% Cr	Large	1330
0.07% V	Small	1300

References

1. Bain, E.C. and Davenport, E.S. *Trans. AIME*, **90**, 117 (1930).
2. Cias, W.W. *Phase Transformation Kinetics and Hardenability of Medium Carbon Alloy Steels*, Climax Molybdenum Co., Greenwich, Connecticut.
3. Atkins, M. *Atlas of Continuous Cooling Transformation Diagrams*, British Steel.
4. Siebert, C.A., Doane, D.V. and Breen, D.H. *The Hardenability of Steels*, ASM, Metals Park, Ohio (1977).
5. Grossman, M.A. *Elements of Hardenability*, ASM, Cleveland (1952).
6. Llewellyn, D.T. and Cook, W.T. *Metals Technology*, December, 517 (1974).
7. deRetana, A.F. and Doane, D.V. *Metal Progress*, September, **100**, 105 (1971).
8. Gladman, T. Private communication.
9. Ueno, M. and Inoue, J. *Trans. ISI Japan*, **13** (3), 210 (1973).
10. Kapadia, B.M., Brown, R.M. and Murphy, W.J. *Trans. Met. Soc. AIME*, **242**, 1689 (1968).
11. Smallman, R.E. *Modern Physical Metallurgy* (Fourth Edition), Butterworths.
12. Grange, R.H. and Baughman, R.W. *Trans ASM*, **48**, 165 (1956).
13. Parrish, G. and Harper, G.S. *Production Gas Carburising*, Pergamon Press.
14. Murray, J.D. *Auto Engineer*, **55**, 186 (1965).
15. Llewellyn, D.T. and Cook, W.T. *Metals Technology*, May, 265 (1977).
16. Wannell, P.H., Blank, J.R. and Naylor, D.J. In *Proc. International Symposium on the Influence of Metallurgy on the Machinability of Steel*, September, ISIJ/ASM, Tokyo (1977).
17. Pickett, M.L. Cristinacce, M. and Naylor, D.J. In *Proc. High Productivity Machining, Materials and Processing*, May, ASM, New Orleans (1985).
18. Irani, R.S. *Metals and Materials*, June, 333 (1987).
19. Naylor, D.J. In *Proc. Integrity of Gas Cylinders: Materials Technology*, NPL, **11** (1985).
20. Oldfield, F.K. In *Proc. Integrity of Gas Cylinders: Materials Technology*, NPL, **1** (1985).
21. Harris, D., Priest, A., Davenport, J., McIntyre, P., Almond, E.A. and Roebuck, B. In *Proc. Integrity of Gas Cylinders: Materials Technology*, NPL, **69** (1985).
22. Zaretsky, E.V. *Effect of Steel Manufacturing Processes on the Quality of Bearing Steels* (ed. Hoo, J.C.C.), ASTM Special Technical Report, **5**, 981 (1988).
23. Bamberger, E.N. In *Proc. Tribology in the 80's* (ed. Loomis, W.F.), NASA CP-2300, Vol. 2, National Aeronautics and Space Administration, Washington, DC, 773 (1983).
24. Johnson, R.F. and Sewell, J. *JISI*, December, 414 (1960).
25. Brooksbank, D. and Andrews, K.W. *JISI*, **210**, 246 (1972).
26. Hampshire, J.M. and King, E. *Effect of Steel Manufacturing Processes on the Quality of Bearing Steels* (ed. Hoo, J.C.C.) ASTM Special Technical Report 981, 61 (1988).

27. Davies, I.G. Clarke, M.A. and Dulieu, D. *Effect of Steel Manufacturing Processes on the Quality of Bearing Steels*, ASTM Special Technical Report 981, 375 (1988).
28. Hoyle, G. *High Speed Steels*, Butterworth (1989).
29. *Source Book on Maraging Steels* (ed. Decker, R.F.), ASM.
30. Spitzig, W.A. In *Source Book on Maraging Steels* (ed. Decker, R.F.), 299.
31. Decker, R.F., Eash, J.T. and Goldman, A.J. In *Source Book on Maraging Steels* (ed. Decker, R.F.), 1.
32. *INCO Databook 1976*, International Nickel Co.
33. Magnee, A., Drapier, J.M., Dumont, J., Coutsouradis, D. and Hadbraken, L. *Cobalt-Containing High Strength Steels*, Cobalt Information Centre, Brussels (1974).
34. Collier, J.G. and Gemmill, M.G. *Metals and Materials*, April, 198 (1986).
35. Wyatt, L.M. *Materials of Construction for Steam Power Plant*, Applied Science Publishers (1976).
36. Reynolds, P.E., Barron, J.M. and Allen, G.B. *The Metallurgist and Materials Technologist*, July, 359 (1978).
37. Viswanathan, R. and Jaffee, R.I., *Trans. ASME*, **105**, October, 286 (1983).
38. Gemmill, M.G. *Metals and Materials*, December, 759 (1985).
39. Boyle, C.J., Curran, R.M., DeForrest, D.R. and Newhouse, D.L. *Proc. ASTM*, **62**, 1156 (1962).
40. Everson, H., Orr, J. and Dulieu, D. In *Proc International Conference on Advances in Material Technology for Fossil Power Plants* (eds Viswanathan, R. and Jaffee, R.I.) (Chicago, 1987), ASM International.
41. Pickering, F.B. and Gladman, T. *ISI Special Report 81* (1961).
42. Gladman, T., McIvor, I.D. and Pickering, F.B. *JISI*, **210**, 916 (1972).
43. Preston, R.R. *Steelresearch 87–88*, British Steel, 57.
44. Hodgson, W.H. and Preston, R.R. *CIM Bulletin*, October, 95 (1988).
45. British Steel *Track Products Brochure*, British Steel.
46. Gladman, T. *Ironmaking and Steelmaking*, **16**, No. 4, 241 (1989).
47. Lagnenborg, R. In *Proc. Fundamentals of Microalloying Forging Steels* (eds Krauss, G. and Banerji, S.K.), TMS of AIME, p. 39 (1987).
48. Naylor, D.J. *Ironmaking and Steelmaking*, **16**, No. 4 (1989).
49. Korchynsky, M. and Paules, J.R. *Microalloyed Forging Steels–A State of the Art Review*, SAE 890801 (1989).
50. Jaiswal, S. and McIvor, I.D. *Ironmaking and Steelmaking*, **16**, No. 1, 49 (1989).

4 Stainless steels

Overview

As chromium is added to steels, the corrosion resistance increases progressively due to the formation of a thin protective film of Cr_2O_3, the so-called *passive layer*. With the addition of about 12% Cr, steels have good resistance to atmospheric corrosion and the popular convention is that this is the minimum level of chromium that must be incorporated in an iron-based material before it can be designated a *stainless steel*. However, of all steel types, the stainless grades are the most diverse and complex in terms of composition, microstructure and mechanical properties. Given this situation, it is not surprising that stainless steels have found a very wide range of application, ranging from the chemical, pharmaceutical and power generation industries on the one hand to less aggressive situations in architecture, domestic appliances and street furniture on the other.

By the late 1800s, iron–chromium alloys were in use throughout the world but without the realization of their potential as corrosion-resistant materials. Harry Brearley, a Sheffield metallurgist, is credited with the discovery of martensitic stainless steels in 1913 when working on the development of improved rifle barrel steels. He found that a steel containing about 0.3% C and 13% Cr was difficult to etch and also remained free from rust in a laboratory environment. Such a steel formed the basis of the cutlery industry in Sheffield and as Type 420 is still used for this purpose to the present day.

During the same period, researchers in Germany were responding to pressures for improved steels for the chemical industry. Up until that time, steels containing high levels of nickel were in use as tarnish-resistant materials but had inadequate resistance to corrosion. Two Krupp employees, Benno Strauss and Eduard Maurer, are credited with the discovery of Cr–Ni austenitic stainless steels and patents on these materials were registered in 1912. However, workers in France and the United States are also cited as independent discoverers of these steels.

During the 1920s and 1930s, rapid developments took place which led to the introduction of most of the popular grades that are still in use today, such as Type 302 (18% Cr, 8% Ni), Type 316 (18% Cr, 12% Ni, 2.5% Mo), Type 410 (12% Cr) and Type 430 (17% Cr). However, even in the 1950s, stainless steels were still regarded as semi-precious metals and were priced accordingly. Up until the 1960s, these steels were still produced in small electric arc furnaces, sometimes of less than 10 tonnes capacity. The process was carried out in a single stage, involving the melting of scrap, nickel and ferro-chrome, with production times in excess of $3\frac{1}{2}$ hours. However, substantial gains were achieved with the installation of larger furnaces with capacities greater than 100 tonnes and the introduction of oxygen refining techniques also increased productivity very substantially. Since the early 1970s, the production of stainless steels has been based on a two-stage process, the first employing a conventional electric arc furnace for the rapid melting of scrap and ferro-alloys but using cheap, high-carbon ferro-chrome as the main source of chromium units. The high-carbon melt is then refined in a second stage, using either an argon–oxygen decarburizer (AOD) or by blowing with oxygen under vacuum (VOD). The AOD process is now employed for over 80% of the world's production of stainless steel and produces 100 tonnes of material in less than one hour. However, in addition to achieving faster production rates, the intimate mixing with special slags results in very efficient desulphurization. Other benefits also accrue from the facility to produce carbon contents of less than 0.01% and hydrogen levels of 2–3 ppm.

Substantial cost savings were also achieved with the adoption of continuous casting in place of ingot casting and these overall gains in production have led to a significant cheapening of stainless steel relative to two of its main competitors, namely plastics and aluminium. For the future, there is the prospect of the direct introduction of cheap chromium ores and their reduction by coal in a converter which would lead to further cost reduction.

In terms of product innovation, perhaps the greatest benefits have been obtained from relatively simple changes, such as the introduction of stainless grades with low carbon contents, i.e. below 0.03% C. This modification has virtually eliminated the risk of intergranular corrosion in unstabilized austenitic grades and has also improved the corrosion performance and ductility of ferritic grades. However, steel users have been reluctant to take advantage of higher strength austenitic steels, such as those based on 0.2% N, which can lead to significant cost reductions through the use of reduced thicknesses in pipework and pressure vessels. This contrasts sharply with the situation outlined earlier in Chapter 3 where the micro-alloy grades are now used extensively in place of plain carbon steel. When these high-nitrogen stainless grades were introduced in the UK in the mid-1960s, their high proof strength values could not be used to full advantage because of limitations imposed by the design codes of the day. Welding problems were also encountered due to the fact that the high nitrogen content led to the formation of a fully austenitic weld metal and susceptibility to solidification cracking. However, these problems have now been resolved and therefore there is the prospect of greater utilization of these materials in the future.

The 1970s saw the introduction of the *low interstitial ferritic* grades, with combined carbon and nitrogen contents of less than 200 ppm. These steels are based on compositions such as 18% Cr 2% Mo and 26% Cr 1% Mo and offered the prospect of being a cheaper alternative to an austenitic grade such as Type 316 (18% Cr, 12% Ni, 2.5% Mo). However, whereas the low interstitial grades exhibit good corrosion resistance, particularly with regard to chloride-induced stress corrosion, they tend to retain the problem of conventional ferritic steels in relation to grain coarsening and loss of toughness after welding. On the other hand, high-alloy steels involving duplex austenite plus ferrite microstructures are now gaining acceptance, because of their higher strength and better resistance to stress corrosion than conventional austenitic grades.

Whereas the consumption of bulk steel products is likely to remain fairly static, stainless steel is still very much in a growth market. This relates to the fact that stainless steel has managed to maintain its traditional image as a decorative material but is now also regarded as an engineering material for use in applications where structural integrity is more important that aesthetic appearance. On the basis of life-cycle costing, stainless steels are also proving to be attractive alternatives to mild steel in structures or components that require frequent painting and maintenance.

Underlying metallurgical principles

As indicated in the *Overview*, stainless steel grades cover a wide range of compositions which results in the generation of a variety of microstructures and mechanical properties. This is clearly illustrated in the following compositions which represent three of the common grades of stainless steel:

Composition	Grade	Microstructure
12% Cr	409	martensitic
17% Cr	430	ferritic
18% Cr 9% Ni	304	austenitic

Reference will be made in this chapter to the composition–structure relationships which show that alloying elements in stainless steels can be divided into two groups, namely those that promote the formation of an austenitic structure at hot rolling or solution treatment temperatures and those that promote the formation of delta ferrite. Chromium is the principal alloying element in stainless steels and this promotes the formation of delta ferrite at high temperature. However, iron can accommodate up to about 13% Cr at a temperature of around 1050°C and still remain completely austenitic at that temperature. On the other hand, the $M_s - M_f$ temperature range of a 12% Cr steel is sufficiently high to allow this material to transform completely to martensite on cooling to ambient temperature. An increase in chromium from 12 to 17% brings about a progressive change from austenite to delta ferrite at high temperature and the ferrite remains unchanged on cooling to ambient temperature. Nickel is a strong austenite-forming element and is added to stainless steels in order to preserve an austenitic structure in the presence of high chromium contents. Thus a steel containing 18% Cr 9% Ni is completely austenitic at a temperature of 1050°C but the overall alloy content now depresses the $M_s - M_f$ temperature range to sub-zero temperatures. Therefore this material retains its austenitic structure on cooling to ambient temperature, providing a relatively low strength but a high level of formability.

Elements such as silicon, molybdenum and titanium also promote the formation of delta ferrite at high temperature, whereas carbon, nitrogen, manganese and copper promote the formation of austenite. Therefore consideration must also be given to the presence of these elements, in addition to the balance between chromium and nickel, in determining the structure of stainless steels at elevated temperatures. However, both austenite- and ferrite-forming elements will depress the $M_s - M_f$ range and influence the microstructure formed on cooling to ambient temperature. Therefore the constitution of stainless steels is governed by:

(i) the balance between austenite- and ferrite-forming elements which controls the structure at hot rolling and solution treatment temperatures,
(ii) the overall alloy content which controls the M_s-M_f temperature range and therefore the structure and properties at ambient temperature.

Austenitic stainless steels can be subjected to severe cold-forming operations, for example in the cold rolling of hot band to strip gauges and also in the production of domestic sinks and tableware from annealed strip. This introduces the topic of *strain-induced martensite*, whereby a material which is austenitic in the solution-treated condition can transform partially or completely to martensite with the application of cold work at ambient temperature. Detailed consideration is given to this topic later in the chapter, but in essence it relates to the stability of the austenitic structure, as influenced by the overall alloy content, and the destabilizing effects due to the magnitude and temperature of cold deformation.

The metallurgy of the 12% Cr martensitic grades is similar to that involved in the engineering grades, although the presence of such a large amount of chromium induces a very high degree of hardenability and these steels are capable of developing a martensitic structure in substantial section sizes, even in the air-cooled condition. However, like their low-alloy counterparts, the 12% Cr grades must be tempered to produce a good combination of strength and ductility/toughness and both types of steel often incorporate additions of molybdenum and vanadium in order to improve the tempering resistance through the formation of stable carbides.

Whereas austenitic stainless steels are used in domestic or architectural applications, where corrosion resistance and aesthetic appeal are the main requirements, they are also employed in pressure vessels where both corrosion resistance and strength are important considerations. As indicated in the *Overview*, solid solution strengthening with additions of nitrogen is the main avenue for the production of higher-strength austenitic stainless steels. Precipitation-strengthening reactions can also be induced in these grades through the precipitation of carbides and intermetallic compounds based on nickel, aluminium and titanium. However, such materials have found little commercial application, due possibly to weldability problems and poor corrosion properties.

Stainless steels resist corrosion through the formation of a thin passive film of Cr_2O_3 and, very broadly, the corrosion resistance of these materials increases with chromium content. However, as illustrated by the previous remarks, marked variations in microstructure are introduced with the addition of alloying elements with a marked effect on mechanical properties. Thus the 12% Cr martensitic grades are capable of developing high levels of strength but with only moderate resistance to corrosion. In contrast, an austenitic grade, based on 18% Cr 9% Ni, has a low strength but a significantly higher resistance to corrosion, the latter being enhanced by the addition of molybdenum. Therefore throughout the broad range of stainless steels, there will be a compromise between corrosion resistance and other properties, such as strength, formability and weldability. Additionally, it is necessary to differentiate between the various types of corrosion in stainless steels, notably:

general corrosion
intergranular corrosion
pitting corrosion
stress corrosion

However, whereas these types of corrosion will be discussed at a later stage, the authors are conscious of the fact that the treatment of corrosion in this text is superficial and reflects their limited knowledge of the subject.

Composition-structure relationships

Iron–chromium alloys

The simplest stainless steels consist of iron–chromium alloys but in fact the binary iron–chromium system can give rise to a wide variety of microstructures with markedly different mechanical properties. The Fe–Cr equilibrium diagram is shown in Figure 4.1 and is characterized by two distinctive features, namely:

1. The presence of sigma phase at about 50% Cr.
2. The restricted austenite phase field, often called the *gamma-loop*.

Sigma phase is an *intermetallic compound*, which is hard and brittle and can be produced in alloys containing substantially less than 50% Cr. It also has an adverse effect on the corrosion resistance of stainless steels and therefore care should be taken to avoid extended exposure in the temperature range 750–820°C which favours its formation.

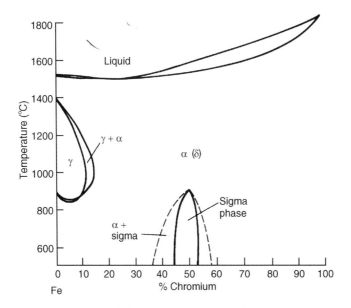

Figure 4.1   *Fe–Cr equilibrium diagram*

From a commercial standpoint, the area of the Fe–Cr diagram of greatest importance is that containing up to about 25% Cr, and a simplified illustration of that region for alloys containing 0.1% C is shown in Figure 4.2. Because chromium is a mild carbide former, many types of stainless steel are solution treated at temperatures significantly higher than those used for low-alloy steels in order to dissolve the chromium carbides. A solution treatment temperature of 1050°C is typical of a variety of stainless steel grades and this will be used as a reference temperature in relation to the microstructure at high temperature. As illustrated in Figure 4.2, 0.1% C steels can accommodate up to about 13.5% Cr at 1050°C and still remain austenitic with a face-centred-cubic structure. As the chromium content of the steels is increased within this range, the hardenability also increases very substantially such that large section sizes can be through-hardened to martensite on cooling to room temperature. For example, a steel containing 12% Cr and 0.12% C will form martensite at the centre of a 100 mm bar on air cooling from 1050°C and the limiting section can be increased to about 500 mm by oil quenching from this temperature. It should also be noted that the M_s–M_f transformation range is depressed significantly with large additions of chromium. However, for most commercial grades of 11–13% Cr steels, the transformation range is above room temperature and therefore the formation of retained austenite is not a major problem.

As the chromium content is increased above about 13%, a significant change takes place in the microstructure at 1050°C as the single-phase austenite region gives way to the *duplex* austenite plus ferrite phase field. Whereas the austenite in this region behaves in a similar way to that within the gamma-loop, i.e. it transforms to martensite on cooling to room temperature, the ferrite formed at

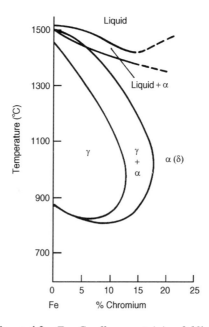

Figure 4.2   *Fe–Cr alloys containing 0.1% C*

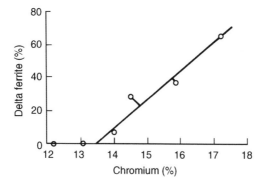

Figure 4.3 *Effect of chromium content on 0.1% carbon steels solution treated at 1050°C (After Irvine et al.[1])*

high temperature undergoes no phase transformation. Although its structure is body-centred-cubic, the high-temperature phase is generally called delta ferrite (δ) in order to differentiate it from alpha ferrite (α), the transformation product from austenite. As illustrated in Figure 4.3, the delta ferrite content increases progressively with further additions of chromium and, in the presence of about 0.1% C, the material becomes completely ferritic with the addition of just over 18% Cr. Therefore, between about 13% and 18% Cr, the hardness of these steels is reduced as the microstructure changes progressively from 100% martensite to 100% delta ferrite. Larger additions of chromium have no further effect on the microstructure, although such materials become increasingly more susceptible to the formation of sigma phase.

Although the changes from $\gamma \rightarrow \gamma + \delta$ and $\gamma + \delta \rightarrow \delta$ occur at chromium levels of about 13.5% and 18.5% respectively, it must be emphasized that this refers to alloys containing a maximum of about 0.1% C and at a reference temperature of around 1050°C. Larger amounts of carbon or the addition of other alloying elements will have a major effect on the microstructure associated with particular levels of chromium. Additionally, for a given composition, an increase in the solution treatment temperature above 1050°C will also increase the amount of delta ferrite at the expense of austenite.

Iron-chromium-nickel alloys

Whereas chromium restricts the formation of austenite, nickel has the opposite effect and, as illustrated in Figure 4.4, the Fe-Ni equilibrium diagram displays an expanded austenite phase field. In the context of stainless steels, chromium is therefore termed a *ferrite former* and nickel an *austenite former*. Thus having created a substantially ferritic microstructure with a large addition of chromium, it is possible to reverse the process and re-establish an austenitic structure by adding a large amount of nickel to a high-chromium steel.

As indicated in Figure 4.3, a steel containing 17% Cr and 0.1% C will have a microstructure of about 65% delta ferrite-35% austenite at a solution treatment temperature of 1050°C. The various changes that then occur with the addition

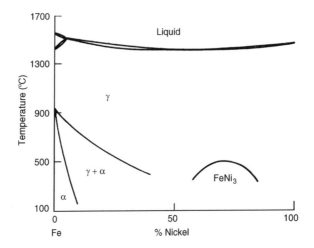

Figure 4.4   *Fe–Ni equilibrium diagram*

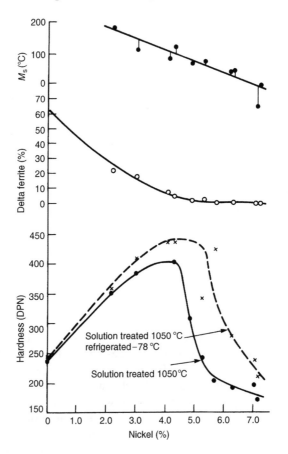

Figure 4.5   *Effect of nickel content on the structure, hardness and M_s temperature of 0.1% C 17% Cr steels (After Irvine* et al.[1]*)*

of nickel to a base steel of this composition are illustrated in Figure 4.5. This shows that the delta ferrite content is steadily reduced and at 1050°C the steels become fully austenitic with the addition of about 5% nickel. On cooling to room temperature, the austenite in these low-nickel steels transforms to martensite and therefore there is initially a progressive increase in hardness with the addition of nickel as martensite replaces delta ferrite. However, the addition of nickel also depresses the $M_s–M_f$ transformation range and at nickel contents greater than about 4% the M_f temperature is depressed below room temperature. Further additions of nickel therefore lead to a decrease in hardness due to incomplete transformation to martensite and the formation of retained austenite. As indicated by the hardness data in Figure 4.5, refrigeration at −78°C causes the retained austenite to transform to martensite over a limited composition range until the M_s temperature coincides with the refrigeration temperature. In commercial 18% Cr 9% Ni austenitic stainless steels, the M_s has been depressed to very low temperatures and little transformation to martensite can be induced, even at the liquid nitrogen temperature of −196°C.

Other alloy additions

Whereas chromium and nickel are the principal alloying elements in stainless steels, other elements may be added for specific purposes and therefore consideration must be given to the effect of these elements on microstructure. Like chromium and nickel, these other alloying elements can be classed as ferrite or austenite formers and their behaviour is illustrated in Figure 4.6 which refers to a base steel containing 17% Cr and 4% Ni. Thus elements such as aluminium, vanadium, molybdenum, silicon and tungsten behave like chromium and promote the formation of delta ferrite. On the other hand, copper, manganese, cobalt, carbon and nitrogen have a similar effect to nickel and promote the formation of austenite. A guide to the potency of the various elements in their role as austenite or ferrite formers[1] is shown in Table 4.1.

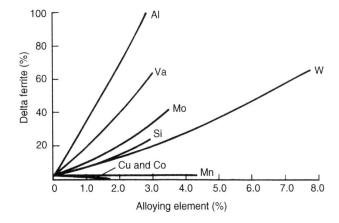

Figure 4.6 *Effect of various alloying elements on the structure of 17% Cr 4% Ni alloys (After Irvine et al.[1])*

Table 4.1

Element	Change in delta ferrite per 1.0 wt%	
N	−200	
C	−180	
Ni	−10	
Co	−6	Austenite
Cu	−3	formers
Mn	−1	
W	+8	
Si	+8	
Mo	+11	Ferrite
Cr	+15	formers
V	+19	
Al	+38	

Thus carbon and nitrogen are particularly powerful austenite formers and the latter is incorporated in certain grades of stainless steel, specifically for this purpose. Elements such as titanium and niobium are also ferrite formers in their own right but have an additional ferrite-promoting effect by virtue of the fact that they are also strong carbide and nitride formers and can therefore eliminate the austenite-forming effects of carbon and nitrogen.

Whereas alloying elements oppose each other in terms of austenite or ferrite formation at elevated temperatures, they act in a similar manner in depressing the martensite transformation range. Andrews[2] has derived the following formula for the calculation of M_s:

$$M_s(°C) = 539 - 423C - 30.4Mn - 17.7Ni - 12.1Cr - 7.5Mo$$

Therefore, in predicting the room temperature microstructure of stainless steels, consideration has to be given to two major effects:

1. The balance between austenite and ferrite formers which dictates the microstructure at elevated temperatures.
2. The overall alloy content which controls the M_s–M_f transformation range and the degree of transformation to martensite at ambient temperature.

A convenient but very approximate method of relating composition and microstructure in stainless steels is by means of the Schaeffler diagram which has been modified by Schneider.[3] This is illustrated in Figure 4.7, which indicates the structures produced in a wide range of compositions after rapid cooling from 1050°C. In this diagram, the elements that behave like chromium in promoting the formation of ferrite are expressed in terms of a *chromium equivalent*:

$$Cr \text{ equivalent} = (Cr) + (2Si) + (1.5Mo) + (5V) + (5.5Al) + (1.75Nb)$$
$$+ (1.5Ti) + (0.75W)$$

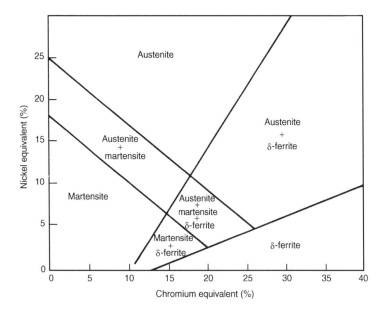

Figure 4.7 *Schaeffler diagram–modified (After Schneider[3])*

In a similar manner, the austenite-forming elements are expressed in terms of a *nickel equivalent*:

$$\text{Ni equivalent} = (\text{Ni}) + (\text{Co}) + (0.5\text{Mn}) + (0.3\text{Cu}) + (25\text{N}) + (30\text{C})$$

all concentrations being expressed as weight percentages.

Commercial grades of stainless steels

From the foregoing remarks, it can be appreciated that stainless steels embrace a wide range of microstructures which are controlled by means of a complex relationship with composition. Although chromium may be the principal alloying element in a stainless steel, the level may give little indication of structure, and steels containing 17% Cr can be martensitic, ferritic or austenitic, depending on heat treatment and the presence of other elements. In discussing the characteristics of stainless steels, it is therefore more convenient to categorize these materials in terms of microstructure rather than composition.

Although most industrialized countries have developed their own national standards for stainless steels, these steels are referred to almost universally by means of the American Iron and Steel Institute (AISI) numbering system. Thus the martensitic stainless steels are classified as the 400 series but, rather confusingly, the 400 series also includes the ferritic grades of stainless steel. The more important grades of austenitic stainless steel are classified in the 300 series. The nominal compositions of the common stainless grades, according to the AISI numbering system, are shown in Table 4.2. In the UK, the familiar specification

Table 4.2   *Nominal composition of stainless steel grades*

AISI Type	Nominal composition
410	12.5% Cr, 0.15% C max.
420	13% Cr, 0.15% C min.
430	17% Cr, 0.12% C max.
434	17% Cr, 1.0% Mo
302	18% Cr, 9% Ni
304	19% Cr, 10% Ni
316	17% Cr, 12% Ni, 2.5% Mo
321	18% Cr, 10.5% Ni, Ti min. $= 5 \times$ C%
347	18% Cr, 11% Ni, Nb min. $= 10 \times$ C%
310	25% Cr, 20% Ni

for stainless flat products was BS 1449: Part 2 *Stainless and heat resisting grades*, but this has now been superseded by the following European specifications:

BS EN 10088-1: 1995	List of stainless steels
BS EN 10088-2: 1995	Technical delivery conditions for sheet/plate and strip for general purposes
BS EN 10088-3: 1995	Technical delivery conditions for semi-finished products, bars, rods and sections for general purposes

The last specification for stainless long products also supersedes those covered earlier in BS 970: Part 1: 1991. As indicated in Table 4.3 BS EN 10088–1 includes 20 ferritic grades, 20 martensitic grades (including precipitation-hardened steels), 37 austenitic grades and 6 austenitic–ferritic (duplex) grades. It also contains useful data on the physical properties of these materials. BS EN 10088–2 and 10088–3 present details of the mechanical properties.

Martensitic stainless steels

A fully martensitic structure can be developed in a stainless steel provided:

1. The balance of alloying elements produces a fully austenitic structure at the solution treatment temperature, e.g. 1050°C.
2. The M_s–M_f temperature range is above room temperature.

As indicated earlier, these conditions are met in the case of a 12% Cr 0.1% C steel and such a grade (Type 410) defines the lower bound of composition in the commercial range of martensitic stainless steels. However, given that the M_s temperature of a 12% Cr 0.1% C steel is substantially above room temperature, there is the facility to make further alloy additions and still maintain the martensitic transformation range above room temperature. Therefore the alloy content can be increased in order to obtain the following improvements in properties:

1. A higher martensitic hardness by means of an increased carbon content.

Table 4.3(a) *BS EN 10088-1: 1995 Stainless Steels*

| Steel designation | | Chemical composition (cast analysis)[1] of ferritic stainless steels | | | | | | | | | | | % by mass |
Name	Number	C max.	Si max.	Mn max.	P max.	S	N max.	Cr	Mo	Nb	Ni	Ti	Others
X2CrNi12	1.4003	0.030	1.00	1.50	0.040	≤0.015	0.030	10.50 to 12.50			0.30 to 1.00		
X2CrTi12	1.4512	0.030	1.00	1.00	0.040	≤0.015		10.50 to 12.50				$6 \times (C+N)$ to 0.65	
X6CrNiTi12	1.4516	0.08	0.70	1.50	0.040	≤0.015		10.50 to 12.50			0.50 to 1.50	0.05 to 0.35	
X6Cr13	1.4000	0.08	1.00	1.00	0.040	≤0.015[2]		12.00 to 14.00					
X6CrAl13	1.4002	0.08	1.00	1.00	0.040	≤0.015[2]		12.00 to 14.00					Al: 0.10 to 0.30
X2CrTi17	1.4520	0.025	0.50	0.50	0.040	≤0.015	0.015	16.00 to 18.00				0.30 to 0.60	
X6Cr17	1.4016	0.08	1.00	1.00	0.040	≤0.015[2]		16.00 to 18.00					
X3CrTi17	1.4510	0.05	1.00	1.00	0.040	≤0.015[2]		16.00 to 18.00				$4 \times (C+N)$ +0.15 to 0.80[3]	
X3CrNb17	1.4511	0.05	1.00	1.00	0.040	≤0.015		16.00 to 18.00		$12 \times C$ to 1.00			
X6CrMo17-1	1.4113	0.08	1.00	1.00	0.040	≤0.015[2]		16.00 to 18.00	0.90 to 1.40				
X6CrMoS17	1.4105	0.08	1.50	1.50	0.040	0.15 to 0.35		16.00 to 18.00	0.20 to 0.60				
X2CrMoTi17-1	1.4513	0.025	1.00	1.00	0.040	≤0.015	0.015	16.00 to 18.00	1.00 to 1.50			0.30 to 0.60	

continued overleaf

Table 4.3(a) (continued)

| Steel designation | | Chemical composition (cast analysis)[1] of ferritic stainless steels | | | | | | | | | | | % by mass |
Name	Number	C max.	Si max.	Mn max.	P max.	S	N max.	Cr	Mo	Nb	Ni	Ti	Others
X2CrMoTi18-2	1.4521	0.025	1.00	1.00	0.040	≤0.015	0.030	17.00 to 20.00	1.80 to 2.50			$4 \times (C+N)$ +0.15 to 0.80[3]	
X2CrMoTiS18-2*	1.4523*	0.030	1.00	0.50	0.040	0.15 to 0.35		17.50 to 19.00	2.00 to 2.50			0.30 to 0.80	(C + N) ≤ 0.040
X6CrNi17-1*	1.4017*	0.08	1.00	1.00	0.040	≤0.015		16.00 to 18.00			1.20 to 1.60		
X6CrMoNb17-1	1.4526	0.08	1.00	1.00	0.040	≤0.015	0.040	16.00 to 18.00	0.80 to 1.40	$7 \times (C+N)$ +0.10 to 1.00			
X2CrNbZr17*	1.4590*	0.030	1.00	1.00	0.040	≤0.015		16.00 to 17.50		0.35 to 0.55			$Zr \geq 7\times$ (C + N) +0.15
X2CrAlTi18-2	1.4605	0.030	1.00	1.00	0.040	≤0.015		17.00 to 18.00				$4 \times (C+N)$ +0.15 to 0.80[3]	Al: 1.70 to 2.10
X2CrTiNb18	1.4509	0.030	1.00	1.00	0.040	≤0.015		17.50 to 18.50		$3 \times C + 0.30$ to 1.00		0.10 to 0.60	
X2CrMoTi29-4	1.4592	0.025	1.00	1.00	0.030	≤0.010	0.045	28.00 to 30.00	3.50 to 4.50			$4 \times (C+N)$ +0.15 to 0.80[3]	

[1] Elements not listed in this table may not be intentionally added to the steel without the agreement of the purchaser except for finishing the cast. All appropriate precautions are to be taken to avoid the addition of such elements from scrap and other materials used in production which would impair mechanical properties and the suitability of the steel.

[2] For bars, rods, sections and the relevant semi-finished products, a maximum content of 0.030% S applies.
For any product to be machined, a controlled sulphur content of 0.015% to 0.030% is recommended and permitted.

[3] The stabilization may be made by use of titanium or niobium or zirconium. According to the atomic number of these elements and the content of carbon and nitrogen, the equivalence shall be the following:

$$Ti \triangleq \tfrac{7}{4} Nb \triangleq \tfrac{7}{4} Zr$$

*Patented steel grade.

Table 4.3(b) BS EN 10088-1: 1995 Stainless Steels

Chemical composition (cast analysis)[1] of martensitic and precipitation hardening stainless steels

Steel designation		C[2]	Si max.	Mn max.	P max.	S	Cr	Cu	Mo	Nb	Ni	Others
Name	Number											% by mass
X12Cr13	1.4006	0.08 to 0.15	1.00	1.50	0.040	≤0.015[3]	11.50 to 13.50				≤0.75	
X12CrS13	1.4005	0.08 to 0.15	1.00	1.50	0.040	0.15 to 0.35	12.00 to 14.00		≤0.60			
X20Cr13	1.4021	0.16 to 0.25	1.00	1.50	0.040	≤0.015[3]	12.00 to 14.00					
X30Cr13	1.4028	0.26 to 0.35	1.00	1.50	0.040	≤0.015[3]	12.00 to 14.00					
X29CrS13	1.4029	0.25 to 0.32	1.00	1.50	0.040	0.15 to 0.25	12.00 to 13.50		≤0.60			
X39Cr13	1.4031	0.36 to 0.42	1.00	1.00	0.040	≤0.015[3]	12.50 to 14.50					
X46Cr13	1.4034	0.43 to 0.50	1.00	1.00	0.040	≤0.015[3]	12.50 to 14.50					
X50CrMoV15	1.4116	0.45 to 0.55	1.00	1.00	0.040	≤0.015[3]	14.00 to 15.00		0.50 to 0.80			V: 0.10 to 0.20
X70CrMo15	1.4109	0.65 to 0.75	0.70	1.00	0.040	≤0.015[3]	14.00 to 16.00		0.40 to 0.80			
X14CrMoS17	1.4104	0.10 to 0.17	1.00	1.50	0.040	0.15 to 0.35	15.50 to 17.50		0.20 to 0.60			
X39CrMo17-1	1.4122	0.33 to 0.45	1.00	1.50	0.040	≤0.015[3]	15.50 to 17.50		0.80 to 1.30		≤1.00	
X105CrMo17	1.4125	0.95 to 1.20	1.00	1.00	0.040	≤0.015[3]	16.00 to 18.00		0.40 to 0.80			

continued overleaf

Table 4.3(b) (continued)

Chemical composition (cast analysis)[1] of martensitic and precipitation hardening stainless steels

| Steel designation | | | | | | | | | | | | % by mass |
Name	Number	C[2]	Si max.	Mn max.	P max.	S	Cr	Cu	Mo	Nb	Ni	Others
X90CrMoV18	1.4112	0.85 to 0.95	1.00	1.00	0.040	≤0.015[3]	17.00 to 19.00		0.90 to 1.30			V: 0.07 to 0.12
X17CrNi16-2	1.4057	0.12 to 0.22	1.00	1.50	0.040	≤0.015[3]	15.00 to 17.00				1.50 to 2.50	
X3CrNiMo13-4	1.4313	≤0.05	0.70	1.50	0.040	≤0.015	12.00 to 14.00		0.30 to 0.70		3.50 to 4.50	N: ≥ 0.020
X4CrNiMo16-5-1	1.4418	≤0.06	0.70	1.50	0.040	≤0.015[3]	15.00 to 17.00		0.80 to 1.50		4.00 to 6.00	N: ≥ 0.020
X5CrNiCuNb16-4	1.4542	≤0.07	0.70	1.50	0.040	≤0.015[3]	15.00 to 17.00	3.00 to 5.00	≤0.60	5 × C to 0.45	3.00 to 5.00	
X7CrNiAl17-7	1.4568	≤0.09	0.70	1.00	0.040	≤0.015	16.00 to 18.00				6.50 to 7.80[4]	Al: 0.70 to 1.50
X8CrNiMoAl15-7-2	1.4532	≤0.10	0.70	1.20	0.040	≤0.015	14.00 to 16.00		2.00 to 3.00		6.50 to 7.80	Al: 0.70 to 1.50
X5CrNiMoCuNb14-5	1.4594	≤0.07	0.70	1.00	0.040	≤0.015	13.00 to 15.00	1.20 to 2.00	1.20 to 2.00	0.15 to 0.60	5.00 to 6.00	

[1]Elements not quoted in this table may not be intentionally added to the steel without the agreement of the purchaser except for finishing the cast. All appropriate precautions are to be taken to avoid the addition of such elements from scrap and other materials used in production which would impair mechanical properties and the suitability of the steel.

[2]Tighter carbon ranges may be agreed at the time of enquiry and order.

[3]For bars, rods, sections and the relevant semi-finished products, a maximum content of 0.030% S applies. For any product to be machined, a controlled sulphur content of 0.015% to 0.030% is recommended and permitted.

[4]For better cold deformability, the upper limit may be increased to 8.30%.

Table 4.3(c) *BS EN 10088-1: 1995 Stainless Steels*

Chemical composition (cast analysis)[1] of austenitic stainless steels

% by mass

Steel designation		C	Si	Mn	P max.	S	N	Cr	Cu	Mo	Nb	Ni	Ti
Name	Number												
X10CrNi18-8	1.4310	0.05 to 0.15	≤2.00	≤2.00	0.045	≤0.015	≤0.11	16.00 to 19.00		≤0.80		6.00 to 9.50	
X2CrNiN18-7	1.4318	≤0.030	≤1.00	≤2.00	0.045	≤0.015	0.10 to 0.20	16.50 to 18.50				6.00 to 8.00	
X2CrNi18-9	1.4307	≤0.030	≤1.00	≤2.00	0.045	≤0.015[2]	≤0.11	17.50 to 19.50				8.00 to 10.00	
X2CrNi19-11	1.4306	≤0.030	≤1.00	≤2.00	0.045	≤0.015[2]	≤0.11	18.00 to 20.00				10.00 to 12.00[3]	
X2CrNiN18-10	1.4311	≤0.030	≤1.00	≤2.00	0.045	≤0.015[2]	0.12 to 0.22	17.00 to 19.50				8.50 to 11.50	
X5CrNi18-10	1.4301	≤0.07	≤1.00	≤2.00	0.045	≤0.015[2]	≤0.11	17.00 to 19.50				8.00 to 10.50	
X8CrNiS18-9	1.4305	≤0.10	≤1.00	≤2.00	0.045	0.015 to 0.35	≤0.11	17.00 to 19.00	≤1.00			8.00 to 10.00	
X6CrNiTi18-10	1.4541	≤0.08	≤1.00	≤2.00	0.045	≤0.015[2]	≤0.11	17.00 to 19.00				9.00 to 12.00[3]	5 × C to 0.70
X6CrNiNb18-10	1.4550	≤0.08	≤1.00	≤2.00	0.045	≤0.015		17.00 to 19.00			10 × C to 1.00	9.00 to 12.00[3]	
X4CrNi18-12	1.4303	≤0.06	≤1.00	≤2.00	0.045	≤0.015[2]	≤0.11	17.00 to 19.00				11.00 to 13.00	
X1CrNi25-21	1.4335	≤0.020	≤0.25	≤2.00	0.025	≤0.010	≤0.11	24.00 to 26.00		≤0.20		20.00 to 22.00	
X2CrNiMo17-12-2	1.4404	≤0.030	≤1.00	≤2.00	0.045	≤0.015[2]	≤0.11	16.50 to 18.50		2.00 to 2.50		10.00 to 13.00[3]	
X2CrNiMoN17-11-2	1.4406	≤0.030	≤1.00	≤2.00	0.045	≤0.015[2]	0.12 to 0.22	16.50 to 18.50		2.00 to 2.50		10.00 to 12.00[3]	

continued overleaf

Table 4.3(c) (continued)

Steel designation		Chemical composition (cast analysis)[1] of austenitic stainless steels — % by mass

Name	Number	C	Si	Mn	P	S	N	Cr	Cu	Mo	Nb	Ni	Ti
					max.								
X5CrNiMo17-12-2	1.4401	≤0.07	≤1.00	≤2.00	0.045	≤0.015[2]	≤0.11	16.50 to 18.50		2.00 to 2.50		10.00 to 13.00	
X1CrNiMoN25-22-2	1.4466	≤0.020	≤0.70	≤2.00	0.025	≤0.010	0.10 to 0.16	24.00 to 26.00		2.00 to 2.50		21.00 to 23.00	
X6CrNiMoTi17-12-2	1.4571	≤0.08	≤1.00	≤2.00	0.045	≤0.015[2]		16.50 to 18.50		2.00 to 2.50		10.50 to 13.50[3]	5 × C to 0.70
X6CrNiMoNb17-12-2	1.4580	≤0.08	≤1.00	≤2.00	0.045	≤0.015		16.50 to 18.50		2.00 to 2.50	10 × C to 1.00	10.50 to 13.50	
X2CrNiMo17-12-3	1.4432	≤0.030	≤1.00	≤2.00	0.045	≤0.015[2]	≤0.11	16.50 to 18.50		2.50 to 3.00		10.50 to 13.00	
X2CrNiMoN17-13-3	1.4429	≤0.030	≤1.00	≤2.00	0.045	≤0.015	0.12 to 0.22	16.50 to 18.50		2.50 to 3.00		11.00 to 14.00[3]	
X3CrNiMo17-13-3	1.4436	≤0.05	≤1.00	≤2.00	0.045	≤0.015[2]	≤0.11	16.50 to 18.50		2.50 to 3.00		10.50 to 13.00[3]	
X2CrNiMo18-14-3	1.4435	≤0.030	≤1.00	≤2.00	0.045	≤0.015[2]	≤0.11	17.00 to 19.00		2.50 to 3.00		12.50 to 15.00	
X2CrNiMoN18-12-4	1.4434	≤0.030	≤1.00	≤2.00	0.045	≤0.015	0.10 to 0.20	16.50 to 19.50		3.00 to 4.00		10.50 to 14.00[3]	
X2CrNiMo18-15-4	1.4438	≤0.030	≤1.00	≤2.00	0.045	≤0.015[2]	≤0.11	17.50 to 19.50		3.00 to 4.00		13.00 to 16.00[3]	
X2CrNiMoN17-13-5	1.4439	≤0.030	≤1.00	≤2.00	0.045	≤0.015	0.12 to 0.22	16.50 to 18.50		4.00 to 5.00		12.50 to 14.50	
X1CrNiSi18-15-4	1.4361	≤0.015	3.70 to 4.50	≤2.00	0.025	≤0.010	≤0.11	16.50 to 18.50		≤0.20		14.00 to 16.00	
X12CrMnNiN17-7-5	1.4372	≤0.15	≤1.00	5.50 to 7.50	0.045	≤0.015	0.05 to 0.25	16.00 to 18.00				3.50 to 5.50	

					max.						
X2CrMnNiN17-7-5	1.4371	≤0.030	≤1.00	6.00 to 8.00	0.045	≤0.015	0.15 to 0.20	16.00 to 17.00			3.50 to 5.50
X12CrMnNiN18-9-5	1.4373	≤0.15	≤1.00	7.50 to 10.50	0.045	≤0.015	0.05 to 0.25	17.00 to 19.00			4.00 to 6.00
X3CrNiCu19-9-2	1.4560	≤0.035	≤1.00	1.50 to 2.00	0.045	≤0.015	≤0.011	18.00 to 19.00	1.50 to 2.00		8.00 to 9.00
X6CrNiCuS18-9-2	1.4570	≤0.08	≤1.00	≤2.00	0.045	0.15 to 0.35	≤0.011	17.00 to 19.00	1.40 to 1.80	≤0.60	8.00 to 10.00
X3CrNiCu18-9-4	1.4567	≤0.04	≤1.00	≤2.00	0.045	≤0.015[2]	≤0.011	17.00 to 19.00	3.00 to 4.00		8.50 to 10.50
X3CrNiCuMo17-11-3-2	1.4578	≤0.04	≤1.00	≤1.00	0.045	≤0.015	≤0.011	16.50 to 17.50	3.00 to 3.50	2.00 to 2.50	10.00 to 11.00
X1NiCrMoCu31-27-4	1.4563	≤0.020	≤0.70	≤2.00	0.030	≤0.010	≤0.011	26.00 to 28.00	0.70 to 1.50	3.00 to 4.00	30.00 to 32.00
X1NiCrMoCu25-20-5	1.4539	≤0.020	≤0.70	≤2.00	0.030	≤0.010	≤0.15	19.00 to 21.00	1.20 to 2.00	4.00 to 5.00	24.00 to 26.00
X1CrNiMoCuN25-25-5	1.4537	≤0.020	≤0.70	≤2.00	0.030	≤0.010	0.17 to 0.25	24.00 to 26.00	1.00 to 2.00	4.70 to 5.70	24.00 to 27.00
X1CrNiMoCuN20-18-7*	1.4547*	≤0.020	≤0.70	≤1.00	0.030	≤0.010	0.18 to 0.25	19.50 to 20.50	0.50 to 1.00	6.00 to 7.00	17.50 to 18.50
X1NiCrMoCuN25-20-7	1.4529	≤0.020	≤0.50	≤1.00	0.030	≤0.010	0.15 to 0.25	19.00 to 21.00	0.50 to 1.50	6.00 to 7.00	24.00 to 26.00

1) Elements not quoted in this table may not be intentionally added to the steel without the agreement of the purchaser except for finishing the cast. All appropriate precautions are to be taken to avoid the addition of such elements from scrap and other materials used in production which would impair mechanical properties and the suitability of the steel.

2) For bars, rods, sections and the relevant semi-finished products, a maximum content of 0.030% S applies. For any product to be machined, a controlled sulphur content of 0.015% to 0.030% is recommended and permitted.

3) Where for special reasons, e.g. hot workability for the fabrication of seamless tubes where it is necessary to minimize the delta ferrite content, or with the aim of low permeability, the maximum Ni content may be increased by the following amounts:

0.50% (m/m): 1.4571
1.00% (m/m): 1.4306, 1.4406, 1.4429, 1.4434, 1.4436, 1.4438, 1.4541, 1.4550
1.50% (m/m): 1.4404,

*Patented steel grade.

Table 4.3(d) *BS EN 10088-1: 1995 Stainless Steels*

Chemical composition (cast analysis)[1] of austenitic–ferritic stainless steels

Steel designation		C max.	Si max.	Mn max.	P max.	S max.	N	Cr	Cu	Mo	Ni	W
Name	*Number*											*% by mass*
X2CrNiN23-4*	1.4362*	0.030	1.00	2.00	0.035	0.015	0.05 to 0.20	22.00 to 24.00	0.10 to 0.60	0.10 to 0.60	3.50 to 5.50	
X3CrNiMoN27-5-2	1.4460	0.05	1.00	2.00	0.035	0.015[2]	0.05 to 0.20	25.00 to 28.00		1.30 to 2.00	4.50 to 6.50	
X2CrNiMoN22-5-3	1.4462	0.030	1.00	2.00	0.035	0.015	0.10 to 0.22	21.00 to 23.00		2.50 to 3.50	4.50 to 6.50	
X2CrNiMoCuN25-6-3	1.4507	0.030	0.70	2.00	0.035	0.015	0.15 to 0.30	24.00 to 26.00	1.00 to 2.50	2.70 to 4.00	5.50 to 7.50	
X2CrNiMoN25-7-4*	1.4410*	0.030	1.00	2.00	0.035	0.015	0.20 to 0.35	24.00 to 26.00		3.00 to 4.50	6.00 to 8.00	
X2CrNiMoCuWN25-7-4	1.4501	0.030	1.00	1.00	0.035	0.015	0.20 to 0.30	24.00 to 26.00	0.50 to 1.00	3.00 to 4.00	6.00 to 8.00	0.50 to 1.00

[1]Elements not listed in this table may not be intentionally added to the steel without the agreement of the purchaser except for finishing the cast. All appropriate precautions are to be taken to avoid the addition of such elements from scrap and other materials used in production which would impair mechanical properties and the suitability of the steel.

[2]For bars, rods, sections and the relevant semi-finished products, a maximum content of 0.030% S applies. For any product to be machined, a controlled sulphur content of 0.015% to 0.030% is recommended and permitted.

*Patented steel grade.

After British Standard BS EN 10088-1: 1995.

2. Improved tempering resistance and toughness via balanced additions of molybdenum, vanadium and nickel.

Type 410 (12% Cr, 0.1% C) is probably the most popular grade in the martensitic series and is used in a wide variety of general engineering applications in both the wrought and cast condition. In Type 420, the carbon content is raised to about 0.3% to provide increased hardness and is typical of the steels used in cutlery. Type 416 and 416Se are free cutting versions of 12% Cr steel and contain additions of 0.15% S min. and 0.15% Se min. respectively in order to provide improved machining characteristics.

During the 1950s, a considerable amount of work was carried out on the development of high-strength 12% Cr steels, particularly in relation to the requirements of the power generation industry for improved steam turbine bolting and blading materials. Two publications from that period by Irvine *et al.*[5] and Irvine and Pickering[6] still serve as classic texts on the metallurgy of high-strength 12% Cr steels. In the solution-treated condition (1050°C AC), a 12% Cr 0.1% C steel develops a tensile strength of about 1300 N/mm^2 and must be tempered in order to achieve a good combination of strength and toughness. Depending on the strength requirement, 12% Cr steels are tempered at temperatures up to about 675°C and the above authors adopted the temperature–time parameter approach to demonstrate the effects of alloying elements on the tempering behaviour. An example is shown in Figure 4.8 for a 12% Cr 0.14% C steel, the tempering parameter being $T(20 + \log t) \times 10^{-3}$ where T is the temperature in °A and t the time in hours. Detailed electron metallographic studies revealed the following changes during the tempering of the steel:

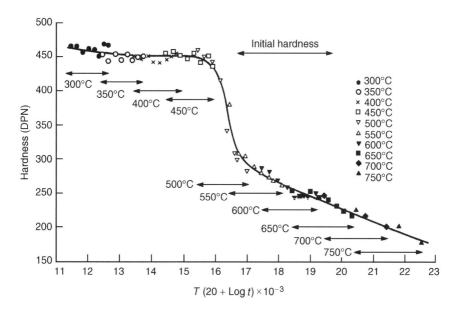

Figure 4.8 *Tempering curve for 0.14% C 12% Cr base steel (After Irvine et al.[5])*

1. In the solution-treated condition, the hardness was 455 HV and there was some evidence for the precipitation of small particles of Fe_3C within the martensitic structure, i.e. slight *autotempering*.
2. After tempering at 350°C, the amount of Fe_3C precipitation had increased, but at 450 HV the loss of hardness was still negligible.
3. Tempering at 450°C caused slight secondary hardening and this was associated with the precipitation of Cr_7C_3 and a small amount of M_2X (based on Cr_2C).
4. A major loss in hardness occurred at 500°C and this was associated with the precipitation of relatively large particles of chromium-rich $M_{23}C_6$ at the martensite plate and prior austenite grain boundaries.
5. Further softening occurred as the tempering temperature was increased and this was associated with the solution of Cr_7C_3 and the growth of $M_{23}C_6$ particles.

On tempering at 600°C, the steel had lost virtually all of the hardness associated with a martensitic structure and there was a particular need to improve the tempering resistance of 12% Cr turbine blading material at temperatures up to at least 650°C. This was concerned with a requirement for good impact properties but brazing operations for the attachment of wear-resistant shields also raised the temperature of the blades locally to a temperature of about 650°C.

As indicated in Chapter 3, additions of molybdenum and vanadium are very effective in improving the tempering resistance of low-alloy martensitic steels and the same is true for their 12% Cr counterparts. This is illustrated in Figure 4.9, which shows the effects of molybdenum and vanadium in a base steel containing 12% Cr, 2% Ni and 0.1% C. In either case, the addition of nickel is made to the steels to counteract the ferrite-forming potential of these elements. In the case of molybdenum (Figure 4.9(b)), a progressive increase in hardness is obtained on tempering at temperatures up to 500°C. This is due to an intensification of the secondary hardening reaction and a fine dispersion of M_2X precipitates persists

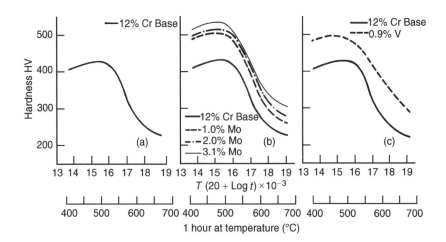

Figure 4.9 *Effect of alloying elements on the tempering characteristics of a 0.1% C 12% Cr steel (After Irvine and Pickering[6])*

at high tempering parameters. The addition of vanadium (Figure 4.9(c)) produces a similar effect and the stabilization of M_2X maintains a high level of hardness, even after tempering at 650°C.

Irvine and Pickering[6] also demonstrated the beneficial effects of carbon, nitrogen and niobium on the tempering resistance of 12% Cr steels. As illustrated in Figure 4.10, carbon and nitrogen increase the hardness throughout the tempering range and in the tempered 650°C 1 h condition, the 0.2% PS values of 12% Cr–Mo–V steels can be expressed by means of the following parameter:

$$0.2\% \ PS = 710 + 772 \ (C + 2N) \ N/mm^2$$

Carbon and nitrogen also have the further benefit of being austenite formers and therefore do not require compensating additions to preserve the required martensitic structure.

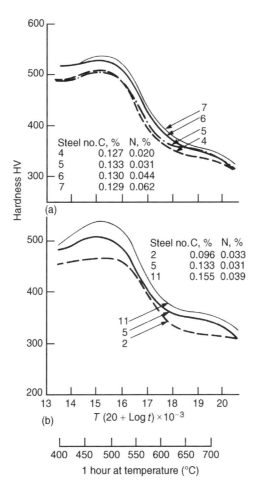

Figure 4.10 *Effect of (a) nitrogen and (b) carbon on the tempering characteristics of a 12% Cr–Mo–V steel (After Irvine and Pickering[6])*

Table 4.4

Property	Base+ 0.015% N	Base+ 0.045% N	Base + 0.37% Nb, 0.013% N	Base + 0.33% Nb, 0.043% N
TS (N/mm^2)	936	1055	981	1084
0.2% PS (N/mm^2)	772	871	815	962
Elongation 4$\sqrt{A}$%	22.8	22.4	18.1	19.1
Charpy V–J at RT	94	47	104	52

The action of niobium in promoting tempering resistance is different to that of the other elements discussed above in that it intensifies the secondary hardening reaction by increasing the lattice parameter of the precipitate relative to that of the ferrite matrix. Thus the effect can be superimposed on other tempering retarding reactions, involving the stabilization of precipitates, in order to gain additional benefit. This is demonstrated in the tensile data given in Table 4.4 for 12% Cr 2.5% Ni 1.5% Mo 0.3% V steels, tempered 650°C 1 h. These data also indicate that nitrogen produces a powerful strengthening effect, albeit with a significant loss in toughness.

As indicated earlier, the carbon content of 12% Cr steels can be increased to high levels in order to promote higher hardness levels but at the expense of toughness and weldability. Notable examples are cutlery steels containing 12% Cr and 0.3% C and stainless razor steels containing 12% Cr and 0.6% C. In the latter steel, substantial amounts of carbon remain out of solution as $M_{23}C_6$ carbides after solution treatment at 1050°C so that the full martensitic hardness associated with 0.6% C is not realized. However, the presence of carbides in the microstructure improves the abrasion resistance of such steels.

Ferritic stainless steels

Ferritic stainless steels can contain up to 30% Cr with additions of other elements such as molybdenum for improved *pitting corrosion* resistance and titanium or niobium for improved resistance to *intergranular corrosion*. These forms of corrosion will be described in a later section. However, the bulk of the requirement for ferritic stainless steels is satisfied by two major grades, namely Type 430 and 434. As shown in Table 4.2, Type 430 is a 17% Cr steel and at the normal solution treatment temperature of 950°C, this steel generally contains a proportion of austenite which transforms to martensite on cooling to room temperature. However, on tempering at 750°C, the martensite breaks down to ferrite and carbide, giving a microstructure which is essentially fully ferritic.

The corrosion resistance of stainless steels in chloride environments is improved substantially with the addition of molybdenum, and Type 434 (17% Cr, 1% Mo) is the most common grade of this type within the ferritic range.

As illustrated later, the corrosion resistance of stainless steels can be seriously impaired by the precipitation of chromium carbides at the grain boundaries. One method of overcoming this problem is to add elements such as titanium and niobium which prevent the formation of chromium carbide. This gives rise to grades such as Type 430 Ti (17% Cr–Ti) and Type 430Nb (17% Cr–Nb).

As discussed earlier, 12% Cr steels are normally martensitic and, as such, tend to have poor forming and welding characteristics. However, the addition of a strong ferrite former to a 12% Cr base steel can produce a fully ferritic microstructure with a marked improvement in the cold-forming and welding behaviour. This effect is achieved in Type 409 (12% Cr–Ti), the titanium also eliminating the problem of chromium carbide formation as well as promoting the ferritic structure. This steel has found extensive application in automobile exhausts in place of plain carbon or aluminized steel.

Type 446 (25% Cr) is the highest chromium grade in the traditional range of ferritic stainless steels and provides the best corrosion and oxidation resistance. Whereas ferritic stainless steels generally possess low toughness, a further embrittling effect can be experienced in steels containing more than 12% Cr when heated to temperatures in the range 400–550°C. The most damaging effect occurs at a temperature of about 475°C and, for this reason, the effect is known as *475°C embrittlement*. The loss of toughness is due to the precipitation of a chromium-rich, α prime phase which becomes more pronounced with increase in chromium content. However, 475°C embrittlement can be removed by reheating to a temperature of about 600°C and cooling rapidly to room temperature.

Austenitic stainless steels

As illustrated in Table 4.2, most of the steels in the AISI 300 series of austenitic steels are based on 18% Cr but with relatively large additions of nickel in order to preserve the austenitic structure. However, as illustrated below, various compositional modifications are employed in order to improve the corrosion resistance of these steels.

Type 304 (18% Cr, 9% Ni) is the most popular grade in the series and is used in a wide variety of applications which require a good combination of corrosion resistance and formability. As discussed later, the stability and work-hardening rate of austenitic stainless steels are related to composition, the leaner alloy grades exhibiting the greater work hardening. Thus a steel such as Type 301 (17% Cr, 7% Ni) work hardens more rapidly than Type 304 (18% Cr, 9% Ni) and, for this reason, Type 301 is often used in applications calling for high abrasion resistance. Steels such as Type 316 (18% Cr, 12% Ni, 2.5% Mo) and 317 (18% Cr, 15% Ni, 3.5% Mo) have greater resistance to corrosion in chloride environments than Type 304 and represent austenitic counterparts of Type 434 (17% Cr, 1% Mo) discussed in the previous section.

Reference was made earlier to the corrosion problems experienced in stainless steels with the formation of chromium carbides. It was indicated that the problem can be overcome in ferritic stainless steels with the addition of titanium and niobium and the same is true in the austenitic grades. This gives rise to Types 321 (18% Cr, 10.5% Ni, Ti) and 347 (18% Cr, 11% Ni, Nb) and, because of their freedom from chromium carbide precipitation and intergranular attack, these grades are often referred to as *stabilized* stainless steels.

Table 4.5   *Compositions of 200 series steels*

	C%	Si%	Mn%	Cr%	Ni%	N%
Type 201	0.1	0.5	6.5	17	4.5	0.25
Type 202	0.1	0.5	8.75	18	5	0.25

An alternative method of preventing the formation of chromium carbide in stainless steels is to reduce the carbon content to a low level. Thus Type 304L (0.03% C max.) is a lower carbon variety of Type 304 and is now used extensively in applications calling for resistance to intergranular attack in the welded condition. In a similar manner, Type 316L is the lower carbon version of Type 316.

Type 310 (25% Cr, 20% Ni) represents the most highly alloyed composition in the popular range of austenitic stainless steels and provides the greatest resistance to corrosion and oxidation.

During the early 1950s, there was a scarcity of nickel in the United States and this led to the development of austenitic stainless steels in which some of the nickel was replaced by alternative austenite-forming elements. The most successful steels of this type were Types 201 and 202, which have the mean compositions shown in Table 4.5.

These steels were developed as alternatives to Types 301 (17% Cr, 7% Ni) and 302 (18% Cr, 8% Ni), in which reductions in the nickel content were compensated by large additions of manganese and nitrogen. The addition of 0.25% N also causes substantial solid solution strengthening and therefore the 200 series steels have high tensile properties. These steels were used extensively for the production of railway carriages in the United States but found little application in the UK. However, they are difficult to produce because of excessive refractory attack and problems in descaling due to their high manganese contents.

Nitrogen additions of about 0.2% are also made to standard grades such as Types 304 and 316 in order to generate high proof stress values. In the UK, these steels are marketed under the *Hyproof* tradename and provide 0.2 PS values of about 330 N/mm^2 compared with 250 N/mm^2 in the standard grades (typically 0.04% N).

The strength of austenitic steels can also be increased by warm working, i.e. by finishing rolling at temperatures below 950°C. As illustrated in Figure 4.11, this results in a major increase in the 0.2% proof stress of standard grades such as Type 304 and 321, but significantly higher levels of strength can be obtained by warm working compositions which are substantially solid solution strengthened with nitrogen. Steels of this type have found limited application in pressure vessels and also for high-strength concrete reinforcement.

Controlled transformation stainless steels

In the 1950s, interest was generated in an entirely new type of stainless steel that was austenitic in the as-delivered, solution-treated condition but which transformed to martensite by means of a simple heat treatment. Such materials

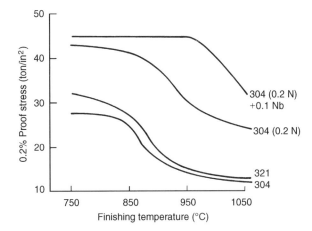

Figure 4.11 *Effect of finishing temperature on the properties of austenitic stainless steel (After McNeely and Llewellyn[7])*

offered the prospect of combining the good forming properties of traditional austenitic stainless steels with the high strength of martensitic grades. The concept required that the composition of these steels was controlled to within fine limits such that the M_s–M_f temperature range was just below room temperature. Following cold-forming operations, the steel could then be transformed to martensite by refrigeration at a temperature such as −78°C (solid CO_2). Such steels became known as *controlled transformation stainless steels* and their behaviour can be illustrated by reference to Figure 4.5. This indicates that a steel containing 0.1% C, 17% Cr and 5.5% Ni is essentially free of delta ferrite and has an M_s temperature near to ambient. In the solution-treated 1050°C AC condition, this composition has a hardness of about 220 HV, but on refrigeration at −78°C, the hardness increases to about 400 HV. This indicates that the steel has been substantially transformed to martensite, although Figure 4.5 shows that hardness values approaching 450 HV are obtained on complete transformation to martensite.

One of the problems encountered in the production of these steels was the very tight control of composition that was required in order to position the transformation range just below room temperature. If the alloy content was excessive, the transformation range was depressed to low temperatures such that transformation to martensite could not be obtained at −78°C. On the other hand, if the alloy content was too low, the high M_s–M_f temperature range meant that substantial transformation to martensite occurred on cooling to room temperature. However, this problem could be eased to some extent by varying the solution treatment temperature and this facility is illustrated schematically in Figure 4.12(a). In austenitic steels containing about 0.1% C and 17% Cr, complete solution of $M_{23}C_6$ carbides can be obtained at a temperature of 1050°C. In such a condition, both chromium and carbon exercise their full potential in depressing the M_s–M_f range. However, if the solution treatment temperature is reduced, then a proportion of $M_{23}C_6$ carbide is left out of solution and the transformation range is raised due to reduced levels of chromium and carbon in solid solution. Thus the use of low solution treatment temperatures provides a means of accommodating casts of steel in which the alloy content was slightly higher than the optimum but, even so, this approach still necessitated very tight control over the solution treatment temperature.

In addition to the use of refrigeration treatments, controlled transformation steels could also be hardened by ageing or *primary tempering* at a temperature of 700°C. This resulted in the precipitation of $M_{23}C_6$ carbide and, as illustrated in Figure 4.12(b), this causes a substantial increase in the M_s–M_f range due to the reduced level of alloying elements in solid solution. Thus on cooling from the tempering treatment, the steel transforms to martensite. In a completely austenitic steel, the precipitation of carbide is essentially restricted to the grain boundaries and this limits the degree to which the transformation range can be raised.

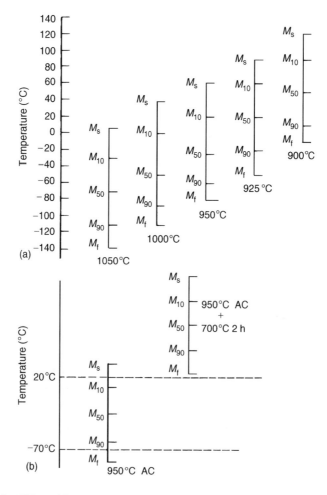

Figure 4.12 *Effect of heat treatment on martensite transformation range: (a) effect of solution treatment; (b) effect of primary tempering (After Irvine et al.[1])*

However, if delta ferrite is present in the microstructure, the ferrite–austenite interfaces provide further sites for carbide precipitation and transformation to martensite is produced more readily. Thus where hardening is to be achieved by primary tempering, rather than refrigeration, the steels should contain 5–10% delta ferrite. However, the primary tempering route to transformation results in the formation of lower carbon martensites and lower strength levels compared with those achieved after transformation by refrigeration. This is illustrated in Table 4.6, which shows the tensile properties obtained in a particular grade of steel after the two forms of hardening. This table also shows the low proof stress value and high tensile strength obtained in the solution-treated 1050°C AC condition, which gives an indication of the high rate of work hardening in these meta-stable grades.

As indicated in Table 4.6, these steels are tempered after the hardening treatments and therefore it is desirable to incorporate alloying elements that will improve the tempering resistance or provide a secondary hardening response. Thus many of the commercial grades that were introduced in the 1950s contained about 2% Mo and elements such as copper, cobalt and aluminium were also added in order to promote secondary hardening reactions in the low-carbon martensite. The effect of copper additions on the secondary hardening behaviour of a 17% Cr 4% Ni steel is shown in Figure 4.13 and examples of the compositions of commercial grades of controlled transformation stainless steels are shown in Table 4.7. Although the range of composition shown in this table is quite wide, the steels embody the following features:

1. A chromium content of 15–17% in order to provide good corrosion resistance.

Table 4.6 *Properties of a 0.1% C 17% Cr 4% Ni 2% Mo 2% Cu steel*

Heat treatment	TS (N/mm^2)	0.2% PS (N/mm^2)	El % 4√A
1050°C 1 h AC	1294	213	27.1
1050°C 1 h AC, 700°C 2 h	1035	689	15.6
1050°C 1 h AC, 700°C 2 h, 450°C 6 h	1161	961	19.6
1000°C 1 h AC, −78°C 2 h	1347	874	15.6
1000°C 1 h AC, −78°C 2 h, 450°C 6 h	1340	1154	22.5

Table 4.7 *Compositions of commercial grades of controlled-transformation stainless steels*

Grade	C%	Mn%	Si%	Cr%	Ni%	Mo%	Cu%	Co%	Al%	Ti%
Armco 17-4PH	0.04	0.5	0.3	16.5	3.5		3.5			
AM350	0.1	0.5	0.3	17	4.2	2.75				
Stainless W	0.07	0.5	0.3	17	7				0.2	0.7
Armco 17-7PH	0.07	0.5	0.3	17	7				1.1	
Armco 17-5PH	0.07	0.5	0.3	15	7	2.5			1.2	
FV 520(S)	0.07	1.5	0.5	16	5.5	2	1.5			0.1
SF 80T	0.08	0.2	0.2	17	4	1	1.2	2		

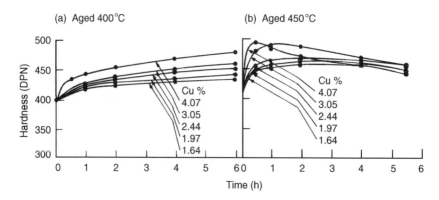

Figure 4.13   *Effect of copper on the secondary hardening response in a 17% Cr 4% Ni steel (initial conditions 950°C AC, −78°C, 1 h) (After Irvine et al.[1])*

2. Sufficient austenite-forming elements to produce a mainly austenitic structure at 1050°C.
3. Alloying elements to promote tempering resistance/secondary hardening reactions.
4. An overall alloy content that produces an M_s–M_f transformation range just below room temperature.

These steels were developed primarily for the aerospace industry but were of limited commercial success due to the fact that:

1. The very tight composition ranges were difficult to achieve in commercial production.
2. The meta-stable nature of the steels results in a very high rate of work hardening which limited some aspects of cold formability.

However, FV 520(S) is still in use in the UK and is used primarily in defence applications such as gun carriages and aircraft.

Steel prices

At the time of reporting (March 1997), stainless steel producers in the UK were experiencing difficult trading conditions due to the strength of sterling and the volatile prices for ferro-alloys. Over the past 12 months, they had experienced price swings for stainless steels of greater than 30% and were therefore very reluctant to quote the current prices for these grades. Because of this situation, the only information that can be presented at this time are the prices of some of the popular grades of strip relative to a value of 1.0 for Type 304 (18% Cr 9% Ni). These data are shown in Figure 4.14 which also includes the relative price for *2205*, the duplex grade, which has gained increased usage over the past few years. Very clearly, the ratios reflect the alloy content of the steels, Type 409 (12% Cr) being the cheapest and Type 310 (25% Cr 20% Ni) the most expensive in this particular selection of grades.

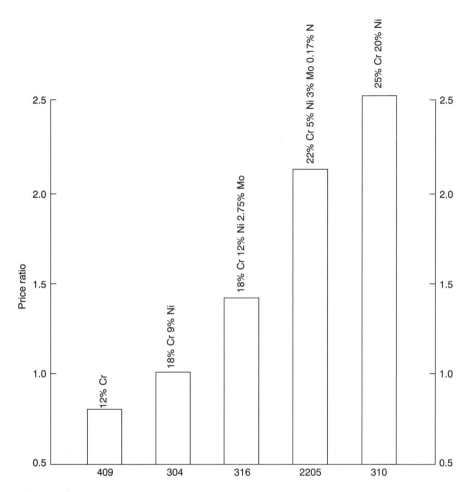

Figure 4.14 *Prices of stainless steel strip relative to a value of 1.0 for Type 304 at March 1997*

Based on the prices quoted in the 2nd edition of this book (1994), Type 304 stainless steel strip was about $3\frac{1}{2}$ times that of cold-reduced, uncoated, mild steel strip. The prices of some high-alloy stainless steels, relative to Type 316L, are shown in Table 4.9 (see p. 332).

Corrosion resistance

Stainless steels owe their corrosion resistance to the formation of the so-called *passive film*. This consists of a layer, only 20–30 Å thick, of hydrated chromium oxide (Cr_2O_3), which is extremely adherent and resistant to chemical attack. If the passive film is damaged by abrasion or scratching, a healing process or *repassivation* occurs almost immediately. In general, the corrosion and oxidation resistance of stainless steels increase with chromium content and the materials are used in a wide range of aggressive media in the chemical and process plant industries. However, under certain conditions, stainless steels are susceptible to highly localized forms of attack in relatively mild environments, rendering them unsuitable for further service. In this context, the main types of localized attack are *intergranular corrosion, pitting corrosion* and *stress corrosion*. However, it should be stressed that these forms of attack are well researched and thoroughly documented and therefore it is now rare for them to lead to premature or unexpected failure in stainless steel components.

Intergranular corrosion

Given favourable temperature conditions, solute atoms can segregate to the grain boundaries, causing enrichment in a particular element or the precipitation of metal compounds. Under highly oxidizing conditions, these effects can cause the grain boundaries of stainless steels to become very reactive, leading to the highly localized form of attack known as *intergranular corrosion*.

Both austenitic and ferritic stainless steel are susceptible to intergranular corrosion and, in either case, the problem is caused by the segregation of carbon to the grain boundaries and the formation of the chromium-rich, $M_{23}C_6$ carbides. The concentration of chromium in these carbides is very much higher than that in the surrounding matrix and, at one time, it was postulated that this resulted in galvanic corrosion between the noble carbides and the more reactive matrix. However, currently the most widely accepted theory for intergranular corrosion in stainless steels is that involving *chromium depletion*. Thus, in forming chromium-rich carbides at the grain boundaries, chromium is drawn out of solid solution, and in areas adjacent to the boundaries, the chromium content becomes severely depleted compared to the bulk chromium concentration of the steel. Such areas are then said to have become *sensitized* in that they no longer contain sufficient chromium to withstand corrosive attack. Corrosion can then proceed along the grain boundaries and a micrograph illustrating this form of attack in an austenitic stainless steel is shown in Figure 4.15.

Two main laboratory tests are carried out for the evaluation of intergranular corrosion, namely the Huey and Strauss tests. The former was developed to determine the performance of stainless steels in nitric acid plant and has been standardized as ASTM A262-70 Practice C. It consists of immersing a sample in boiling 65% HNO_3 for five periods, each of 48 hours, using fresh acid solution in each period. The samples are weighed after each period and the weight loss is generally converted to a corrosion rate in terms of mm/year. The geometry of the test piece has to be carefully controlled to avoid excessive 'end grain' effects, which will be described later, and the type of acid used must also be controlled very carefully in order to obtain reproducible results.

The Strauss test was introduced to determine the pickling behaviour of stainless steels and is now standardized as ASTM A262-70 Practice E. This involves exposure to boiling 15.7% H_2SO_4 + 5.7% $CuSO_4$ solution with the test specimen in contact with metallic copper which increases the severity of attack. The

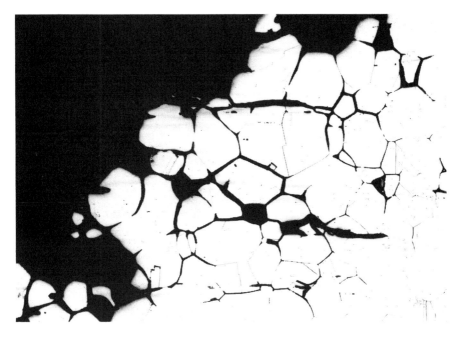

Figure 4.15 *Intergranular corrosion in Type 304 stainless steel*

performance of the steel is then judged in terms of the presence or absence of cracks or fissures after bending.

A third test, often designated the oxalic acid test, has been standardized as ASTM A262–70 Practice A. In this test, a polished sample of austenitic stainless steel is etched electrolytically in a 10% solution of oxalic acid at room temperature. In the absence of chromium carbide precipitates, the grain boundaries of the etched sample exhibit a *step* between adjacent grains compared with a *ditch* in material which has experienced intergranular attack. The test also recognizes a dual, intermediate condition in which some intergranular corrosion has occurred but no single grain is surrounded completely by ditches.

At temperatures below about 850°C, the solubility of carbon in an austenitic stainless steel falls below 0.03% and exposure in the temperature range 450–800°C can result in the precipitation of $M_{23}C_6$ carbide. This can occur during:

1. Slow cooling through the sensitization temperature range following solution treatment at 1050°C or after welding.
2. Stress relieving after welding or service exposure in the critical temperature range.

Slow cooling from welding and subsequent intergranular attack gave rise to the term *weld decay* before the mechanism was fully understood. The weld decay area is generally a band of material, some distance from the weld, which has been exposed to the temperature range favouring the precipitation of chromium carbide.

From the foregoing remarks, it will be obvious that a reduction in the carbon content will reduce the susceptibility to intergranular corrosion but, at one time, major expense was incurred in reducing the carbon content of an austenitic stainless steel to below 0.06%. However, as indicated earlier, the problem was controlled initially by adding elements such as titanium or niobium, which are stronger carbide formers than chromium, and therefore TiC or NbC are formed rather than the damaging chromium-rich $M_{23}C_6$. The addition of these elements is said to *stabilize* stainless steels against intergranular attack and gives rise to the standard grades–Type 321 (Ti-stabilized) and Type 347 (Nb-stabilized). These elements also form nitrides and the additions made to stainless steels are slightly in excess of those required by stoichiometry for complete precipitation of carbon and nitrogen, namely, $Ti = 5 \times (C + N)\%$ and $Nb = 10 \times (C + N)\%$.

Although the problem of intergranular corrosion is controlled by the addition of stabilizing elements, Types 321 and 347 are not completely immune to this form of corrosion since they are susceptible to *knife-line attack*. During welding, the heat-affected zone is raised to temperatures above 1150°C and this can result in the partial dissolution of TiC and NbC. Carbon is therefore taken into solution in a narrow region adjacent to the weld and can be available for the formation of chromium carbide on cooling through the sensitization range of 450–800°C. The susceptible region may only be a few grains wide but can give rise to a thin line of intergranular attack. Hence the term *knife-line attack*.

The problem of intergranular corrosion in austenitic stainless steels can also be overcome by reducing the carbon content to low levels. Although steels containing 0.03% C max. (*L grades*) have been available since the 1950s, their use was restricted very severely in the early days due to high production costs. This involved the use of low-carbon ferro-chrome, which was expensive. However, the introduction of AOD refining has cheapened the production of the *L-grades* very considerably and these steels are now used extensively in applications formerly satisfied by the stabilized grades. However, the presence of TiC and NbC particles gives rise to some dispersion strengthening and Types 321 and 347 are still used in applications where advantage can be taken of their higher strength.

According to Sedriks,[8] molybdenum has an adverse effect on intergranular corrosion as assessed in the Huey test and a beneficial effect in the Strauss test. It is suggested that the adverse effect in the former may be concerned with the fact that nitric acid attacks areas other than chromium-depleted regions, e.g. areas associated with solute segregation or the early formation of sigma phase. At one time, small additions of boron were made to austenitic stainless steels in order to improve hot workability and the high-temperature creep properties. However, such additions are very deleterious to the performance in the Huey test. It has been suggested that boron additions lead to the formation of M_2B and $M_{23}(CB)_6$ precipitates at the grain boundaries, both of which give rise to chromium-depletion effects. Performance in the Huey test is also improved by reducing the phosphorus content of austenitic stainless steels to low levels, i.e. less than 0.01% P.

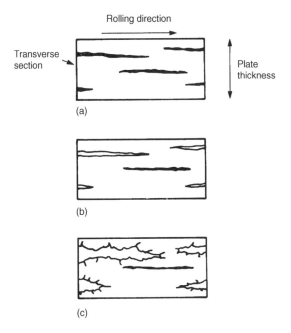

Figure 4.16 *Schematic illustration of end grain attack from outcropping inclusions: (a) elongated inclusions outcropping on transverse sections; (b) outcropping inclusions dissolved in nitric acid; (c) heavy intergranular attack from crevices*

In addition to segregation and precipitation effects at grain boundaries, the intergranular corrosion behaviour of austenitic stainless steels can also be influenced very markedly by the presence of elongated particles or clusters of second phases. These can take the form of sulphides or other plastic inclusions but adverse effects can also be induced in niobium-stabilized Type 347 steel due to the presence of stringers of coarse niobium carbonitrides. The mechanism is shown schematically in Figure 4.16. In the Huey test, outcropping inclusions appear to dissolve quickly, creating long narrow passages from the end faces into the body of the samples. These passages do not provide ready access for the ingress of new acid and it is thought that this gives rise to the formation and concentration of hexavalent Cr^{6+} ions. This leads to particularly aggressive corrosion conditions within the passages and intergranular attack proceeds very quickly along the adjacent grain boundaries. The effect is known as *end grain attack* and can lead to weight losses far in excess of that experienced in cleaner steels or those containing shorter inclusions. In chemical plant, end grain corrosion can be experienced in forgings or in tube-to-tube welds where end faces are exposed due to ovality effects or differences in tube diameter. In such cases, the exposed end grain faces should be covered by capping welds.

Because carbon has a very low solid solubility in ferritic stainless steels, it is extremely difficult to prevent the formation of chromium carbides in these grades. Thus when carbon is taken into solution by heating to temperatures above 925°C, carbide precipitation may occur even when the material is water quenched from the solution treatment temperature. However, intergranular corrosion will only proceed in sensitized material in aggressive, oxidizing media and, ordinarily, ferritic stainless grades such as Types 430 and 434 are not exposed to such environments. Even so, care should be taken to minimize the risk of intergranular corrosion.

Ferritic stainless steels sensitize more rapidly and at lower temperatures than their austenitic counterparts, the fastest reaction occurring at a temperature of about 600°C. However, by holding at a temperature of about 800°C, or by cooling slowly through the temperature range 700–900°C, the risk of intergranular attack can be eliminated. Following the initial, damaging precipitation of chromium carbides, these treatments allow sufficient time for chromium to diffuse into the depleted zones and thereby eliminate the sensitization effects.

The addition of titanium also reduces the risk of intergranular attack in ferritic stainless steels. However, whereas the reduction of carbon to below 0.03% is effective in preventing sensitization in austenitic grades (L grades), significantly lower levels are required in ferritic stainless steels. Nitrogen is also damaging, causing chromium depletion via the formation of chromium nitride. Therefore, the total (carbon + nitrogen) must be reduced to below 0.01% in order to prevent sensitization in a 17% Cr ferritic steel. The development of *low interstitial ferritic steels* will be discussed later.

Pitting corrosion

As its name suggests, pitting is a highly localized form of corrosion which, in its initial form, results in the formation of shallow holes or pits in the surface of the component. However, the pits can propagate at a fast rate, resulting in *pin-holing* or complete perforation in the wall of the component. Therefore pitting can be completely destructive in terms of further useful life when only a very small amount of metal has been attacked by corrosion. In stainless steels, pitting corrosion generally takes place in the presence of chloride ions and it is widely held that the initiation stage is associated with attack on non-metallic inclusions. However, other microstructural features may also play a part. The formation of a pit in an austenitic stainless steel is shown in Figure 4.17.

Fontana and Greene[9] have stated that pits usually grow in the direction of gravity, i.e. downwards from horizontal surfaces, and only rarely do they proceed in an upward direction. These authors have formulated a model for pitting in terms of an autocatalytic process for stainless steels in aerated sodium chloride solution. The sequence of events is as follows:

1. Anodic dissolution takes place at the bottom of the pit:

$$M \rightarrow M^+ + e^-$$

2. A cathodic reaction takes place on adjacent surface:

$$O_2 + 2H_2O + 4e^- \rightarrow 4OH^-$$

Figure 4.17   *Pitting corrosion in Type 304 stainless steel*

3. The build up of M^+ ions within the pit causes the Cl^- ions to migrate to the pit in order to preserve electrical neutrality.
4. The soluble metal chloride hydrolyses to form hydroxide and free acid:

$$M^+Cl^- + H_2O \rightarrow MOH + H^+Cl^-$$

Thus acid is produced by the reaction which decreases the pH to a low level whereas the bulk solution remains neutral.

Traditionally, the susceptibility of stainless steels to pitting corrosion was evaluated by immersion tests in acidified ferric chloride solutions (ASTM Practice G48–76 and Practice G46–76). Such tests involve the measurement of pit density, size and depth. However, laboratory tests on pitting behaviour are now more generally based on electrochemical techniques.

The resistance to pitting increases with chromium content but major benefit is obtained from the addition of molybdenum in stainless steels, e.g. Type 316 (18% Cr, 12% Ni, 2.5% Mo). The addition of nitrogen is also beneficial and the combined effects of nitrogen plus molybdenum will be discussed later in this chapter. The potential pitting resistance of stainless steels is often expressed in terms of a pitting index:

$$\text{Pitting index} = Cr\% + 3.3Mo\% + 16N\%$$

In terms of microstructure, MnS inclusions are important sites for pit initiation but other features such as delta ferrite, alpha prime and sigma phase can also promote pitting corrosion.

rosion cracking

ress corrosion cracking (SCC) is a form of failure induced by the conjoint action of tensile stresses and particular types of corrosive environments. The stresses can be either applied or residual and cracking takes place in a direction normal to the tensile stresses, often at stress levels below the yield strength of the material. A micrograph illustrating SCC in an austenitic stainless steel is shown in Figure 4.18. Cracking can take place in either a transgranular or intergranular manner and can proceed to the point where the remaining material can no longer support the applied stress and fracture then takes place.

SCC occurs in chloride, caustic or oxygen-rich solutions but the majority of failures in austenitic stainless steel take place in chloride-bearing environments. Chloride ions may be present in the process stream but can also be introduced accidentally through the incomplete removal of sterilizing agents, such as hypochlorite solutions.

It is now generally acknowledged that SCC is initiated by the formation of a pit or crevice at anodic sites on the surface, e.g. outcropping inclusions which disrupt the passive film. The reactions controlling pitting corrosion were described in the previous section. A number of mechanisms has been proposed for crack propagation but the surviving theories are based on anodic dissolution and hydrogen embrittlement mechanisms. In the former, cracking propagates by local anodic dissolution at the crack tip, passivation at that point being prevented by plastic deformation. Alternatively, the principal role of plastic deformation

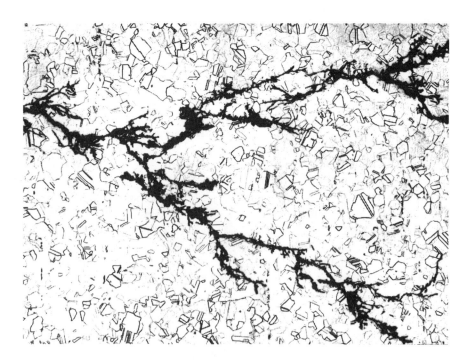

Figure 4.18 *Stress corrosion cracking in Type 304 stainless steel*

is to accelerate the dissolution process. In the hydrogen embrittlement theory, hydrogen is generated due to the cathodic reaction ($H^+ + e^- \rightarrow H$) and migrates to the crack tip where it is absorbed within the metal lattice, causing mechanical rupture. This extends the crack and further anodic dissolution then occurs on the newly exposed surfaces.

Although laboratory tests are carried out in a variety of salt solutions, that involving boiling 42% magnesium chloride solution is probably the most popular. The test specimens may be U-bend specimens which are clamped so as to provide a large tensile stress on the outside of the bend or else tensile specimens are employed, loaded to perhaps 0.8 times the 0.2% PS of the material.

Major research effort has been devoted to the effects of composition on the stress corrosion behaviour of stainless steels. One of the most important elements is nickel and the celebrated Copson curve[10] is shown in Figure 4.19. This relates to stainless steels containing 18–20% Cr in a magnesium chloride solution boiling at 154°C. This figure indicates a minimum resistance to SCC at nickel contents in the range 8–10%, i.e. typical of those in standard austenitic steels such as Types 302 and 304. A reduction in nickel from these levels leads to an improvement in the SCC behaviour but of course this is associated with the progressive replacement of austenite by delta ferrite in the microstructure. Alternatively, the resistance to SCC can be improved by increasing the nickel content to levels of 40% and above. The beneficial effect of nickel at levels above 8–10% has been attributed to an increase in the *stacking fault energy* of the materials but nickel must also exert an ennobling effect which inhibits anodic dissolution.

Silicon has a beneficial effect on the SCC behaviour in magnesium chloride solutions but has virtually no effect in high-temperature solutions of sodium chloride.

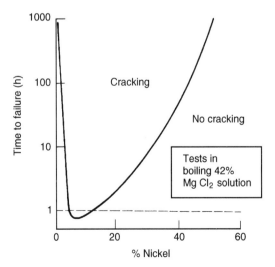

Figure 4.19 *Effect of nickel on the stress corrosion behaviour of austenitic stainless steels (After Copson[10])*

Although clearly beneficial in the formation of passive films and in suppressing pitting corrosion, both chromium and molybdenum appear to have a variable effect on the SCC behaviour of stainless steels. These effects may well be linked to austenite stability and a reduction in stacking fault energy which promotes crack propagation.

Temperature is a very important parameter in the SCC behaviour of stainless steels and is rarely observed at temperatures below about 60°C.

High-alloy stainless steels

As indicated at the beginning of this chapter, most of the popular grades of stainless steel were formulated in the 1920s and 1930s and have undergone little change to the present day. However, in the intervening period, a large amount of research effort has been devoted to the development of new grades but relatively few of these compositions have enjoyed major commercial exploitation.

High-alloy austenitic/duplex grades

From the basic Type 304 (18% Cr, 9% Ni) composition, a number of developments have taken place which were designed to improve a particular property. These include:

1. The addition of up to 0.5% N to improve the strength (solid solution strengthening) or to compensate for reductions in the nickel content (e.g. Types 201 and 202).
2. The addition of elements such as aluminium, titanium or phosphorus to produce precipitation-strengthening reactions.
3. The inclusion of high levels of nickel and molybdenum and moderate levels of copper and nitrogen to improve the resistance to stress corrosion or reducing acids.

The properties of steels in the first category were discussed briefly on p. 316. In contrast to the situation in low-carbon strip and structural grades, very little use has been made of *Hiproof*-type compositions which contain about 0.2% N and provide 0.2% PS values of about 330 N/mm^2 (cf. 250 N/mm^2 in standard grades). Having overcome the welding problems that were encountered when these steels were first introduced, it is surprising that they are not yet being used in large quantities. Whereas the 200 series have enjoyed commercial success in the United States, they have always been regarded as inferior substitutes to the traditional 300 steels in the UK and, as stated earlier in the text, they also present problems in production.

The metallurgy of precipitation-strengthened austenitic stainless steels was the subject of detailed research work in the 1950s and 1960s.[11] Depending upon the precipitation system and heat treatment, such steels can develop 0.2% PS values up to 750 N/mm^2, i.e. three times that produced in Type 304 in the solution-treated condition. However, such grades were produced in very limited quantities, probably for the following reasons:

1. The lack of facilities and expense involved in carrying out the ageing treatments – typically 700–750°C for 24 hours.
2. The reduction in corrosion resistance produced by the precipitation of carbides, phosphides or intermetallic compounds.
3. The loss of strength in the heat-affected zones of welds.
4. The reluctance of plant designers to consider the very large reductions in plate thickness etc. that could be provided by these high-strength steels.

As illustrated in Table 4.8, a large number of standard and proprietary grades have been developed with the objective of increasing the resistance to reducing acids, pitting and crevice corrosion. Many of the steels can be regarded as simple extensions of Type 316 (18% Cr, 12% Ni, 2.5% Mo) with higher molybdenum contents and with increased levels of nickel to preserve a fully austenitic structure. Some of these steels also contain up to 0.25% N which increases the strength and augments the chromium and molybdenum contents in improving the pitting resistance, according to the following pitting index formula:

$$\text{Pitting index} = Cr + 3.3\% \text{ Mo} + 16\% \text{ N}$$

Copper additions of up to 2% feature in some of the compositions which improve the resistance to sulphuric acid.

Many of the compositions listed in Table 4.8 are used in flue gas desulphurization plant where acid condensates, contaminated with chlorides, produce severe corrosive conditions. Steels such as 94L are also used in the food industry and also for heat exchangers, particularly where salt water is involved. This grade

Table 4.8 *Compositions of high-alloy austenitic and duplex stainless steels*

Alloy designation	Chemical composition (wt %)						
	C	*Cr*	*Mo*	*Ni*	*Cu*	*N*	*Others*
316L	0.03	17	2.25	12	–	–	–
316LM	0.03	17	2.75	12	–	–	–
317L	0.03	18	.5	15	–	–	–
317LN	0.03	18	4.5	13	–	0.15	
317LM	0.03	18	4.5	15	–	–	–
317LMN	0.04	18	6	14	–	0.15	2.1Nb + Ta,
94L	0.02	20	3	25		–	2.5Co, 1W
254SMo	0.02	20	6.5	18	2	0.12	5Co, 1.5W
Incoloy 825	0.03	22		42		–	3.65Nb + Ta
Hastelloy G	0.05	22	7	44	1.9	–	3W
			9			–	1.25Co, 4W
Hastelloy G-3	0.015	22	13	41	–	–	–
Inconel 625	0.05	22	16	61	–	–	0.75W
Hastelloy C-22	0.015	22	3		–	–	
Hastelloy C-276	0.02	25	3.5		0.75	–	
2205	0.03	22	3	5.5		0.15	
Zeron 100	0.03	25	3.5	7	0.75	0.25	0.75W

After Lane and Needham.[12]

Table 4.9   *Prices for high-alloy grades relative to Type 316L*

Alloy types	Price factor
316L	1
2205	1.1
317L	1.2
317LM	1.3
94L	2
High-nickel alloys	2.3–3
High-molybdenum alloys ($\geq$6% Mo)	2.7–3.8
Nickel-base alloys	3.3–9
Titanium	7.6–8.1

contains 25% Ni and will therefore provide better resistance to stress corrosion than traditional grades such as Type 316.

The last two compositions in Table 4.8 differ from the rest in that the high chromium and molybdenum contents are not complemented with a large addition of nickel and this results in a duplex, austenite plus ferrite structure. Such steels have good resistance to both chloride and sulphide stress corrosion and the duplex structure also provides high proof stress values, e.g. 480 N/mm^2. This combination of properties is used to advantage in the petroleum industry, particularly for the handling of crude oil which is contaminated with hydrogen sulphide. Typical applications include transfer pipelines, downhole tubing and liners and also topside process equipment, such as heat exchangers and seawater transport piping.[13]

British Steel[13] has presented the information in Table 4.9 on the prices of some of the high-alloy grades and nickel superalloys relative to Type 316L.

High-alloy ferritic grades

As indicated earlier, standard austenitic stainless steels such as Type 304 (18% Cr, 9% Ni) are highly susceptible to chloride stress corrosion and the nickel content must be raised to about 40% in order to provide immunity from this form of attack. Such materials are expensive and the addition of nickel alone provides little benefit in terms of resistance to pitting and crevice corrosion. On the other hand, standard ferritic grades such as Type 430 (17% Cr) and 446 (25% Cr) are resistant to stress corrosion but are prone to intergranular corrosion and have poor toughness, even in moderate plate thicknesses. However, as early as 1951,[14] it was shown that these deficiencies could be overcome by reducing the interstitial content of ferritic stainless steels to low levels. However, for many years, this information remained little more than laboratory-based data until the development of steelmaking processes that enabled the materials to become commercially viable. In particular, the Airco-Temescal Division of Airco Reduction Co. Inc. in the United States introduced a vacuum melting process with electron beam stirring which could achieve total carbon and nitrogen contents of the order of 0.01%.[15] In addition, an LD converter coupled with a Standard-Messo decarburizer was developed in Germany to achieve total interstitial contents of about 0.025%.[16]

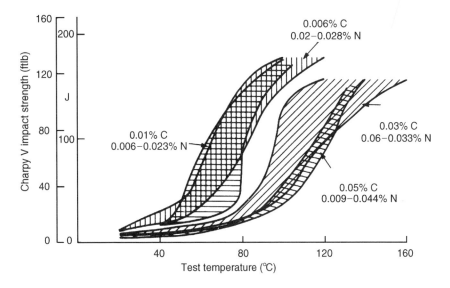

Figure 4.20 *Effect of carbon and nitrogen on the impact properties of 18% Cr 2% Mo steel (After Hooper* et al.[17] *)*

However, the argon–oxygen decarburization (AOD) process has also been used for the production of these grades. Thus in the mid-1960s major commercial interest was stimulated in low interstitial ferritic stainless steels which offered the potential of cheaper alternatives to austenitic grades, particularly in relation to resistance to stress corrosion.

The influence of carbon and nitrogen on the impact properties of an 18% Cr 2% Mo base steel was investigated by Hooper *et al.*[17] and the data are shown in Figure 4.20. From this figure, it is apparent that a reduction in carbon content to 0.01% produces a marked improvement in the impact transition behaviour, whereas a change in nitrogen content has little effect. However, the grain size of these steels was about ASTM 0 and even at low interstitial levels, the impact transition temperature is still above room temperature. Therefore, in order to produce attractive impact properties, the grain size must be reduced, i.e. to a level below ASTM 8.

In addition to a reduction in carbon and nitrogen contents, the new range of ferritic steels contained higher levels of chromium and additions of molybdenum and nickel. Streicher[18] has reviewed the development of these grades and the types of composition that have been produced in commercial quantities are shown in Table 4.10. Although nitrogen has a relatively small effect on the impact properties, its low solubility in ferritic grades means that chromium nitride forms very readily and contributes to the susceptibility to intergranular attack. Thus many of the high-alloy ferritic grades must be stabilized with titanium or niobium, even when the carbon and nitrogen has been reduced to very low levels.

As illustrated in Table 4.11, most of these grades are resistant to stress corrosion cracking in $MgCl_2$ and NaCl solutions and they also display good resistance to pitting and crevice corrosion. The inclusion of nickel in steels such as 28% Cr

Table 4.10 *Commercial grades of low interstitial ferritic stainless steels (After Streicher[18])*

	Alloy	Limits for carbon, nitrogen and stabilizers (%)	Melting and refining processes
I	Fe-18 Cr-2 Mo-Ti	C – 0.0250 max. N – 0.0250 max. C + N < 0.030 desirable Ti + Nb = 0.20 + 4 (C + N) min. = 0.80 max.	Argon–oxygen decarburization or vacuum–oxygen decarburization
II	Fe-26 Cr-1 Mo	C – 0.0050 max. N – 0.0150 max. Nb = 13 to 29 (N)	Electron beam hearth refining or vacuum induction melting
II-A	Fe-26 Cr-1 Mo-Ti	C – 0.0400 max. N – 0.0400 max. 0.2 to 1.0 Ti C + N = 0.050 typical	Argon–oxygen decarburization
III	Fe-28 Cr-2 Mo	C + N ≤ 0.0100 Desirable C + N ≤ 0.050	Vacuum melting followed by arc remelting
III-A	Fe-28 Cr-2 Mo-4 Ni–Nb	C – 0.0150 max. N – 0.0350 max. C + N ≤ 0.0400 with Nb ≥ 12 (C + N) + 0.2	Vacuum–oxygen decarburization Argon–oxygen decarburization
IV	Fe-29 Cr-4 Mo	C – 0.0100 max. N – 0.0200 max. C + N = 0.0250 max.	Vacuum induction melting or electron beam refining
IV-A	Fe-29 Cr-4 Mo-2 Ni	Same	Vacuum induction melting

2% Mo 4% Ni–Nb and 29% Cr 4% Mo 2% Ni is beneficial in improving the corrosion resistance in reducing acids. It is also claimed[18] that nickel improves the toughness of ferritic stainless steels.

Low interstitial ferritic stainless steels have enjoyed some commercial success in Sweden and the United States but failed to make much impression in the UK. However, the author recalls one trial, involving an 18% Cr 2% Mo steel, for the manufacture of a brewery vessel. This trial proved to be unsuccessful due to fabrication problems and poor impact properties in the heat-affected zones of welds.

Welding of stainless steels

A very high proportion of stainless steels is welded in the fabrication of pressure vessels, storage tanks, chemical plant and domestic appliances. In each case, the welds are required to be of high integrity and to provide corrosion resistance or

Table 4.11 *The properties of low interstitial ferritic stainless steels (After Streicher[18])*

Property	Alloy					
	18Cr-2Mo-Ti	26Cr-1Mo (High purity)	w26Cr-1Mo-Ti	28Cr-2Mo-4Ni-Nb	29Cr-4Mo	29Cr-4Mo-2Ni
Stress Corrosion Cracking						
MgCl$_2$ (155 C, 310 F)	R*‡	R*	R*95	F	R*	R
NaCl (103 C, 217 F)	R	R	R	R	R	R
Pitting and Crevice Corrosion						
KMnO$_4$-NaCl						
Room Temperature	F	R	R	R	R	R
50 C (120 F)	–	R	R	R	R	R
90 C (195 F)	–	F	F	F	R	R
Ferric Chloride						
Room Temperature	F	F	F	R	R	R
50 C (120 F)	–	–	–	F	R	R
Corrosion in Boiling Acids[1]						
Nitric–65%	F†	R	F†	R	R	R
Formic–45%	F	R	R	R	R	R
Oxalic–10%	F	F	F	R	R	R
Sulfuric–10%	F	F	F	R	F	R
Hydrochloric–1%	F	F	F	R	F	R
Transition Temperature[2]	+25 to +75 C (+75 to 165 F)	−62 C (−80 F)	+40 C (+105 F)	−5 C (+25 F)	+16 C (+60 F)	−7 C (+20 F)
Refining Processes	AOD or VOD	EB or VIM	AOD	VOD or AOD	VIM or EB	VIM

R = Resistant
F = Fails by type of corrosion shown
AOD = Argon–Oxygen Decarburization
VOD = Vacuum–Oxygen Decarburization
VIM = Vacuum–Induction Melting
EB = Electron Beam Refining

[1] R indicates passive or self-repassivating with rates <0.2 mm/yr 0.008 in./yr
[2] Full size Charpy V-notch specimens
* Copper and nickel residuals must be kept low in these alloys to resist cracking in this solution
† Not recommended for oxidizing solutions
‡ Data from Climax Molybdenum Co. publication, '18Cr-2Mo Ferritic Stainless Steel'

mechanical properties similar to that of parent material. Therefore good weld-ability is a particularly attractive feature of stainless steels and this contributes very significantly to the wide usage and versatility of these materials. The welding of stainless steels has been reviewed extensively by Castro and de Cadenet[19] and the topic has been updated more recently by Gooch.[20]

Martensitic stainless steels

Because of their high alloy content, 12% Cr steels and other martensitic stainless grades exhibit a high level of hardenability and are capable of transforming to martensite in large section sizes. Therefore these steels are susceptible to cold cracking in the weld metal and HAZ in a similar manner to low-alloy marten-sitic grades. As with their low-alloy counterparts, the problem is exacerbated by the presence of hydrogen, and the risk of cracking increases as the carbon content and hardness of the steels are increased. For this reason, the welding of 12% Cr martensitic grades is generally restricted to compositions containing a maximum of about 0.25% C. However, the problem of hydrogen cracking is well understood and can be readily overcome in martensitic stainless steels provided the normal precautions are taken. Thus in gas-shielded processes, pure argon should be used rather than argon-hydrogen mixtures, and in manual metal arc welding, low-hydrogen basic electrodes should be employed. The thorough baking of electrodes is also recommended in order to remove the last traces of moisture.

Like other highly hardenable grades, 12% Cr steels are generally preheated to reduce the risk of cracking. Preheating temperatures of up to 250°C are employed which ensure that the weld metal and HAZ cool slowly through the $M_s - M_f$ temperature range, thereby facilitating stress relaxation and reducing thermal stresses. On the other hand, martensitic stainless steel containing less than 0.1% C and with nickel additions of up to 4% are often welded without preheat. However, it is still imperative that low levels of hydrogen are produced in the weld metal with matching fillers so as to avoid weld metal cracking. The use of Cr–Ni austenitic filler metal can also remove the need for preheating, although the strength of the joint will be lower than that achieved with martensitic weld metals.

After welding, 12% Cr steels are subjected to heat treatment in order to provide an adequate balance between strength and toughness in the weld zone. Because these steels have a low $M_s - M_f$ temperature range, austenite may well be present immediately after the completion of welding and it is essential that the assembly is cooled to room temperature in order to complete the transformation before applying post-weld heat treatment. If adequate cooling is not carried out, austenite will be present in the microstructure during the subsequent tempering treatment and this may well transform to untempered martensite on cooling to room temper-ature. Post-weld heat treatments are generally carried out in the range 600–750°C.

Martensitic stainless steels can be welded using all the conventional fusion processes, providing precautions are taken to minimize the level of hydrogen in the weld.

Austenitic stainless steels

Austenitic stainless steels are readily welded, given their high level of toughness and freedom from transformation to martensite. Therefore they are not prone to the cold-cracking problems encountered in martensitic stainless steels and require neither preheating nor post-weld heat treatment. Because of these characteristics, austenitic fillers are often used for joining dissimilar steels or, as mentioned earlier, in the welding of brittle martensitic stainless grades. However, precautions must be taken to avoid other problems in austenitic stainless steels, namely *solidification cracking* and *sensitization*.

Solidification cracking takes place in the weld metal as it is about to solidify. The problem is due to the generation of high contraction stresses in austenitic stainless steels because of the high thermal expansion characteristics of these materials. These stresses pull the solidified crystals apart when they are still surrounded by thin films of liquid metal, giving rise to interdendritic cracking. However, it was discovered at an early stage that welds with a fully austenitic structure were particularly susceptible to solidification cracking and that the problem could be overcome with the introduction of a small amount of delta ferrite. The amount of ferrite required to eliminate cracking depends upon the degree of restraint imposed upon the joint and also on the composition of the steel. However, a level of about 5% ferrite is generally adequate. Therefore in autogenous welds, the compositions of commercial grades such as Types 304, 316, 321 and 347 are balanced such that they are free of delta ferrite in the solution-treated condition but generate about 5% ferrite in the weld metal due to the thermal excursion into the liquid state. Similarly, in the joining of thick sections of austenitic stainless steels, filler metals of the correct composition are employed which develop the required microstructure.

The prediction of the microstructure in the weld metal of stainless steels can be carried out using diagrams prepared by Schaeffler[21] and De Long.[22] The latter was developed later and has the advantage that it takes account of the nitrogen content of the steel. As indicated earlier, nitrogen is a powerful austenite-forming element and can therefore have a significant effect on the microstructure. The De Long diagram is shown in Figure 4.21 and the chromium and nickel equivalents are calculated in the following manner:

$$\text{Cr equivalent} = \%\text{Cr} + \text{Mo} + 1.5\text{Si} + 0.5\text{Nb}$$

$$\text{Ni equivalent} = \%\text{Ni} + 30\text{C} + 30\text{N} + 0.5\text{Mn}$$

Solidification cracking is promoted by the presence of certain elements which segregate to the remaining liquid during the solidification process, producing interdendritic films of low melting point. Elements such as nickel, silicon, sulphur and phosphorus increase the susceptibility to cracking whereas chromium, nitrogen and manganese reduce the cracking tendency. With this knowledge, filler metals can be designed which are completely austenitic and yet resistant to solidification cracking. Such fillers are low in sulphur, phosphorus and silicon but often contain 7–10% Mn. One particular use of these *zero ferrite* electrodes is in the welding of grades such as Type 310 (25% Cr 20% Ni) in applications

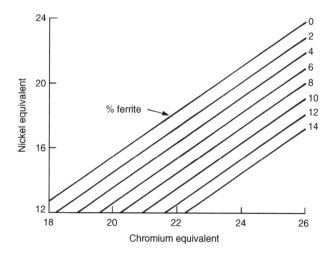

Figure 4.21   *De Long diagram for prediction of microstructure in stainless steel welds (After De Long[22])*

calling for low magnetic permeability, which precludes the introduction of delta ferrite.

The problem of chromium carbide precipitation at grain boundaries and sensitization to intergranular attack was discussed earlier in the section dealing with the corrosion behaviour of stainless steels. However, this problem can now be avoided with the use of stabilized grades such as Types 321 and 347 or with the L grades where the carbon is restricted to 0.03% max.

Ferritic stainless steels

Again, the absence of transformation in ferritic grades eliminates the potential for cold cracking and, due to their moderate thermal expansion characteristics, these steels are generally free from solidification cracking. On the other hand, ferritic stainless steels are prone to grain growth and this leads to low levels of toughness in the weld, particularly in thicker sections. However, sound autogenous welds are produced in Type 430 (17% Cr) steel in thin gauges in applications such as domestic sinks, washing machines and dish washers. Where thicknesses of several millimetres are to be welded, it is preferable to use an austenitic filler such as Type 316 which produces weld metal of high toughness.

Ferritic stainless steels are susceptible to hydrogen cracking, and in gas shielded processes, pure argon rather than argon–hydrogen mixtures should be used.

Variable weld penetration

With the introduction of automatic welding processes, such as orbital TIG for tube joining, major problems were encountered with stainless steels due to variable weld penetration. Thus marked differences in the depth of penetration were observed when steels of nominally the same composition were welded under identical conditions. Alternatively, there was a tendency for the weld to deflect to one side of the joint when steels from different casts were welded together. For this reason, the problem is also known as *cast-to-cast* variability. Although problems of this kind had been observed previously, and also in materials other than stainless steels, they could be accommodated in manual welding by adjustments to power or welding speed. Whereas factors such as arc–metal interactions may be contributory, it is now generally agreed that variable peneration is due to composition-induced, surface energy effects.

Llewellyn *et al.*[23] investigated the problem in commercial tube and plate samples of Type 304L and found a reasonable correlation between current to penetrate and sulphur content. This effect is shown in Figure 4.22 and regression analysis on the pooled data for tube and plate materials yielded the following relationship:

$$\text{Current to penetrate (amps)} = 47 - 686 \ (\%S)$$

which explained 60% of the variance. No other factors were statistically significant at the 95% confidence level. When similar work was carried out on laboratory casts of Type 304L, sulphur had no effect. However, these materials had oxygen contents of 190–330 ppm compared to 20–70 ppm in the commercial steels. Following the work of Heiple and Roper,[24] it was anticipated that both sulphur and oxygen would be beneficial in promoting good weldability. However, the lack of a sulphur effect in the laboratory casts of Type 304L may well have been due to the overriding effect of oxygen in these materials.

Changes in surface energy across the temperature gradients of the weld pool surface give rise to *Marangoni convection* which has a marked effect on the direction and velocity of fluid flow. Thus direction of fluid flow can be changed by surface active impurities which reverse the negative temperature dependence of surface energy for pure materials to a positive temperature dependence when

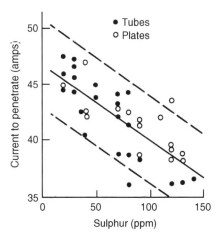

Figure 4.22 *Effect of sulphur on the current required to give penetration of a 2-mm-thick sheet or tube wall with a 5 second arc duration (After Llewellyn* et al.[23]*)*

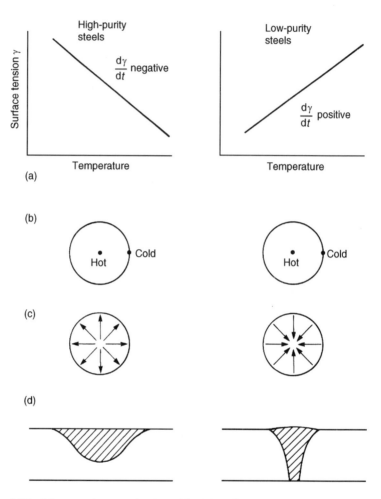

Figure 4.23 *Marangoni convection in welds: (a) surface energy dependence; (b) temperature distribution in welds; (c) fluid flow on surface; (d) bad and good weld penetration*

impurities are present. The effects are illustrated schematically in Figure 4.23(a). The temperature distribution in the weld pool is similar in both 'good' and 'bad' materials (Figure 4.23(b)) but the different relationships between surface tension and temperature in the two materials cause a marked difference in fluid flow (Figure 4.23(c)). Thus in high-purity steels, the surface liquid flow is from the centre to the outside, giving an upward axial flow and poor penetration (Figure 4.23(d)). On the other hand, in steels containing surface active impurities, the flow is towards the centre of the liquid pool, giving downward axial flow and good penetration.

Leinonen[25] has also demonstrated the beneficial effect of sulphur on the TIG weldability of Type 304 steels. He showed that the maximum welding speed that could achieve through-penetration in a steel containing 0.13% S was 74% higher than that for a 0.003% S steel, when using 100% argon shielding gas. The addition of hydrogen and helium to the shielding gas enabled the welding speed to be increased substantially but, in both cases, the differential in weldability between the high- and low-sulphur steels was still preserved.

In order to achieve good weld penetration, fabricators may well specify a minimum of 0.01% S but high levels of sulphur impair the corrosion resistance of stainless steels. Therefore the sulphur content needs to be controlled to a narrow range to ensure good weldability on the one hand and the avoidance of pitting corrosion on the other.

Cold working of stainless steels

The cold-working characteristics of stainless steels are important in relation to:

1. The conversion of hot band to cold-reduced gauges.
2. The production of components by cold forming.

Austenitic stainless steels are characterized by high rates of work hardening which limit the amount of cold deformation that can be undertaken before an annealing treatment is required. However, the work-hardening behaviour is influenced markedly by chemical composition and can often dictate the grade of steel that is used for a particular application.

Role of alloying elements

The austenite to martensite transformation in a material can be discussed in terms of thermodynamics and this aspect has been covered in detail by Kaufman and Cohen.[26] In Figure 4.24, F^{γ} and $F^{\alpha'}$ represent the chemical free energies of austenite and martensite which vary with temperature in the manner shown. At temperature T_0, the two phases have equal free energies, but above T_0, austenite is thermodynamically stable relative to martensite, and below T_0, martensite is the more stable phase. On cooling from above T_0, the austenite to martensite reaction might be expected to occur at T_0 but, in fact, this reaction does not occur until the temperature is lowered to M_s, which in iron-base alloys may be of the order of 200°C below T_0. At M_s, the austenite has derived sufficient energy from undercooling to overcome the barriers of nucleation and the interfacial strains that inhibit the growth of the bcc nucleus in its foreign fcc environment. On heating from a low temperature, the reverse reaction at A_s may also require superheat and A_s is generally as far above T_0 as M_s is below. Alloying elements depress the transformation from austenite to martensite and in very highly alloyed stainless grades, the theoretical T_0 temperature will be below the absolute temperature. However, in the leaner alloy grades, T_0 is above and M_s is below room temperature.

When stress is applied to an austenitic stainless steel during cold working, the mechanical energy interacts with the thermodynamics of the martensite reaction, as demonstrated by Patel and Cohen.[27] Thus in Figure 4.25, F^{γ} and $F^{\alpha'}$ represent the conditions in the unstressed system and $F^{\gamma'}$ and $F^{\alpha'}$ are the relative

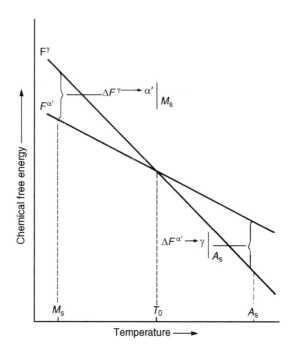

Figure 4.24   *Thermodynamics of the martensite reaction (After Kaufman and Cohen[26])*

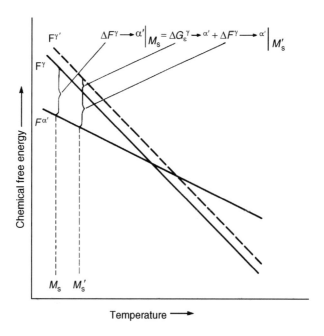

Figure 4.25   *Effect of applied stress on the martensite reaction (After Patel and Cohen[27])*

conditions in the stressed system, where F^γ and $F^{\gamma'}$ are displaced by an amount $\Delta G_\varepsilon^{\gamma \to \alpha'}$, the mechanical energy due to the applied stress. Therefore M_s in the unstrained system can be moved to a higher temperature, M'_s, where the sum of the mechanical and thermodynamic energies $(\Delta G_\varepsilon^{\gamma \to \alpha'} + \Delta F^{\gamma \to \alpha'}/M'_s)$ is equal to $\Delta F^{\gamma \to \alpha'}/M_s$, the critical undercooling energy at M_s. Thus cold working at temperatures above M_s may result in the formation of *strain-induced martensite*, depending upon the chemical composition of the steel and the temperature of working.

In addition to their influence on the martensite reaction, alloying elements can also affect the work-hardening characteristics of austenitic stainless steels through their effect on the *stacking fault energy* of the system. Stacking faults are planar imperfections in the normal stacking of the fcc lattice which increase the rate of work hardening by hindering the movement of dislocations. Elements such as carbon, manganese and cobalt facilitate the formation of stacking faults whereas nickel and copper raise the stacking fault energy which inhibits their formation. The effects of these elements on the true stress–true strain behaviour of 18% Cr 13% Ni-base steels is shown in Figure 4.26. This base composition has sufficient stability to withstand the formation of strain-induced martensite at ambient temperature and therefore any changes that take place in the work-hardening behaviour can be attributed to stacking fault effects. Thus cobalt and manganese raise the rate of work hardening whereas nickel and copper decrease the rate. The effect of chromium on the stacking fault energy was examined in steels containing 18% Ni and, as illustrated in Figure 4.27, it would appear that chromium reduces the stacking fault energy and thereby increases the rate of work hardening.

Work hardening of commercial grades

The effects of rolling at ambient temperature on the properties of commercial grades of austenitic stainless steel are shown in Figure 4.28 and 4.29. Type 301

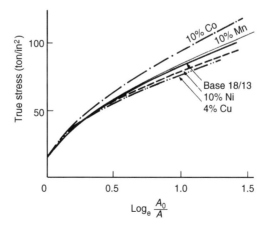

Figure 4.26 *True stress–true strain in 18% Cr 13% Ni steel with alloy additions (After Llewellyn and Murray[28])*

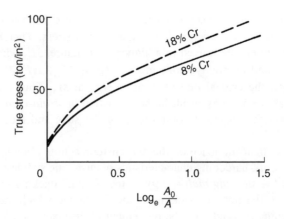

Figure 4.27 *True stress–true strain in 18% Ni base steels with different chromium levels (After Llewellyn and Murray[28])*

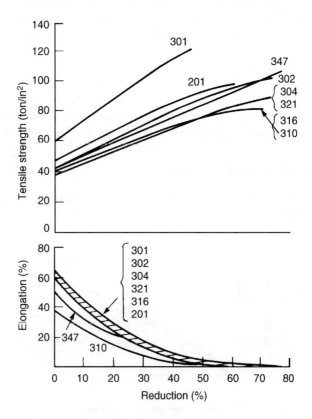

Figure 4.28 *Effect of cold rolling on the tensile strength and elongation of austenitic stainless steels (After Llewellyn and Murray[28])*

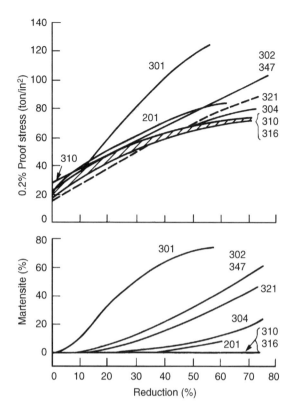

Figure 4.29 *Effect of cold rolling on the 0.2% proof stress and martensite content of austenitic stainless steels (After Llewellyn and Murray[28])*

(17% Cr, 7% Ni) shows the highest rate of work hardening and this can be related to the fact that it has the lowest alloy content in the range examined. Therefore it has the lowest stability and forms the greatest amount of strain-induced martensite during cold rolling. Conversely, Type 316 (18% Cr, 12% Ni, 2.5% Mo) and Type 310 (25% Cr, 20% Ni) are the most highly alloyed steels in the series and these show the lowest rates of work hardening because their alloy content is sufficient to suppress the formation of strain-induced martensite at ambient temperature. However, if rolling is carried out at a lower temperature, e.g. −78°C, then transformation to martensite can be induced in Type 316 steel and this results in a higher rate of work hardening than that observed at room temperature. This is illustrated in Figure 4.30.

Optimization of cold-forming properties

A considerable amount of work has been carried out on the optimization of composition in austenitic stainless steels in order to promote high levels of stretch formability in applications such as the production of sink bowls. Up until the early 1970s, the standard Type 304 grade (nominally 18% Cr, 10% Ni) was used in

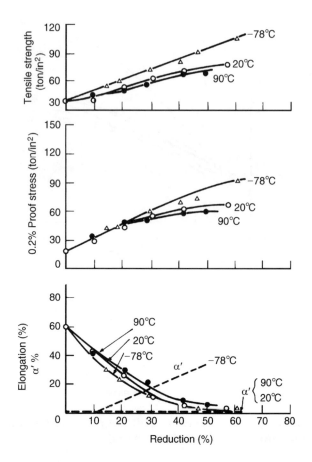

Figure 4.30   *Effect of the temperature of rolling on the properties and martensite content of Type 316 steel (After Llewellyn and Murray[28])*

the UK for this application but the degree of precision required in the pressing operation, for the consistent production of high-quality bowls, was uneconomical. This related to the fact that the 18% Cr, 10% Ni composition is relatively stable at ambient temperature and the rate of work hardening was insufficient to compensate for the reduction in thickness that accompanies the stretching operation. Failure therefore occurred before the required degree of stretching (uniform strain) could be achieved. However, in very lean compositions, which form large amounts of strain-induced martensite at low strain, the rate of work hardening reaches a maximum at an early stage and decreases when the reaction is complete. Workers in British Steel[29] therefore focused their attention on identifying the compositions that would provide the optimum rate of work hardening and uniform strain via the formation of controlled amounts of strain-induced martensite. The results of this work are shown in Figure 4.31 which illustrates the interdependence of composition, stability and uniform strain. In this figure, the lower diagram shows the relationship between composition and the amounts

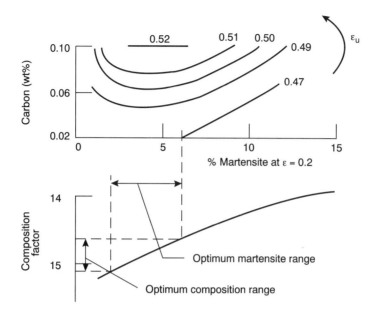

Figure 4.31   *Effect of composition factor and carbon content on the formation of martensite at a true strain of 0.2 and on the uniform true strain (ε_u) (After Gladman et al.[29])*

of martensite induced at a true strain of 0.2, the composition being expressed through the following factor:

$$CF = 18\ (\%C) + 13\ (\%N) + 0.90\ (\%Ni) + 0.27\ (\%Cr) + 0.47\ (\%Mn)$$
$$+ 0.53\ (\%Mo) + 0.97\ (\%Cu)$$

By controlling the CF value to within the limits shown, the level of martensite is controlled to between 3 and 6% at a true strain of 0.2 and, as shown in the upper diagram, this results in the optimum uniform strain at various carbon levels. It will also be noted that uniform strain increases with carbon content. Ordinarily, it would be anticipated that an increase in carbon would reduce the rate of work hardening by suppressing the formation of martensite. However, for a given composition factor, and therefore a defined level of austenite stability, it is proposed that an increase in carbon leads to an increase in work-hardening rate through its strengthening effect on the strain-induced martensite.

Austenitic stainless steels are also used in fastener applications where they undergo severe deformation during cold-heading operations. In this context, a low rate of work hardening is required and the grades involved must have sufficient stability to resist the formation of strain-induced martensite. Beyond this basic requirement, the performance can be enhanced by the selection of alloying elements that provide both austenite stability and a high stacking fault energy.

In stable compositions, the rate of work hardening is dictated solely by stacking fault energy and the effects of alloying elements in an 18% Cr, 13% Ni base were described earlier (Figure 4.26). Having eliminated stability effects, the addition of copper has a significantly greater effect than nickel in reducing the rate of work hardening, but the strengthening effect of adding 10% Mn was very small compared with that produced by a similar level of cobalt.

In the United States, Armco 18–9 LW (18% Cr, 9% Ni, 4% Cu) was developed for the production of cold-headed fasteners and grades based on 18% Cr, 12% Ni, 2% Cu have been used in the UK. In addition to fastener applications, materials with a low yield stress and low rates of work hardening are also required in cold-drawing operations. These characteristics therefore contrast sharply with the compositions used in stretch forming operations which are designed to work harden and accommodate the reduction in cross-section via the formation of controlled amounts of strain-induced martensite. Deep-drawing grades are therefore based on compositions with a low level of interstitial elements, sufficient alloy content to preserve a stable structure and ideally with a high stacking fault energy. A coarse grain size is also beneficial, consistent with the avoidance of the *orange peel* effect. Texture considerations are also important in deep drawing performance and high *r* values are associated with strong {111} and/or {110} planes parallel to the surface of the sheet.

Mechanical properties at elevated and sub-zero temperatures

Although austenitic stainless steels are used primarily because of their high corrosion resistance, they also possess extremely good mechanical properties over a wide temperature range. Unlike ferritic materials, austenitic stainless steels do not exhibit a ductile–brittle transition and maintain a high level of toughness at liquid gas temperatures. On the other hand, they also exhibit good creep rupture strength at temperatures above 600°C, where ferritic and martensitic steels undergo microstructural degradation. The standard grades of austenitic stainless steel are therefore used in cryogenic applications, such as liquid gas storage vessels and missiles, and at elevated temperatures in chemical and power plant applications. These aspects are discussed on pp. 351–356.

Tensile properties

The tensile properties of the standard grades over the temperature range −200 to 800°C have been examined by Sanderson and Llewellyn[30] and the data are summarized in Figure 4.32. This indicates that the tensile strengths of the majority of the grades fall within a fairly narrow scatter band at temperatures above about 100°C. As the temperature falls below this level, there is a marked increase in tensile strength, the lower alloy steels such as Types 302 and 304 achieving higher strengths than the highly alloyed grades such as Types 309 and 310. However, this situation tends to be reversed in the case of the 0.2% proof stress values, the more highly alloyed grades such as Types 316, 309 and 310 providing higher strengths than Types 302, 304, 321 and 347. This behaviour is related to the stability of the austenitic structure in these materials and the relative ease with which they undergo transformation to strain-induced martensite.

At temperatures above 100°C, solid solution strengthening is the main criterion controlling strength and therefore the more highly alloyed steels, such as

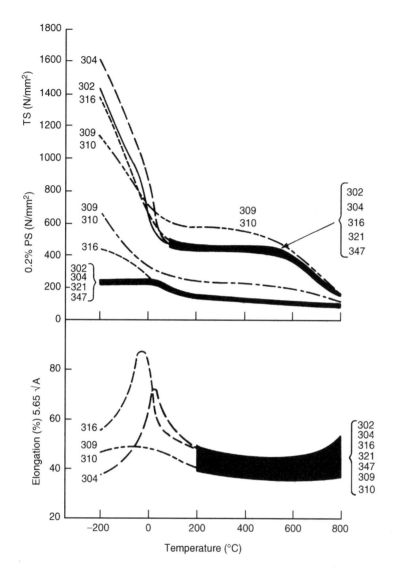

Figure 4.32 *Tensile properties of standard austenitic stainless steels: −200 to 800°C (After Sanderson and Llewellyn[30])*

Types 309 (25% Cr, 15% Ni) and 310 (25% Cr, 20% Ni), possess the highest strengths. However, at temperatures below 100°C, deformation in the tensile test takes place by means of:

1. Slip/dislocation movement.
2. Transformation to strain-induced martensite.

Thus the more lowly alloyed grades such as Type 304 (18% Cr, 9% Ni) undergo strain-induced transformation to martensite with relatively little strain

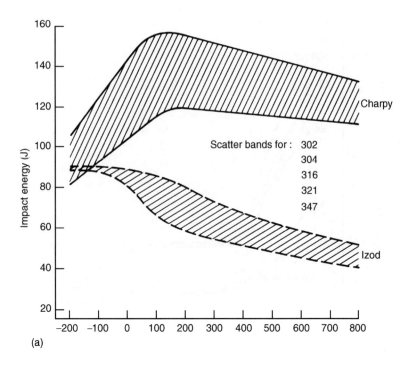

Scatter bands for : 302
                    304
                    316
                    321
                    347

(a)

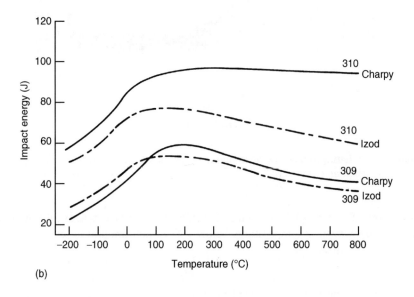

(b)

Figure 4.33   *Impact properties of standard austenitic stainless steels: −200 to 800°C (After Llewellyn and Sanderson[30])*

at temperatures below about 50°C, resulting in low proof stress values. However, this behaviour also leads to a very high rate of work hardening and produces high levels of tensile strength. Conversely, steels such as Types 309 and 310 are stable enough to resist the formation of strain-induced martensite and their proof stress values increase significantly below 100°C, following the characteristic of fcc alloys in showing a marked increase in flow stress with decreasing temperature. However, the suppression of strain-induced martensite leads to low rates of work hardening and therefore to relatively low levels of tensile strength.

As indicated in Figure 4.32, the elongation values of these materials are also strongly correlated with alloy content at temperatures below about 100°C. Thus stable grades such as Types 309 and 310 exhibit a modest peak in ductility at temperatures of the order of −50°C whereas the less stable grades such as Types 304 and 316 show sharply defined maximum values at temperatures of about +20 and −20°C respectively. The reason for this behaviour is complex but can be explained in terms of the temperature dependence of the flow stress and work hardening rate.[30]

Impact properties

The impact properties of the standard austenitic grades over a similar range of temperature are shown in Figure 4.33 (a) and (b). Again with the exception of Types 309 and 310, the materials fall within reasonably well defined scatter bands but with a very clear differentiation between Charpy and Izod values. In the Charpy test, the energy values decrease at temperatures below about 100°C, whereas in the Izod test the values decrease progressively as the temperature is increased. However, these variations are largely a function of the operating conditions in the testing machines used for this work, and the main point to emerge is the excellent toughness provided by these steels at extremely low temperatures.

Steels for boilers and pressure vessels

Carbon, alloy and stainless steels are used in boilers operating at elevated temperatures and also in pressure and containment vessels operating at room and sub-zero temperatures. Whereas the metallurgy and applications of these various types of steel have been discussed in discrete chapters of this text, it was considered more appropriate to deal with this general topic under a single heading at this stage. Thus, although this chapter is concerned primarily with stainless steels, this particular section will also deal with the use of carbon and alloy steels in boilers and pressure vessels.

Steel specifications

Pressure vessels are manufactured from plates, sections, tubes and forgings and steel specifications for these and other product forms have been formulated specifically for use in pressure vessels. A summary of the relevant BS and

Table 4.12 *Specifications for boiler and pressure vessel steels*

Specification	Product form	Title
BS EN 10028	Plates	Part 2: Non-alloy and alloy steels with specified elevated temperature properties Part 3: Weldable fine grain steels, normalized Part 4: Nickel alloy steels with specified low temperature properties
BS 1501		Part 3: Corrosion and heat-resisting steels
BS 1502	Sections and bars	Steels for fired and unfired pressure vessels
BS 1503	Forgings	Steel forgings for pressure purposes
BS EN 10213	Castings	Steel castings for pressure vessel purposes
BS 3059	Tubes	Steel boiler and superheat tubes Part 1: Low-tensile carbon steel tubes without specified elevated temperature properties Part 2: Carbon, alloy and austenitic stainless tubes with specified elevated temperature properties
BS 3601	Pipes and tubes	Steel pipes and tubes for pressure purposes: carbon steel with specified room temperature properties
BS 3602	Pipes and tubes	Steel pipes and tubes for pressure purposes: carbon and carbon manganese steels with specified elevated temperature properties
BS 3603	Pipes and tubes	Steel pipes and tubes for pressure purposes: carbon and low-alloy steel with specified low-temperature properties
BS 3604	Pipes and tubes	Steel pipes and tubes for pressure purposes: ferritic alloy steel with specified elevated temperature properties
BS 3605	Pipes and tubes	Seamless and welded austenitic steel pipes and tubes for pressure purposes
BS 3606	Tubes	Steel tubes for heat exchangers

European specifications in this sector is given in Table 4.12. Thus, whereas particular grades of structural steel are specified for use in bridges and buildings, similar compositions for pressure vessels are also identified in other specifications, e.g. BS EN 10028. The first reaction to this situation might well be to suggest that this represents an unnecessary and confusing duplication but, in fact, the two types of specification deal with the property data that are important to the specific end use. Whereas both types of specification cover tensile and impact properties at ambient and sub-zero temperatures, pressure vessel codes may also specify the tensile and stress rupture properties at elevated temperatures.

Pressure vessel codes

In an earlier chapter, an introduction was given to the British Standard Specification relating to the construction of bridges (BS 5400) and steelwork in buildings (BS 5950). In a similar vein, specifications have also been prepared for the construction of boilers and pressure vessels, e.g.:

1. BS 1113: 1992 Design and manufacture of water tube steam generation plant.
2. BS 5500: 1997 Unfired fusion welded pressure vessels.

Both specifications have a materials section which defines the steels that can be used and the associated design stress values. An example is shown in Table 4.13 which has been reproduced from BS 1113. The steels identified in this table are from BS 3059 (Steel boiler and superheat tubes) and cover a range of compositions from 0.3% Mo up to 12% Cr–Mo–V. Against each steel grade, a series of design stress values (f) is provided for various operating temperatures up to a maximum of 620°C in this particular table. As illustrated shortly, these design stresses are calculated on the basis of short-term tensile properties at low to moderate temperatures, but at higher operating temperatures design is based on creep rupture strength. Where the latter is the case, the values in Table 4.13 are shown in italics.

Thus specifications such as BS 1113 and BS 5500 provide design engineers with clearly defined data on the allowable design stress values at different operating temperatures for the various steel specifications and product forms identified in the previous section.

Steels for elevated-temperature applications

Prior to dealing with the derivation of design stress values, the following terminology must be defined:

- R_m min. TS at room temperature
- $R_{e(T)}$ min. 0.2% PS (1.0% PS for austenitic steels) at temperature T
- S_{Rt} mean value of stress to cause rupture in time t at temperature T
- f_E design strength based on short-term tensile strength properties
- f_F design strength based on creep rupture characteristics

In BS 1113 and BS 5500, the allowable design stresses at elevated temperatures are calculated in the following manner:

1. C, C–Mn and low-alloy steels

BS 1113

$$>250°C \quad f_E = \frac{R_{e(T)}}{1.5} \text{ or } \frac{R_m}{2.7}$$

BS 5500

$$\geq150°C \quad f_E = \frac{R_{e(T)}}{1.5} \text{ or } \frac{R_m}{2.35}$$

whichever gives the lower value.

Table 4.13 *Design stress values for tube steels according to BS 1113 (Design and manufacture of water tube steam generating plant)*

Type–grade and method of manufacture	Rm (N/mm²)	Re (N/mm²)	Thickness (mm)	250	300	350	400	440	450	460	470	480	490	500	510	520	530	540	550	560	570	580	590	600	610	620	Design lifetime (h)
				| Values of f for design temperature (°C) not exceeding |																							
BS 3059: Part 2 – Alloy steel																											
243 0.3Mo S1, S2 ERW and CEW	480	275		149	128	120	117		115		113	112	95	78	62	52	41	32									100 000
												107	88	70	57	46	37	30									150 000
												100	81	65	53	42	34	28									200 000
620 1Cr½Mo S1, S2 ERW and CEW	460	180		129	129	129	120		116			114	113	112	93	76	62	52	42	33	27						100 000
													113	102	83	67	55	44	35	29	25						150 000
													113	94	76	61	49	40	32	26	22						200 000
													107	88	70	57	45	37	30	25	20						250 000
622–490 2¼CrMo S1, S2	490	275		157	153	149	145		137	135	133	131	118	105	94	82	72	61	53	45	39	34	29	26			100 000
											133	122	108	97	85	73	63	56	48	42	36	31	27	23			150 000
											130	117	104	92	79	68	59	52	45	38	33	28	25	22			200 000
											126	113	100	87	75	65	57	49	42	36	32	27	23	20			250 000
629–590 9 Cr 1 Mo S1, S2	590	400		219	217	215	211	208	207	195	176	159	144	129	115	103	91	80	68	58	48	38	31	26			100 000
									207	188	169	152	137	123	110	97	85	74	62	52	42	34	28	23			150 000
									204	183	165	148	132	118	105	93	81	69	58	48	38	31	25	22			200 000
									199	179	159	144	129	115	102	89	78	66	55	45	35	28	24	21			250 000
762 12CrMoV S1, S2	720	470		241	234	230	225		215			205	201	191	173	155	138	122	107	93	80	68	58	48	40	33	100 000
													200	184	168	152	135	115	98	85	72	62	52	44	37	30	150 000
													197	180	164	146	128	110	94	80	68	58	49	41	34	28	200 000
													192	176	160	142	124	105	90	77	65	55	46	38	32	25	250 000

After BS 1113: 1989.

2. Austenitic stainless steels

BS 1113

$$>250°C \ f_E = \frac{R_{e(T)}}{1.35} \text{ or } \frac{R_m}{2.7}$$

BS 5500

$$\geq 150°C \ f_E = \frac{R_{e(T)}}{1.35} \text{ or } \frac{R_m}{2.5}$$

whichever gives the lower value.

3. For time-dependent design stresses (creep range)

BS 1113 and BS 5500

$$f_F - \frac{S_{Rt}}{1.3}$$

These rules give rise to the type of design curve shown in Figure 4.34 and which is a graphical representation of the sort of information that is presented in Table 4.13. As illustrated schematically in this figure, there is a marked change in the design stress values at the *cross-over* temperature where the values based on short-term tensile properties give way to those based on the time-dependent creep rupture strength.

The actual design stress values for a series of ferritic grades and austenitic stainless steels are shown in Figure 4.35(a) and (b) respectively. The steels concerned are in tube form (to BS 3059) and the design values in the creep range relate to rupture in 100 000 hours. As illustrated in Figure 4.35(a), the allowable stresses increase progressively as the alloy content of the ferritic grades is increased. Whereas the C–Mn steel exhibits a cross-over temperature of 430°C, the alloy steels undergo this change in the range 450–500°C. In the austenitic series, Figure 4.35(b), the design stress values increase progressively in the order:

- Type 304 – 18% Cr, 9% Ni
- Type 316 – 18% Cr, 12% Ni, 2.75% Mo
- Esshete 1250 – 16% Cr, 10% Ni, 8% Mn–Mo–V–Nb

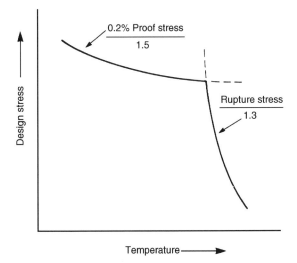

Figure 4.34 *Design strength values for low-alloy steels according to BS 1113*

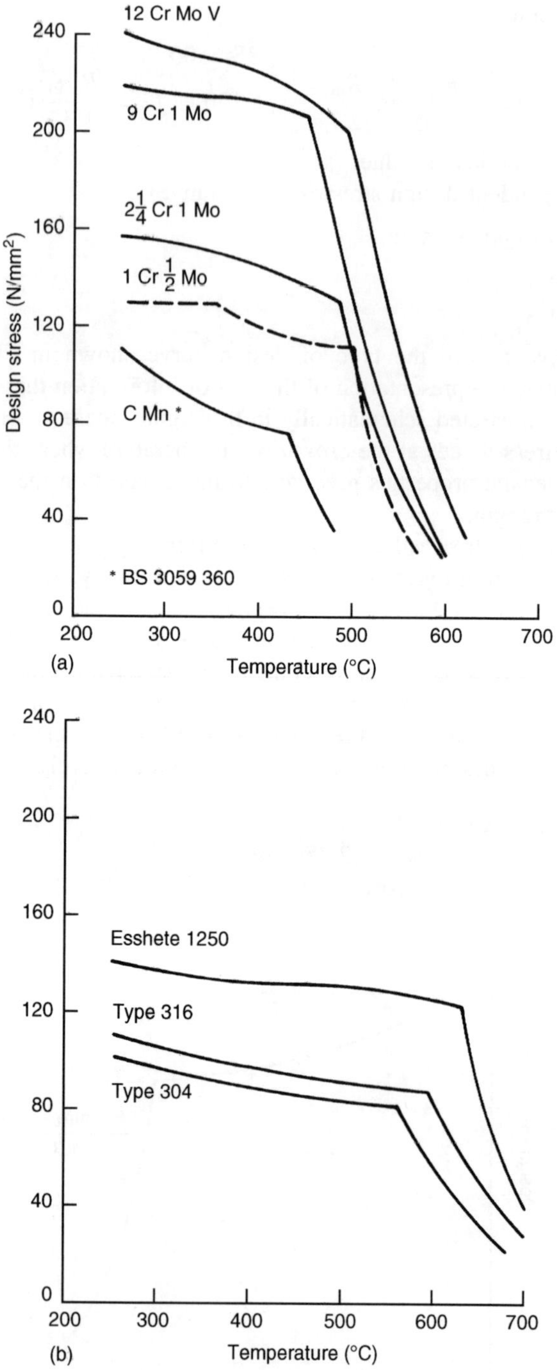

Figure 4.35 *Design stress values according to BS 1113: 1992 (Design and manufacture of water tube steam generating plant); (a) C–Mn and alloy steels; (b) austenitic stainless steels*

The last composition was developed specifically for use in power plant boilers and will be discussed later on pp. 363–365. However, the main feature to note with these steels is that the design stress values are relatively low at temperatures up to about 500°C compared with the ferritic grades, whereas the cross-over between short-term tensile properties and time-dependent properties is extended to considerably higher temperatures. As discussed later, Esshete 1250 offers a significant advantage over standard grades because of the facility to design on the basis of proof strength at operating temperatures up to 630°C.

From the data presented in Figure 4.35, it can be appreciated that the economic selection of steels for service at elevated temperature would dictate the use of C–Mn or low-alloy grades at temperatures up to 500–525°C. However, at higher temperatures, use must be made of austenitic stainless grades which exhibit better properties, particularly in the creep rupture range.

Steels for low-temperature applications

As illustrated on pp. 350–351 austenitic stainless steels possess excellent impact properties at extremely low temperatures, which is a characteristic of face-centred-cubic metals. Therefore stainless steels, together with alloys of aluminium, copper and nickel, are used extensively for the construction of equipment operating at temperatures below −200°C. However, there are many structures which operate at only moderately low temperatures which do not require the high toughness levels of these materials and which are satisfied by the use of less expensive steels. This situation is demonstrated very clearly by Wigley[31] in Figure 4.36, which identifies the boiling temperatures for a wide range of gases and the materials that are used for the associated structures and containment vessels. Thus for temperatures down to about −50°C, C–Mn steels (to BS 4360) may be adequate and a range of nickel steels ($2\frac{1}{4}$, $3\frac{1}{2}$ and 9% Ni) can be used progressively down to temperatures of the order of −200°C. Wigley states that 9% Ni steel is the only ferritic steel which is permitted for use at liquid nitrogen temperatures (−196°C). It is also economical for the construction of storage tanks for liquid argon (−186°C), oxygen (−183°C) and methane (−161°C).

The specifications, composition ranges and properties of these C–Mn and nickel steels are summarized in Table 4.14. In this table, Wigley makes the point that although specific impact test temperatures are laid down in the relevant standards, this does not guarantee that the steel will be satisfactory for use at those particular temperatures. This aspect is particularly pertinent if the materials are used in thick sections. Annex D of BS 5500: 1997 therefore specifies particular requirements for ferritic steels for use in vessels operating below 0°C. These take the form of diagrams providing relationships between the lowest operating temperature (design reference temperature) and impact test temperature for different thicknesses of material. These relationships are shown in Figures 4.37(a) and (b) for as-welded and post-weld heat-treated conditions respectively. Given the beneficial effect of post-weld stress-relieving treatments, the impact test requirements are less demanding than those required in the as-welded condition. For example, 20 mm material operating at a minimum design temperature

Table 4.14 *The compositions and mechanical properties of ferritic steels for use at low temperatures*

Type	Condition	Designation	Composition							Mechanical properties at 20°C			
			C (%)	Mn (%)	Si (%)	Mo (%)	Nb (%)	Ni (%)	Cr (%)	Yield strength (MPa)	Tensile strength (MPa)	Elongation (gauge length 5.65 √S_0) (%)	Charpy V-notch impact test temperature[a] (°C)
C–Mn semi-killed	As rolled	BS 4360: grade 40C / grade 43C	0.22 / 0.22	1.6 / 1.6						260 / 280	400–480 / 430–510	25 / 22	0 / 0
C–Mn–Nb semi-killed	Normalized	BS 4360: grade 40D / grade 43D / grade 50C	0.19 / 0.19 / 0.24	1.6 / 1.6 / 1.6	0–0.55		0.1 / 0.1 / 0.1			260 / 280 / 355	400–480 / 430–510 / 490–620	25 / 22 / 20	−20 / −20 / −15
C–Mn–Nb silicon killed	Normalized	BS 4360: grade 50D	0.22	1.6	0.1–0.55		0.1			355	490–620	20	−30
C–Mn silicon killed, fine grain	Normalized	BS 4360: grade 40E / grade 43E / grade 55E	0.19 / 0.19 / 0.26	1.6 / 1.6 / 1.7	0.1–0.55 / 0.1–0.55 / 0–0.65		0.1			260 / 280 / 450	400–480 / 430–510 / 550–700	25 / 22 / 19	−50 / −50 / −50
Ni–Cr–Mo	Quenched and tempered	BSC QT 445 ASTM A517 Gd F (USS T1)	0.1–0.2	0.6–1	0.15–0.35	0.4–0.6		0.7–1	0.4–0.65	690	795–930	18	−45
2¼% Ni	Normalized	ASTM A203 Gd A	0.17–0.23	0.7–0.8	0.15–0.3			2.1–2.5		255	450	25	−60
3½% Ni	Normalized and tempered	BS 1501/503 ASTM A203 Gd D	0.17–0.2	0.7–0.8	0.15–0.3			3.25–3.75		255	450	22	−100
9% Ni	Double normalized and tempered	BS 1501/509								515	690	22	−196
	As welded with Ni–Cr–Fe (Inconel 92) electrodes	ASTM Code Case 1308 and A353	0.13	0.9	0.13–0.32			8.4–9.6		480	655	–	−196

[a]The impact test temperature is that laid down by the relevant standards for quality control purposes. This does not guarantee that the steel, especially if used in thick sections, will be completely satisfactory for use at low temperatures.

After Wigley.[31]

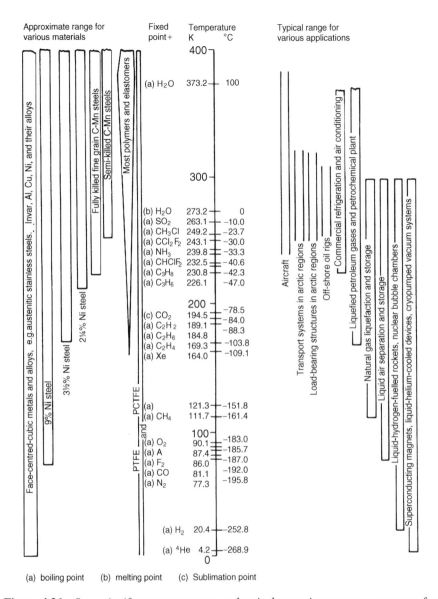

Figure 4.36 *Some significant temperatures and typical operating temperature ranges for various structures and constructional materials (After Wigley[31])*

Table 4.15

Specified min. tensile strength (N/mm²)	Min. impact energy (J) at material test temperature			
	10 mm × 10 mm	10 mm × 7.5 mm	10 mm × 5 mm	10 mm × 2.5 mm
<450	27	22	19	10
≥450	40	32	28	15

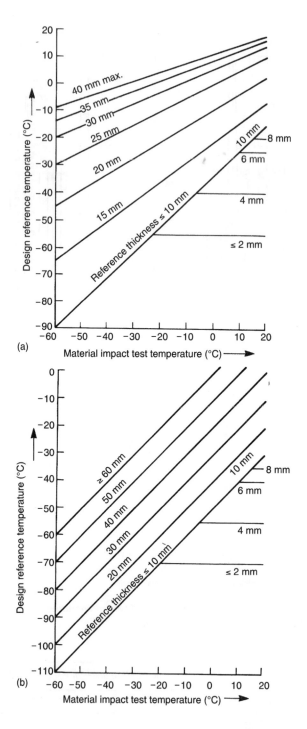

Figure 4.37 *Relationship between minimum operating temperature and impact test temperature for various thicknesses of ferritic steels: (a) as-welded condition; (b) post-weld heat-treated condition (From Annex D, BS 5500: 1997)*

of −40°C would need to provide a minimum Charpy V value at a test temperature of −50°C in the as-welded condition compared with a temperature of 0°C
in the post-weld heat-treated condition. The minimum Charpy V value depends
upon the strength of steel employed as indicated in Table 4.15.

Steels in fossil-fired power plants

Since the early 1970s, major events have taken place which have had a profound
effect on the choice of fuels for power generation. First, the oil crises of 1973
and 1979 curtailed the use of oil for steam-raising purposes but, currently, this
fuel accounts for about 8% of power generation. At a later stage, the fire at Three
Mile Island in the United States and the devastating *melt-down* at Chernobyl had
a very significant effect in limiting further commitment to nuclear power. More
recently, the privatization of the electricity industry in the UK has prompted the
dash for gas and in 1993, natural gas accounted for 9.2% of power generation.
As a result, the use of coal is declining but, currently, coal makes up 53% of
the power requirements. However, with proven reserves for more than 300 years,
coal-fired power stations are likely to feature prominently well into the twenty-
first century.

Boiler layout and operation

The selection of steels for the various parts of fossil-fired boilers will be appreciated more readily by first providing an outline of boiler construction and the
steam−water circuit. A schematic cross-section through a boiler is shown in
Figure 4.38 The boiler has a large rectangular combustion chamber in which
high-temperature gases are generated by burning coal or oil, providing maximum
temperatures of up to 1600 and 2000°C respectively. The walls of the furnace
section are formed with the *evaporator* tubes which are arranged in closely
packed, vertical rows. Each tube has two longitudinal fins at 180° and the fins
on adjacent tubes are welded together to form an airtight wall, except for entry
ports for burners or exits for flue gas.[32] Although the temperatures of the furnace
gases are very high, the cooling effect of the steam/water mixture generally limits
the outer metal *(fire-side)* temperature of the evaporator tubing to about 450°C.
On the other hand, significantly higher metal temperatures can be generated in
some circumstances, e.g. when a flame impinges directly onto the surface of the
evaporator tubes.

The atmosphere in the combustion chamber of fossil-fired boilers can be
extremely aggressive, leading to severe attack and premature replacement of
the boiler tubes. This arises from the chlorine content of coals whereas sulphur
and vanadium are the damaging elements in fuel oils. Under normal oxidizing
conditions, a protective oxide scale is formed on the surface of the tubing which
limits the corrosion rate. However, reducing conditions promote complex reactions between sulphur, chlorine, carbon and oxygen, leading to the formation
of non-protecting scales. Particularly damaging conditions are produced when
unburnt particles of coal impinge on the side-wall of the evaporator tubes, the

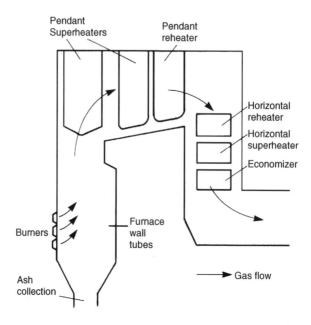

Figure 4.38   *Schematic illustration of coal-fired boiler*

generation of carbon monoxide promoting sulphidation and the formation of hydrogen chloride.[33]

At the top of the combustion zone, the hot gases are turned through an angle of 90° and over banks of *superheater* tubes which are suspended from the boiler roof. In this region of the boiler, the steam is heated progressively to maximum temperatures of 540°C in the case of oil-fired boilers and 565°C in coal-fired stations. These temperature limits in UK power stations are dictated very largely by the aggressive environmental conditions described above and the need to maintain an adequate balance between operating efficiency and excessive corrosion/oxidation of the boiler tubing. In the superheater section, the first few rows of tubing are subjected to very arduous conditions due to the impingement of hot gases and fly ash.

Beyond the superheater section of the boiler, the hot gases turn through a second angle of 90° and over the *reheater* tubes. Being in the same part of the boiler, the temperatures in the superheater and reheater are similar but, due to the fact that they operate on different steam circuits, the pressure of steam in the reheater is significantly lower than that in the superheater.

Before leaving the boiler, the gases pass over a final bank of tubes that constitute the *economizer*. At this stage, the gas temperature has fallen to about 300°C and the economizer operates at a steam temperature of about 250°C.

The flow of steam/water through the boiler is shown in Figure 4.39. Feedwater is first preheated in the economizer before passing into the evaporator tubing in the furnace walls. From the various outlets in the evaporator, the steam/water mixture then passes into a large collection vessel, known as the *steam drum*. In this vessel, the steam is separated from the water, the steam being passed to the

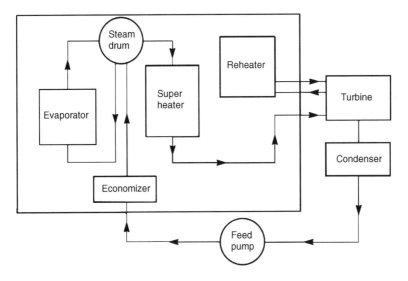

Figure 4.39 *Water–steam flow in coal-fired boiler*

superheater and the water being directed to the feedwater circuit. On reaching a temperature of 540–565°C and a pressure of 170 bar in the superheater, the steam is conveyed from the boiler to the turbine via *headers, steam pipes* and valves. The headers are very large pressure vessels, with many stub tube connections, and act as collection units for steam from the various tubes in the superheater. A typical unit in a large power station would measure 500 mm diameter × 40 mm wall thickness × 12 m long.

As indicated in Figure 4.39, part of the steam leaving the turbine is diverted to the reheater section of the boiler. At this stage, the steam has lost a substantial amount of pressure but, after reheating, the steam is then passed again into the turbine via a header system.

The conditions described above relate to traditional coal or oil-fired stations but in recent years, the combined cycle gas turbine (CCGT) has been introduced in which the maximum steam temperature is only of the order of 400–450°C.

Steel selection

From the foregoing remarks, it can be appreciated that boiler tube materials are subjected to a wide range of temperatures and operating conditions and economic selection dictates that a variety of steel grades is used to satisfy these conditions. However, one statistic worthy of note is that in the 660 MW boilers at Drax, there are 480 km of tubing with 30 000 butt and fillet welds.

Evaporator tubing

As indicated earlier, the fire-side temperature of evaporator tubing is generally of the order of 450°C and C–Mn steel, typically BS 3059 Grade 440, is used in this region. However, in areas of the furnace wall where the operating conditions are particularly troublesome, it is sometimes necessary to use tubing which has a surface cladding of stainless steel in order to achieve reasonable life. These so-called *co-extruded* tubes will be discussed later.

Superheater/reheater tubing

Depending upon the position in the boiler and steam temperature, the tubing in various stages of the superheater will be made from the following steel compositions:

Increasing	1% Cr, $\frac{1}{2}$% Mo
temperature	$2\frac{1}{4}$%, Cr, 1% Mo
resistance	18% Cr, 12% Ni, $2\frac{1}{2}$% Mo (Type 316)
	16% Cr, 10% Ni, 8% Mn–Mo–V–Nb
	(Esshete 1250)

Esshete 1250 was developed by the former United Steel Companies Ltd in the 1960s specifically as a high-temperature superheater material and the number indicates the maximum operating temperature in degrees fahrenheit (675°C). As indicated in Figure 4.35, the proof strength of the steel at 600°C is 40% higher than that of Type 316, whereas its rupture strength is approximately 95% greater than that of the standard grade. This provides a design cross-over temperature from proof to rupture strength of 630°C in the case of Esshete 1250 compared with 590°C for Type 316. This enables the use of substantially thinner tubing in Esshete 1250 compared to other austenitic steels, which results in a considerable cost saving.[34]

Reheater tubing operates at the same temperature as superheater tubing and this usually dictates the use of an austenitic stainless steel, such as Type 316, in coal-fired stations. However, $2\frac{1}{4}$% Cr 1% Mo steel can be used in oil-fired boilers or in coal-fired stations where the design temperature has been reduced deliberately to allow the use of ferritic steels. In this respect, steels such as 9% Cr 1% Mo can also be considered as candidate materials.

Steam headers

Although operating at high temperatures and pressure, steam headers are not subjected to the corrosive conditions encountered by superheater or reheater tubing and therefore headers are generally made from $\frac{1}{2}$% Cr–Mo–V, 1% Cr–Mo or $2\frac{1}{4}$% Cr 1% Mo steel. For steam temperatures in excess of 565°C, highly alloyed ferritic or austenitic steels would be required, as discussed in the next section dealing with steam pipe.

Steam pipe

Steam pipes carry large volumes of steam from the headers to the turbine and, like headers, they are not subjected to aggressive environments. In the UK, $\frac{1}{2}$% Cr–Mo–V steel is used extensively for steam temperatures up to 565°C and therefore the material is operating in its creep range.

Steam pipe is heavy-walled material and, according to Wyatt,[32] $\frac{1}{2}$% Cr–Mo–V piping operating at a temperature of 565°C would have a wall thickness of

6.3 cm. At higher temperatures, greater wall thicknesses would be required or else stronger steels would be selected. These would include 9% Cr–Mo–V–Nb (T91) and 12% Cr–Mo–V (X20) which could be used in thinner sections, thereby reducing the cost of the pipework. At the super-critical Drakelow 'C' station, operating at 599°C/250 bar, both Type 316 and Esshete 1250 have been used for steam pipes.

Other components

The steam drum in the steam–water circuit between the evaporator and super-heater is fabricated from C–Mn steel plates and presents no problems in material selection.

The economizer tubing, situated near the exhaust end of the boiler, operates in a gas temperature of 400°C and again a C–Mn steel is adequate for these conditions.

Feedwater pipes are made in C–Mn–Nb steel (BS 3602 Grade 490 Nb) and operate at temperatures up to 250°C. Previously, a 0.4% C–Mn steel was used for this purpose but BS 3602 Grade 490 Nb (0.23% C max.) is more weldable and has a higher proof strength than the traditional medium-carbon steel.

Co-extruded tubing

As indicated earlier, C–Mn or low-alloy steels, clad with a stainless steel or superalloys, are used in parts of the boiler where metal loss is excessive due to high-temperature corrosion. The experience of these materials in UK power stations is the subject of a detailed paper by Flatley *et al.*[35]

The tubes are manufactured by conventional hot extrusion and in essence comprise a corrosion-resistant outer layer over a conventional, load-bearing inner layer. Co-extruded tubing involving Type 310 (25% Cr, 2% Ni)/carbon steel was first introduced into the furnace wall of a 500 MW boiler in 1974 and gave a corrosion benefit of three to ten times, depending upon the severity of the location. This practice has now become widespread in UK coal-fired stations. The tube dimensions have varied from 76.2 mm OD × 8.3 mm total wall thickness (4.3 mm outer 310 on 4 mm C steel) to 50.8 mm OD × 6 mm total wall (3 mm outer on 3 mm inner).[35]

In the superheater and reheater stages of the boiler, the combination of materials has generally involved Type 310 heat-resistant steel over Esshete 1250, the high rupture strength steel. Co-extruded tubing of this type has been installed in various power stations in the UK, providing a benefit of 2.5 or more over monobloc Esshete 1250.[35] To avoid the possibility of intergranular attack due to the migration of carbon at the service temperature, later installations have incorporated niobium-stabilized, Type 310 material.

Flue gas desulphurization equipment

Prior to the privatization of the electricity industry, the UK Government announced plans for the installation of flue gas desulphurization (FGD) equipment on all new coal-fired power stations and also for the retrofitting of such plant on certain existing coal-fired stations. Following privatization, PowerGen have completed the installation of FGD at Ratcliffe and National Power have FGD operational at Drax power station. The main function of FGD plant is to remove at least 90% of the SO_2 from boiler exhaust gases and, ideally, the process should also generate a saleable by-product.

Most FGD plants operate on the basis of a wet scrubbing process in which SO_2 is obsorbed in a suspension of lime (CaO) or limestone ($CaCO_3$) in water:

$$CaCO_3 + SO_2 \rightarrow CaSO_3 + CO_2$$

In the above reaction, calcium sulphite ($CaSO_3$) forms as a sludge and having no commercial value would have to be disposed of to landfill. However, by incorporating an oxidation stage in the process, the calcium sulphite is converted to gypsum:

$$CaSO_3 + O + 2H_2O \rightarrow CaSO_4.2H_2O$$

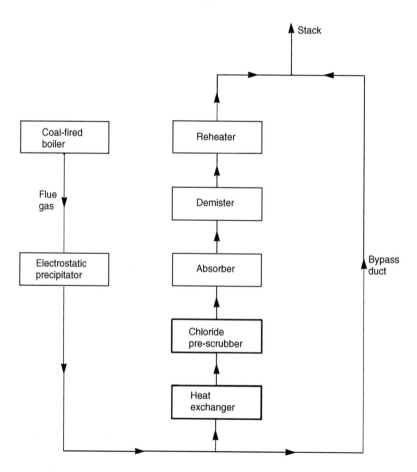

Figure 4.40 *Schematic layout for flue gas desulphurization plant*

Gypsum is more environment-friendly than the calcium sulphite sludge and also represents a commercially useful product in the building industry.

The various process stages in a typical FGD are shown schematically in Figure 4.40. The flue gas leaves the boilers at a temperature of about 130°C and passes through an electrostatic precipitator for the removal of solid particles, e.g. fly ash. The gas then passes to a heat exchanger where it is cooled to about 90°C before entering a *chloride prescrubber*. UK coals contain high levels of chlorine (typically 0.25 wt%) and chloride ions are removed from the flue gas by a spray containing calcium chloride, limestone or dilute hydrochloric acid. The gas stream leaves the chloride prescrubber at a temperature of about 50°C and then enters the SO_2 *absorber* section.

As indicated earlier, the SO_2 is reacted with alkalis in this section to form gypsum. On leaving the absorber, the gas passes on to a mist eliminator for the removal of entrained droplets and then via a *reheater* to the stack.

Lane and Needham[12] have reported on the temperature and chemical conditions in the various stages of the FGD process and these are shown in Table 4.16. Prior to the heat exchanger, the gas is hot and dry and therefore the operating conditions are not aggressive. Therefore materials such as mild steel and Cor-Ten *(weathering steel)* perform adequately up to this stage.

Table 4.16 *Operating conditions in flue gas desulphurization plant*

Zone	Temperature (°C)	Chloride ion content (ppm)	Acidity (pH)	Comments
Electrostatic precipitator	120–150	–	–	Dry gas
Inlet duct	120–150	–	–	Dry gas
Reheater (heat extraction)	88	–	Low	Below dewpoint of flue gas plus entrained particulate matter not removed in ESP
Inlet/quench zone	88–49	High	Low	Highly abrasive, scale deposits and accumulation of chloride ions
Prescrubber	49	Up to 350 000	1.5–20	Conditions will be process dependent
Absorber	49	5–30 000	4.5–6.5	Conditions will be process dependent
Outlet duct	49–80	Variable	Low	Temperature below dewpoint, slurry carryover and mixing with raw gas can lead to scale deposits and high chloride levels

After Lane and Needham.[12]

Table 4.17   *Materials used in absorber towers of FGD plant*

Alloy designation	Chloride ion content (ppm)	pH	Plant
316L	1000	7.5	Cane Run 5, La Cygne 1, Cane Run 4, Colstrip 1 and 2, Duck Creek 1, Coal Creek 1 and 2, Jeffrey 1 and 2, Lawrence 4 and 5 (new) and Green River
316LM	1700	6	Dallman 3, Jim Bridger 2 and Laramie River 1 and 2
317L	Not specified		Muscatine 9
317LM	7000	6.5	Schahfer 17 and 18, Colstrip 3 and 4, Four Corners 4 and 5 and Jim Bridger 4
94L	Few thousand	6	Big Bend 4
Hastelloy G	1600	–	Cholla 1
Inconel 625	10 000	2	San Juan 1, 2, 3 and 4 and Somerset 1

After Lane and Needham.[12]

In the heat exchanger, the conditions become more aggressive as the temperature falls below the dewpoint of the flue gas and grades such as 316L and 317LM are used in this region.

As indicated in Table 4.16, the conditions can vary significantly in the prescrubber and absorber and the choice of grade will depend very much on the particular combination of chloride ion concentration and acidity (pH). A summary of the stainless grades used in the absorber section is shown in Table 4.17. The compositions of these grades are included in Table 4.8.

Nuclear fuel reprocessing plant

The nuclear fuel reprocessing industry is a major user of stainless steels in the form of plates, tubes and forgings and the selection of steels in this sector has been reviewed very thoroughly by Shaw and Elliott.[36] Whereas the choice of materials for chemical plant is generally governed solely by the nature of the process stream, due regard must also be paid in the nuclear industry to the hazards of radiation which have a major influence on the opportunity for equipment repair and maintenance. Thus materials operating in areas of high radioactivity in nuclear reprocessing plant, where access is difficult or even impossible, are designated *primary plant materials*, whereas those operating in more accessible areas with low levels of radiation are defined as *secondary plant materials*.

Reprocessing is based upon the dissolution of irradiated nuclear fuels in hot nitric acid, followed by solvent extraction and evaporation stages. The first reprocessing plant at Sellafield in Cumbria was commissioned in 1952 and was constructed in two main types of stabilized stainless steel, namely 18% Cr 13% Ni–Nb (similar to Type 347) for primary plant and 18% Cr 9% Ni–Ti (Type 321) for secondary plant. Early work had indicated that the niobium–stabilized steel corroded at less than half the rate of Type 321 in boiling nitric acid and this was ascribed to the fact that the former had a completely austenitic structure whereas Type 321 contained some delta ferrite. However, the above authors state that it was shown subsequently that delta ferrite had no significant effect on the corrosion behaviour and that the inferior performance of Type 321 was due to the presence of TiC particles which are readily dissolved in hot nitric acid.

The use of 18% Cr 13% Ni–Nb and Type 321 remained in force until the mid-1970s, when several factors brought about a re-appraisal of steel selection. It is stated that these included the fact that some forgings in 18% Cr 13% Ni–Nb were badly corroded and that Type 321 was prone to knife-line attack in the heat-affected zone of welds. Welding problems had been experienced with the fully austenitic, niobium-stabilized steel and this material was also prone to end grain attack due to outcropping stringers of coarse NbC carbides. However, perhaps the greatest impetus for the re-appraisal of material selection stemmed from the fact that the nitric acid manufacturing industry had ceased to use the stabilized grades of stainless steels, having experienced improved manufacturing and operating performance with the low-carbon grades such as 304L and 310L.

Following extensive testing in nitric acid-based liquors and vapours, BNFL elected to replace 18% Cr 13% Ni–Nb by a low-carbon grade of 18% Cr 10% Ni steel. Initially, this material was called Nitric Acid Grade 304L but was subsequently designated by BNFL as *NAG 18/10L*. The carbon content of this grade is restricted to 0.025% max. compared with 0.03% max. in 304L and restriction on the phosphorus content (0.018% max.) also ensures very low levels of sensitization and intergranular attack. In addition, the steelmaking practice for this grade of steel has to be tightly controlled in order to produce a very clean steel so as to minimize the inclusion content and the tendency for end grain corrosion. Large quantities of *NAG 18/10L* have been used in the construction of the Thermal Oxide Reprocessing Plant (THORP) at Sellafield, and Shaw and Elliott report that the steel has given excellent welding performance. Whereas 9.3% of welds showed signs of cracking in 18% Cr 13% Ni–Nb, this has been reduced to 0.2% in *NAG 18/10L*.

Stainless steels are also used in the nuclear power and fuel reprocessing industries for the storage and transportation of spent fuel elements. For such applications, the materials must provide substantial neutron absorption, and this characteristic is provided by the inclusion of up to 1% B in a base steel containing 18% Cr and 10% Ni. The production and properties of such material, designated *Hybor 304L*, have been described by King and Wilkinson.[37] The addition of large amounts of boron to a stainless steel results in the formation of an austenite–$(FeCr)_2$ B eutectic which reduces the hot workability of the material. However, steels containing up to 1% B have been successfully rolled to plate and welded satisfactorily by the TIG, MIG and MMA processes. As illustrated in Figure 4.41, the addition of boron also produces a substantial dispersion strengthening effect, coupled with significant loss of toughness and ductility. The above authors therefore conclude that the conflicting interests of the steelmaker, fabricator and the nuclear industry are best served by steels with boron contents in the range 0.5–1%.

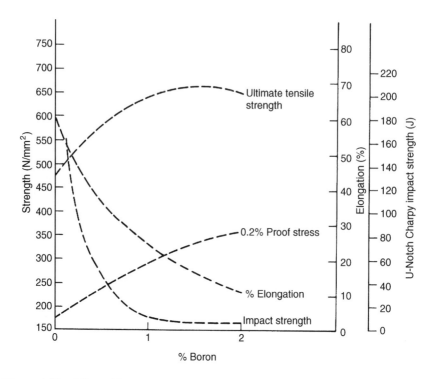

Figure 4.41   *Effect of boron on the mechanical properties of Type 304 steel (After King and Wilkinson*[37]*)*

Corrosion/abrasion-resistant grades

Large quantities of wear-resistant materials are used for the handling of bulk solids in industries such as steelmaking, coal mining, quarrying and power generation. Many of the requirements are satisfied by quenched and tempered low-alloy steels, with hardness values in the range 360–500 BHN, and provide long life and satisfactory performance under a wide range of operating conditions. However, under wet corrosive conditions, corrosion losses can contribute significantly to the wear process. Additionally, corrosion in low-alloy steel linings can increase friction and prevent the free flow of material in chutes and bunkers, causing metering and blockage problems. Under such conditions, stainless steels become attractive alternatives to low-alloy wear-resistant grades.

One interesting example of the use of stainless steel in this field is as a composite material bonded to a thick resilient rubber backing. Both ferritic (Type 409) and austenitic (Type 304) stainless steels are used in such applications, the rubber backing reducing the amount of steel required and also the level of noise generated. Ferritic stainless steel has been used very successfully in such composites for lining coal silos in British Steel plant and many of the installations have handled over 2 million tonnes of coal.[38]

Austenitic stainless steels are generally used in situations where the conditions are very corrosive but the high rate of work hardening in these materials also provides good resistance to wear. Cook *et al.*[39] have reported on the performance of austenitic steels in a number of installations handling coal, coke and ore and, as well as providing long life compared with low-alloy steels or high-density polythene, the authors also comment favourably on the flow-promoting characteristics of Type 304 steel.

For many applications, a 12% Cr steel provides adequate corrosion resistance and the high martensitic hardness of such material also offers comparable abrasion resistance to the traditional quenched and tempered low-alloy steels. Many of the 12% Cr steels used for this type of work are based on Type 420 steel (12% Cr 0.14–0.2%C) and provide hardness levels of 400–450 BHN in section sizes up to 40 mm thick in the as-rolled condition. *Hyflow 420R* is an example of a proprietary 12% Cr steel that is marketed as a corrosion/abrasion-resisting material and provides the mechanical properties shown in Table 4.18 in the as-rolled condition. Thus low-carbon material of this type can provide a good combination of strength and toughness without resort to costly heat treatments.

Metcalfe *et al.*[40] have described the development of 8–12% Cr steels for corrosion/abrasion-resistant applications in South African gold mines. The rock in which the gold-bearing ore is embedded is extremely hard and causes severe abrasion problems. In addition, large quantities of water are used to allay dust and cool machinery. In turn, the water becomes contaminated with acids and salts from the working environment and, at air temperatures of about 30°C, this leads to very humid and corrosive operating conditions. The compositions of the steels developed for use in such situations are shown in Table 4.19.

After oil quenching from 1100°C and tempering at 200°C, the first three steels in Table 4.19 develop microstructures that are predominantly martensitic with hardness levels in the range 471–583 HV. Steel 1210 involves an entirely different approach in that the high manganese and nitrogen contents produce an austenitic microstructure. However, the austenite is unstable and transforms to strain-induced martensite under mechanical action in service.

Table 4.18 *Mechanical Properties of Hyflow 420R*

TS (N/mm^2)	0.2% PS (N/mm^2)	El (%)	2 mm CVN (J)		HV
			Long.	Trans.	
1520	1050	16	55	16	450

Table 4.19 *Corrosion/abrasion resistant steels*

Code	DIN designation	Composition %					
		C	Mn	Cr	Ni	Al	N
825	X 25 CrNi 8.3	0.25		8	3		
102A	X 20 CrNiAl 10.1	0.2	1	10		0.5	
122	X 20 CrMn 12.1	0.2	1	12			
1210	X 5 CrMnN 12.10	<0.05	10	12			0.17

Automotive exhausts/catalytic converters

Motor car owners were conditioned to the fact that a traditional mild steel exhaust would need to be replaced after a typical service life of only 18 months, whereas engine and transmission components are expected to last in excess of 10 years/100 000 miles. The use of aluminized mild steel in place of uncoated material is cost-effective but, even so, the average life of the exhaust is only extended to about 30 months. On the basis of life-cycle costing, the most effective material for exhausts is a 12% Cr steel which will give a life of four to five years at a cost of about 1.5 times that of a mild steel system.

During the 1970s, British Steel directed major effort to the optimization of high-chromium steels for automotive exhausts following a major survey that showed that corrosion was responsible for about 80% of the failures that occurred in mild steel exhausts. Two types of corrosion are operative in automotive exhausts, namely:

1. That taking place on the outside of systems due to the action of water/solids thrown up from road surfaces and, more particularly, from the use of deicing salts in winter periods.
2. That occurring on the inside of exhausts due to the action of corrosive exhaust gas condensate which forms in the cooler parts of the system.

Information was available which showed that Type 302 (18% Cr, 8% Ni) austenitic stainless steel would be very effective in overcoming these corrosion problems but it was considered that this material would prove too expensive to achieve a major market conversion from mild steel systems. Additionally, experience in the United States had indicated that a 12% Cr *muffler* grade steel might provide adequate corrosion resistance.

On the basis of preliminary laboratory tests, samples of steels containing up to 15% Cr were exposed on both the inside and outside of an exhaust system for $8\frac{1}{2}$ months, including the winter period, and during this time the car covered 16 000 miles.[41] This trial confirmed that a very significant improvement in corrosion resistance over mild steel could be achieved in steels containing 10–12% Cr and therefore further work was focused on steels within this chromium range. This culminated in the development of *Hyform 409*, a variant of the standard titanium-stabilized Type 409 grade, with the following typical composition:[42]

0.02% C, 0.6% Si, 0.3% Mn, 11.4% Cr, 0.4% Ti

Because the manufacture of exhausts involves major welding operations, a stabilized grade was required in order to eliminate any possibility of intergranular corrosion. However, as indicated earlier, the addition of titanium to a 12% Cr steel also ensures that the microstructure is ferritic, thereby providing a much more formable and weldable material than the 12% Cr martensitic grades.

Hyform 409 can be welded satisfactorily using most techniques, including MIG, TIG and HF. When MIG welding is employed, the preferred filler metal is

Type 316L. It has been established that exhausts in Hyform 409 can be manufactured on the same equipment as that used for mild steel, which is an important factor in maintaining the cost of stainless steel exhausts at a relatively low level.

Whereas 80% of failures in mild steel exhausts were attributable to corrosion, it was found that the remaining 20% were due to fatigue. Having improved the corrosion performance very substantially, fatigue therefore becomes the predominant mode of failure in stainless steels. It was established that the limiting fatigue stresses of Hyform 409 and mild steel were virtually identical. The thermal conductivity and expansion coefficients were also very similar, indicating that the 12% Cr material could replace mild steel without increasing the risk of thermal fatigue. Therefore to ensure that a stainless steel exhaust system does not undergo premature failure due to fatigue, careful consideration must be given to design details and standards of manufacture rather than to the intrinsic material properties. Extensive road trials have shown that attention to these details can provide exhausts with excellent fatigue characteristics, thereby ensuring the long-term integrity of stainless steel systems.

Whereas some car manufacturers fit complete 12% Cr exhaust systems as original equipment, others have opted for hybrid systems involving a combination of mild steel, aluminized mild steel and Hyform 409. In such systems, the 12% Cr steel is restricted to the areas that constitute the greatest risk of failure by corrosion, i.e. the rear silencer boxes.

The exhaust manifold is generally made in cast iron and the material is perfectly satisfactory in this application. However, cast iron manifolds are relatively heavy components, and for weight reduction/improved fuel economy, consideration has been given to their replacement with manifolds fabricated from stainless steel strip. Type 304 has been used for this purpose and also some of the 17% Cr steels that will be discussed in the following paragraphs on catalytic converters.

Catalytic converters have been fitted to cars in the United States since 1974 in order to reduce the level of toxic products such as CO, NO_x and unburnt hydrocarbons in the exhaust gases. These devices are also being introduced gradually in Europe and represent a major potential market for stainless steel. Typically, a platinum catalyst mounted on a ceramic substrate is contained in a stainless steel case, comprising a cylindrical shell flanked by inlet and outlet cones. In early models, the operating temperatures were low and casing materials reached a maximum temperature of 550–600°C. Given the experience with the 12% Cr *muffler* grade in the United States, steels such as Type 409 were evaluated as candidate materials for the early converters and proved to be perfectly satisfactory. However, as the requirements of legislation have tightened, the operating temperatures in the converter have been increased up to 900°C, which imposes severe demands on casing materials. These include good elevated-temperature strength to withstand the stresses produced by exhaust gas pressure and also by the weight of the device itself. Additionally, the materials are required to have good resistance to oxidation and scaling, not only for long life, but also to avoid blockage and malfunction of the converter. Type 409 has a maximum operating temperature of about 700°C and therefore more highly alloyed steels are required for modern converter systems. Whereas austenitic grades such as Type 304 (18% Cr, 9% Ni) have excellent creep rupture strength, coupled with

Table 4.20   *Steels for catalytic converters*

Grade	Cr%	Al%	Nb%	Ti%	Zr%
409	11.5			0.3	
430Nb	17		0.7		
18CrNb	18.6		0.6	0.3	
Armco 12SR	12	1	0.6	0.3	
Uginox FK	17		$12 \times C$		0.4

good scaling resistance at temperatures up to 900°C, they are expensive and their high thermal expansion characteristics could also introduce large thermal stresses and cyclic fatigue in constrained parts of the component. The lower cost and thermal expansion characteristics of ferritic steels therefore offer greater attraction as candidate materials for high-temperature converters.

The nominal compositions used in current catalytic converters are compared with Type 409 in Table 4.20.

Thus the majority of the grades rely on higher chromium contents in order to achieve improved scaling resistance compared with Type 409. However, both silicon and aluminium are also beneficial in this respect and a high aluminium content is included in Armco 12SR. Each of the above steels also contains substantial amounts of strong carbide-forming elements which stabilize the materials against intergranular corrosion. However, it is interesting to note that these steels employ dual stabilization, involving niobium plus either titanium or zirconium. Whereas the latter are the more powerful stabilizing elements and also ensure a martensitic-free weld, the inclusion of niobium ensures better impact properties in the HAZ. The creep resistance of ferritic stainless steels is also improved significantly by the precipitation of Fe_2Nb (*Laves* phase). In addition to its function as a stabilizing element, zirconium is also beneficial in improving the oxidation resistance of these steels. Therefore significant benefits are obtained from the inclusion of more than one stabilizing element in ferritic stainless steels for operation at high temperatures.

Because of the very large number of cars involved and the substantial amount of material in each converter, it is reported that General Motors is now the largest consumer of stainless steel in the United States.

Architectural applications

With its aesthetic appeal, excellent resistance to atmospheric corrosion and low maintenance requirements, stainless steel offers major attractions as an architectural material. However, whereas stainless steel has featured prominently in the architectural sector in Japan and the United States, it is only recently that it has begun to enjoy similar success in the UK and Europe. One of the earliest architectural uses of stainless steel was in the cladding of the top section of the 320 m-high Chrysler building in New York in 1930. After a period of 60 years, an ASTM examination has shown that the stainless steel on this building is still in excellent condition. Of particular note in recent years have been the construction of the Lloyd's building and Canary Wharf tower in London, where extensive use has been made of stainless steel cladding. In addition, stainless steel is now being used more frequently in doors, entrance halls, portals, stairways and roofing and also in street furniture such as lighting columns and telephone boxes.

As illustrated in Figure 4.42, stainless steels are now available in a range of textured finishes which increase the options open to designers and architects. Fashion and preference have also turned away from the traditional, highly reflective finishes in stainless steel in favour of the No. 4 *brushed* finish. However, this finish is produced by mechanical abrasion which can introduce fissures and crevices into the surface which can act as initiation sites for pitting corrosion. Therefore, in producing the No. 4 finish, great care has to be taken to produce a cleanly cut surface and this is generally achieved with silicon carbide belts rather than alumina abrasive belts. For architectural applications, major attention must also be given to strip flatness and to the complete removal of oxide particles from hot-band material prior to cold rolling. Small particles of scale which remain on the surface can introduce a galvanic corrosion effect which leads to rusting during atmospheric exposure.

British Steel has operated atmospheric corrosion test sites in the UK since the 1960s in various rural, industrial and marine locations. A wide range of steels and finishes are being investigated, but in the stainless steel range, attention has been focused on the following grades:

Type 304	18% Cr	10% Ni	
Type 315	18% Cr	10% Ni	1.5% Mo
Type 316	18% Cr	12% Ni	2.5% Mo
Type 317	18% Cr	12% Ni	3.5% Mo

The stainless steel samples were examined thoroughly for evidence of pitting corrosion after 15 years exposure and the performance has been reported by

Figure 4.42   *Textured finishes in stainless steel strip (Courtesy of British Steel Stainless)*

Stone *et al.*[43] and also by Needham.[44] As indicated in Figure 4.43, the marine location is the most aggressive because of the high chloride content in the atmosphere. However, as anticipated, the corrosion behaviour of these steels correlates strongly with the molybdenum content. This is illustrated by the following pitting data in Tables 4.21 and 4.22 which were obtained after 15 years exposure in a marine environment.

Thus both the maximum pit depth and the pit density decrease progressively with molybdenum content. It is also apparent that washing, either by the natural action of the rain in unsheltered conditions or by manual washing in sheltered areas, exerts a beneficial effect in reducing pitting attack. Thus the design of structures is important and, ideally, the material should be unsheltered such that rain water will remove airborne salts and other damaging debris.

The information derived from this work has confirmed that Type 316 is the most appropriate grade of stainless steel for architectural applications, particularly where the atmosphere is relatively high in chloride ions. The long-term data obtained in marine locations were also instrumental in convincing operators of offshore oil and gas platforms of the benefits of using stainless steels for topside

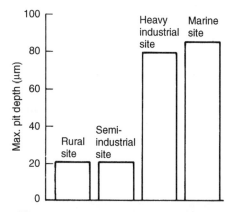

Figure 4.43 *Atmospheric corrosion in Type 304 steel after 15 years exposure (unwashed samples) (After Needham*[44]*)*

Table 4.21

Type	Pit density (pits/cm²)			
	Unsheltered		*Sheltered*	
	Washed	*Unwashed*	*Washed*	*Unwashed*
304	3060	3172	3798	4214
315	647	728	789	890
316	364	377	419	450
317	37	46	52	67

After Stone *et al.*[43]

Table 4.22

Type	Maximum pit density (μm)			
	Unsheltered		Sheltered	
	Washed	Unwashed	Washed	Unwashed
304	55	85	88	102
315	37	39	50	72
316	25	24	46	70
317	22	18	25	24

After Stone *et al.*[43]

Figure 4.44 *North Sea oil rig with accommodation module clad in Type 316 stainless steel (Courtesy of British Steel Stainless)*

architecture in place of traditional painted mild steel. An example is shown in Figure 4.44. In the North Sea, the environment is particularly aggressive and the high corrosion resistance of Type 316 is attractive. However, fire resistance is a very important requirement which rules out the use of organic materials and also raises questions as to the integrity of low melting point metals such as aluminium. In the event of fire, stainless steel is also preferable to painted mild steel because of its higher strength at elevated temperature and the elimination of the smoke or toxic hazards associated with burning paint.

Because no corrosion allowance is required, stainless steel module walls can be up to 50% thinner than similar components in mild steel, resulting in significant

weight savings. Whereas Type 316 stainless is considerably more expensive, the
cost differential at *float out* is less than 15% compared with mild steel fabrications. A cost analysis[43] indicates that this initial differential is eliminated after
the first five years of operation due to the need for the first repaint on mild steel
at that time.

Based on this work, Type 316 stainless steel has now been used for wall
cladding on accommodation modules, ventilation louvres and other architectural
features on more than half the oil fields in the North Sea.

References

1. Irvine, K.J., Llewellyn, D.T. and Pickering, F.B. *JISI*, July, 218 (1959).
2. Andrews, K.W. *Atlas of Continuous Cooling Transformation Diagrams*, British Steel.
3. Schneider, H. *Foundry Trade Journal*, **108**, 562 (1960).
4. *Stainless Steels Specifications*, 2nd edn, British Steel Stainless.
5. Irvine, K.J., Crowe, D.J. and Pickering, F.B. *JISI*, **195**, 386 (1960).
6. Irvine, K.J. and Pickering, F.B. *ISI Special Report 86*, 34 (1964).
7. McNeely, V.J. and Llewellyn, D.T. *Sheet Metal Industries*, **49**, No. 1, 17 (1972).
8. Sedriks, A.J. *Corrosion of Stainless Steels*, John Wiley & Sons, New York (1979).
9. Fontana, M.G. and Greene, N.D. *Corrosion Engineering*, 2nd edn, McGraw-Hill (1978).
10. Copson, H.R. In *Proc. 1st Int. Cong. Met. Corr.* (London, 1962), Butter-worths, p. 328 (1962).
11. Irvine, K.J., Llewellyn, D.T. and Pickering, F.B. *JISI*, October (1961).
12. Lane, K.A.G. and Needham, N.G. *Materials for Flue Gas Desulphurisation Plant No. 1 – Stainless Alloys and Coated Steels*, British Steel Stainless.
13. *Steelresearch 84–85*, British Steel, 23.
14. Binder, W.R. and Spendelow, H.O. *Trans. ASM*, **43**, 759 (1951).
15. Gregory, E. and Knoth, R.J. *Metal Progress*, January, 114 (1970).
16. *Stahl und Eisen*, **88**, 153 (1968).
17. Hooper, R.A.E., Llewellyn, D.T. and McNeely, V.J. *Sheet Metal Industries*, January, **49**, No. 1, 26 (1972).
18. Streicher, M.A. In *Proc. Stainless Steels '77*, Climax Molybdenum Co.
19. Castro, R. and de Cadenet, J.J. *Welding Metallurgy of Stainless and Heat Resisting Steels*, Cambridge University Press.
20. Gooch, T.G. In *Proc. Stainless Steels '87* (York, 1987), The Institute of Metals, p. 53.
21. Schaeffler, L.A. *Metal Progress*, **56**, 680 (1949).
22. DeLong, W.T. *Metal Progress*, **77**, 98 (1960).
23. Llewellyn, D.T., Bower, E.N. and Gladman, T. In *Proc. Stainless Steels '87* (York, 1987), The Institute of Metals, p. 62.
24. Heiple, C.R. and Roper, J.R. *Welding J.*, **60**, 74 (1981).

25. Leinonen, J.I. In *Proc. Stainless Steels '87* (York, 1987), The Institute of Metals, p. 74.
26. Kaufman, L. and Cohen, M. *Progress in Metal Physics*, Vol. 1, Pergamon Press.
27. Patel, J.R. and Cohen, M. *Acta Met.*, **1**, 531 (1953).
28. Llewellyn, D.T. and Murray, J.D. *ISI Special Report 86*, 197 (1964).
29. Gladman, T., Hammond, J. and Marsh, F.W. *Sheet Metal Industries*, **51**, 219 (1974).
30. Sanderson, G.P. and Llewellyn, D.T. *JISI*, August (1969).
31. Wigley, D.A. *Materials for Low Temperature Use*, Engineering Design Guides, Oxford University Press.
32. Wyatt, L.M. *Materials of Construction for Steam Power Plant*, Applied Science Publishers (1976).
33. Cutler, A.J.B., Flatley, A. and Hay, K.A. *The Metallurgist and Materials Technologist*, **13**, No. 2, 69 (1981).
34. Townsend, R. In *Proc. International Conference on Advances in Materials Technology for Fossil Power Plants* (Chicago, 1987) (eds Viswanathan, R. and Jaffee, R.I.), ASM International.
35. Flatley, T., Latham, E.P. and Morris, C.W. In *Proc. Advances in Materials Technology for Fossil Power Plants* (Chicago, 1987) (eds Viswanathan, R. and Jaffee, R.I.), ASM International.
36. Shaw, R.D. and Elliott, D. In *Proc. Stainless Steels '84* (Göteborg, 1984), The Institute of Metals, p. 395.
37. King, K.J. and Wilkinson, J. In *Proc. Stainless Steels '87* (Göteborg, 1984), The Institute of Metals, p. 368.
38. Guy, D.J. and Peace, J. In *Proc. Steels in Mining and Minerals Handling Applications* (London, 1990), The Institute of Metals, p. 58.
39. Cook, W.T., Peace, J. and Fletcher, J.R. In *Proc. Stainless Steels '87* (York, 1987), The Institute of Metals, p. 307.
40. Metcalfe, B., Whittaker, W.M. and Lenel, U.R. In *Proc. Stainless Steels '87* (York, 1987), The Institute of Metals, p. 300.
41. *Steelresearch 75*, British Steel, p. 43 (1975).
42. Hooper, R.A.E., Shemwell, K. and Hudson, R.M. *Automotive Engineer*, February (1985).
43. Stone, P.G., Hudson, R.M. and Johns, D.R. In *Proc. Stainless Steels '84* (Göteborg, 1984), The Institute of Metals, p. 478.
44. Needham, N.G. *Steelresearch 87–88*, British Steel, p. 49 (1988).

Index